Cycles and Rays

NATO ASI Series

Advanced Science Institutes Series

A Series presenting the results of activities sponsored by the NATO Science Committee, which aims at the dissemination of advanced scientific and technological knowledge, with a view to strengthening links between scientific communities.

The Series is published by an international board of publishers in conjunction with the NATO Scientific Affairs Division

A Life Sciences **B Physics**	Plenum Publishing Corporation London and New York
C Mathematical **and Physical Sciences** **D Behavioural and Social Sciences** **E Applied Sciences**	Kluwer Academic Publishers Dordrecht, Boston and London
F Computer and Systems Sciences **G Ecological Sciences** **H Cell Biology**	Springer-Verlag Berlin, Heidelberg, New York, London, Paris and Tokyo

Series C: Mathematical and Physical Sciences - Vol. 301

Cycles and Rays

edited by

Geňa Hahn
Département d'Informatique et de Recherche Opérationnelle,
Université de Montréal,
Montréal, Québec, Canada

Gert Sabidussi
Département de Mathématiques et de Statistique,
Université de Montréal,
Montréal, Québec, Canada

and

Robert E. Woodrow
Department of Mathematics and Statistics,
University of Calgary,
Calgary, Alberta, Canada

Kluwer Academic Publishers

Dordrecht / Boston / London

Published in cooperation with NATO Scientific Affairs Division

0372 07 67
MATH-STAT,

Proceedings of the NATO Advanced Research Workshop on
Cycles and Rays: Basic Structures in Finite and Infinite Graphs
Montréal, Canada
May 3–9, 1987

Library of Congress Cataloging in Publication Data

Cycles and rays / edited by Geňa Hahn, Gert Sabidussi, Robert E.
Woodrow.
        p.   cm. -- (NATO ASI series. Series C, Mathematical and
physical sciences ; vol. 301)
    "Published in cooperation with NATO Scientific Affairs Division."
    ISBN 0-7923-0597-3 (alk. paper)
    1. Paths and cycles (Graph theory)--Congresses. 2. Rays (Graph
theory)--Congresses.   I. Hahn, Geňa, 1949-   . II. Sabidussi,
Gert. III. Woodrow, Robert E., 1948-   . IV. North Atlantic Treaty
Organization. Scientific Affairs Division.   V. Series: NATO ASI
series. Series C, Mathematical and physical sciences ; no. 301.
QA166.22.C9 1990
511'.5--dc20   89-26724

ISBN 0-7923-0597-3

Published by Kluwer Academic Publishers,
P.O. Box 17, 3300 AA Dordrecht, The Netherlands.

Kluwer Academic Publishers incorporates the publishing programmes of
D. Reidel, Martinus Nijhoff, Dr W. Junk and MTP Press.

Sold and distributed in the U.S.A. and Canada
by Kluwer Academic Publishers,
101 Philip Drive, Norwell, MA 02061, U.S.A.

In all other countries, sold and distributed
by Kluwer Academic Publishers Group,
P.O. Box 322, 3300 AH Dordrecht, The Netherlands.

Printed on acid-free paper

TABLE OF CONTENTS

PREFACE

What is the "archetypal" image that comes to mind when one thinks of an infinite graph? What with a finite graph - when it is thought of as opposed to an infinite one? What structural elements are typical for either - by their presence or absence - yet provide a common ground for both?

In planning the workshop on "Cycles and Rays" it had been intended from the outset to bring infinite graphs to the fore as much as possible. There never had been a graph theoretical meeting in which infinite graphs were more than "also rans", let alone one in which they were a central theme. In part, this is a matter of fashion, inasmuch as they are perceived as not readily lending themselves to applications, in part it is a matter of psychology stemming from the insecurity that many graph theorists feel in the face of set theory - on which infinite graph theory relies to a considerable extent. The result is that by and large, infinite graph theorists know what is happening in finite graphs but not conversely. Lack of knowledge about infinite graph theory can also be found in authoritative sources. For example, a recent edition (1987) of a major mathematical encyclopaedia[1] proposes to "... restrict [itself] to finite graphs, since only they give a typical theory". If anything, the reverse is true, and needless to say, the graph theoretical world knows better. One may wonder, however, by how much.

This volume shows that in spite of the relatively small number of practitioners, infinite graph theory is an active and varied field. It shows that the reputation of being tainted with set theory may well be unfounded: only two of the seven major papers on infinite graphs use anything beyond the basic set theoretic baggage of any mathematician. It also testifies to an inner coherence that is much greater than in finite graphs. There is quite literally a common thread running through most of infinite graph theory: only rarely does one encounter a situation where rays (infinite paths) do not play a significant part. It is difficult to find a single structure in finite graph theory which enjoys the same degree of fundamentality - in the sense of being ubiquitous. Cycles are perhaps the best candidates. There is one context where rays and cycles meet on an equal footing - as manifestations of the same principle, if such an expresssion be permitted - thereby providing at least an aesthetic justification for mentioning them in the same breath: namely, as building blocks in problems of decomposition and multiple covering. Several papers in the present volume deal with problems of this kind.

The thirteen one-hour survey lectures given at the Workshop were evenly divided between the two main topics. On the side of the Cycles, the core was provided by four lectures on transition systems, compatibility and decomposition problems in eulerian graphs and associated isotropic systems (Bouchet, Jackson, Jaeger), as well as hamiltonian cycle decompositions (Bermond). Two other main lectures dealt with variations on the double

---

[1] Encyclopedic Dictionary of Mathematics, 2nd ed., MIT Press, Cambridge, Mass., and London (1987), vol. 2, p. 693.

cover conjecture (Bondy), and connections between cyclicity and perfect graphs (Berge) . On the infinite side there was a presentation of compactness arguments for the extension of finitary properties to infinite graphs (Thomassen), an account of infinite graphs with primitive automorphism group (Watkins), and of simplicial decompositions of infinite graphs (Halin). One lecture each was devoted to connections between infinite graphs and topology (Polat) and analysis (Picardello), dealing respectively with the space of ends of a graph and with harmonic functions on a graph. One lecture (Hajnal), involving coloured graphs, illustrated the set theoretic side.

In addition to the 13 invited speakers there were 30 participants coming from a dozen different countries. The meeting was intentionally kept small in order to achieve maximum thematic coherence. All participants, with the exception of a few graduate students, presented papers. It is worth mentioning that they carefully respected the theme of the conference. To all, a most heartfelt Thank you! for their enthusiasm and the quality of their contributions which made this Workshop a success.

Most of the funding for the Workshop came from NATO through its ARW-Programme. An additional grant was obtained from the Natural Sciences and Engineering Research Council of Canada. The meeting took place at the Université de Montréal with the help of the Département de mathématiques et de statistique and the Centre de recherches mathématiques. It is with great pleasure that we express our gratitude to all these institutions.

These Proceedings have been a long time in the making, and we apologize to all concerned. Most of the delay was due to our scanty experience with TEX. It would have been greater still but for the expert help of Michel Toulouse whom we owe sincere thanks.

Finally, we wish to express our gratitude to the members of the Committee of the NATO ARW-Programme as well as - and especially so - to its Director, Dr. Craig Sinclair. His advice and help in the months preceding the Workshop have been invaluable, and his patience thereafter most reassuring.

August 1989   The Editors

PARTICIPANTS

Ron AHARONI
Department of Mathematics
Technion
32000 Haifa
Israel

Thomas ANDREAE
Institut für Mathematik II (WE 2)
FU Berlin
Arnimallee 3
D-1000 Berlin 33
FRG

Hans-Jürgen BANDELT/KE
Faculty of Economic and Business Administration
Main Bldg.
Rijksuniversitet Limburg
Postbus 616
6200 MD Maastricht
The Netherlands

Claude BERGE
10, rue Galvani
75017 Paris
France

Jean-Claude BERMOND
Informatique CRNS
3, rue Albert Einstein
Sophia Antipolis
06560 Valbonne
France

J. Adrian BONDY
Combinatorics & Optimization
University of Waterloo
Waterloo, Ontario, N2L 3G1
Canada

André BOUCHET
Dép. de Mathématiques et d'Informatique
Université du Maine
Route de Laval
B.P. 535
72017 Le Mans Cedex
France

Jean-Michel BROCHET
Institut de Mathématiques-Informatique
Lab. Algèbre Ordinale
Université Claude Bernard (Lyon I)
69622 Villeurbanne Cedex
France

Christine CHARRETTON
Institut de Mathématiques-Informatique
Lab. Algèbre Ordinale
Université Claude Bernard (Lyon I)
69622 Villeurbanne Cedex
France

Genghua FAN
Combinatorics & Optimization
University of Waterloo
Waterloo, Ontario, N2L 3G1
Canada

Joan FEIGENBAUM
AT&T Bell Laboratories
600 Montain Avenue
Murray Hill, NJ 07974
USA

Herbert FLEISCHNER
Institut für Informationsverarbeitung
Akademie der Wissenschaften
A1010 Wien
Sonnenfelsgasse 19
Austria

Stephane FOLDES
GERAD
Ecole des Hautes Etudes Commerciales
5255 ave Decelles
Montréal, Qué., H3T 1V6
Canada

Roland FRAÏSSÉ
Université de Provence I
U.E.R. de Mathématiques
3, Place Victor Hugo
13331 Marseille
France

Luis A. GODDYN
Combinatorics & Optimization
University of Waterloo
Waterloo, Ontario, N2L 3G1
Canada

Lucien HADDAD
Physical Science Division
Scarborough College
University of Toronto
Scarborough, Ontario, M1C 1A4
Canada

Geňa HAHN
Dép. d'Informatique et de Recherche Opérationnelle
Université de Montréal
C.P. 6128, Succ. A
Montréal, Qué., H3C 3J7
Canada

András HAJNAL
Mathematical institute
Hungarian Academy of Sciences
1053 Budapest V.
Reáltanoda u. 13-15
Hungary

Rudolf HALIN
Mathematisches Seminar
Universität Hamburg
Bundesstrasse 55
2000 Hamburg 13
FRG

Simone HAZAN
Department of Mathematics
University of California at Berkeley
Berkeley, CA 94720
U.S.A.

Pavol HELL
School of Computing Science
Simon Fraser University
Burnaby, B.C., V5A 1S6
Canada

Anthony HILTON
Department of Mathematics
University of Reading
P.O. Box 220
Whiteknights, Reading RG6 2AX
United Kingdom

Petr HORÁK
Katedra Matematiky
SvF SVST
Radlinskeho 11
813 68 Bratislava
Czechoslovakia

Bill JACKSON
Mathematical Sciences Department
Goldsmiths' College
London SE14 6NW
United Kingdom

François JAEGER
LSD, Institut IMAG
BP 68
38042 St-Martin-d'Hères Cedex
France

Tommy JENSEN
Combinatorics & Optimization
University of Waterloo
Waterloo, Ontario, N2L 3G1
Canada

Heinz Adolf JUNG
Fachbereich 3- Mathematik
TU Berlin
Strasse des 17. Juni 135
1000 Berlin 12
FRG

François LAVIOLETTE
Dép. de Mathématiques et de Statistique
Université de Montréal
C.P. 6128, Succ. A
Montréal, Qué., H3C 3J7
Canada

Charles H.C. LITTLE
Department of Mathematics and Statistics
Massey University
Palmerston North
New Zealand

Wolfgang MADER
Universität Hannover
Institut für Mathematik
Welfengarten 1
D-3000 Hannover 1
FRG

Eric MILNER
Department of Mathematics and Statistics
University of Calgary
Calgary, Alberta, T2N 1N4
Canada

Massimo PICARDELLO
Dipartimento di Matematica Pura e Applicata
Università dell'Aquila
67100 L'Aquila
Italy

Norbert POLAT
Département de Mathématiques
Bâtiment 101
Université Claude Bernard (Lyon I)
43, Boulevard du 11 Novembre 1918
69622 Villeurbanne Cedex
France

Maurice POUZET
Institut de Mathématiques-Informatique
Lab. Algèbre Ordinale
Université Claude Bernard (Lyon I)
69622 Villeurbanne Cedex
France

Neil ROBERTSON
Department of Mathematics
Ohio State University
Columbus, OH 43210
U.S.A.

Alexander ROSA
Department of Mathematical Sciences
McMaster University
Hamilton, Ontario, L8S 4K1
Canada

Ivo ROSENBERG
Dép. de Mathématiques et de Statistique
Université de Montréal
C.P. 6128, Succ. A
Montréal, Qué., H3C 3J7
Canada

Gert SABIDUSSI
Dép. de Mathématiques et de Statistique
Université de Montréal
C.P. 6128, Succ. A
Montréal, Qué., H3C 3J7
Canada

Norbert SAUER
Department of Mathematics and Statistics
University of Calgary
Calgary, Alberta, T2N 1N4
Canada

Raffaele SCAPELLATO
Dipartimento di Matematica
Via dell' Università 12
43100 Parma
Italy

Alejandro SCHÄFFER
Department of Computing Science
Rice University
P.O. Box 1892
Houston, TX 77251
U.S.A.

Karen SEYFFARTH
Combinatorics & Optimization
University of Waterloo
Waterloo, Ontario, N2L 3G1
Canada

John SHEEHAN
Department of Mathematical Science
University of Aberdeen
Aberdeen AB9 2TY
United Kingdom

Dominque SOTTEAU
CNRS, UA410 Informatique
Bâtiment 490
Université Paris-Sud
91405 Orsay
France

Teo STURM
Department of Mathematics
University of Natal
King George V Avenue
Durban 4001
South Africa

Claude TARDIF
Dép. de Mathématiques et de Statistique
Université de Montréal
C.P. 6128, Succ. A
Montréal, Qué., H3C 3J7
Canada

Carsten THOMASSEN
Mathematical Institute
Technical University of Denmark
2800 Lyngby
Denmark

Bjarne TOFT
Department of Mathematics and Computer Science
Odense University
5230 Odense M
Denmark

Preben Dahl VESTERGAARD
Institut of Electronic Systems
Aalborg University
Strandvejen 19
9000 Aalborg
Denmark

Mark E. WATKINS
Department of Mathematics
Syracuse University
Syracuse, New York 13244-1150
U.S.A.

W. WOESS
Dipartimento di Matematica
Università di Milano
Via C. Saldini 50
I-20133 Milano
Italy

Robert E. WOODROW
Department of Mathematics and Statistics
University of Calgary
Calgary, Alberta, T2N 1N4
Canada

CONTRIBUTORS

Ron AHARONI
Department of Mathematics
Technion
32000 Haifa
Israel

Brian ALSPACH
Department of Mathematics and Statistics
Simon Fraser University
Burnaby, B.C., V5A 1S6
Canada

G.D. BADENHORST
Department of Mathematics
University of Natal
King George V Avenue
Durban 4001
South Africa

Jean-Claude BERMOND
Informatique CRNS
3, rue Albert Einstein
Sophia Antipolis
06560 Valbonne
France

J. Adrian BONDY
Combinatorics & Optimization
University of Waterloo
Waterloo, Ontario, N2L 3G1
Canada

André BOUCHET
Dép. de Mathématiques et d'Informatique
Université du Maine
Route de Laval
B.P. 535
72017 Le Mans Cedex
France

Jean-Michel BROCHET
Institut de Mathématiques-Informatique
Lab. Algèbre Ordinale
Université Claude Bernard (Lyon I)
69622 Villeurbanne Cedex
France

N. CHAKROUN
CNRS, UA410 Informatique
Bâtiment 490
Université Paris-Sud
91405 Orsay
France

M. EL-ZAHAR
Department of Mathematics and Statistics
University of Calgary
Calgary, Alberta, T2N 1N4
Canada

Genghua FAN
Combinatorics & Optimization
University of Waterloo
Waterloo, Ontario, N2L 3G1
Canada

Herbert FLEISCHNER
Institut für Informationsverarbeitung
Akademie der Wissenschaften
A1010 Wien
Sonnenfelsgasse 19
Austria

Anthony HILTON
Department of Mathematics
University of Reading
P.O. Box 220
Whiteknights, Reading RG6 2AX
United Kingdom

Wilfried IMRICH
Institut für Mathematik und Angewandte Geometrie
Montanuniversität Leoben
A-8700 Leoben
Austria

Bill JACKSON
Mathematical Sciences Department
Goldsmiths' College
London SE14 6NW
United Kingdom

François JAEGER
LSD, Institut IMAG
BP 68
38042 St-Martin-d'Hères Cedex
France

S. KRSTIC

Matematički Institut
University of Belgrade
YU-11000 Beograd
Yugoslavia

Wolfgang MADER
Universität Hannover
Institut für Mathematik
Welfengarten 1
D-3000 Hannover 1
FRG

Eric MILNER
Department of Mathematics and Statistics
University of Calgary
Calgary, Alberta, T2N 1N4
Canada

Massimo PICARDELLO
Dipartimento di Matematica Pura e Applicata
Università dell'Aquila
67100 L'Aquila
Italy

Norbert POLAT
Département de Mathématiques
Bâtiment 101
Université Claude Bernard (Lyon I)
43, Boulevard du 11 Novembre 1918
69622 Villeurbanne Cedex
France

Maurice POUZET
Institut de Mathématiques-Informatique
Lab. Algèbre Ordinale
Université Claude Bernard (Lyon I)
69622 Villeurbanne Cedex
France

C. A. RODGER
Department of Algebra, Combinatorics and Analysis
Mathematical Annex
Auburn University
Auburn, Alabama, 36849
U.S.A.

Gert SABIDUSSI
Dép. de Mathématiques et de Statistique
Université de Montréal
C.P. 6128, Succ. A
Montréal, Qué., H3C 3J7
Canada

Norbert SAUER
Department of Mathematics and Statistics
University of Calgary
Calgary, Alberta, T2N 1N4
Canada

Dominque SOTTEAU
CNRS, UA410 Informatique
Bâtiment 490
Université Paris-Sud
91405 Orsay
France

Teo STURM
Department of Mathematics
University of Natal
King George V Avenue
Durban 4001
South Africa

E. C. TURNER
Department of Mathematics and Statistics
State University of New York at Albany
Albany, NY 12222
USA

Mark E. WATKINS
Department of Mathematics
Syracuse University
Syracuse, New York 13244-1150
U.S.A.

W. WOESS
Dipartimento di Matematica
Università di Milano
Via C. Saldini 50
I-20133 Milano
Italy

LINKABILITY IN COUNTABLE-LIKE WEBS

R. AHARONI
Department of Mathematics and Statistics
The University of Calgary
Calgary, Alberta, T2N 1N4
Canada

Abstract

A *web* is a triple (G, A, B), where G is a directed graph and $A, B \subseteq V(G)$. It is called *linkable* if there exists a family of disjoint paths from all of A into B. It is *countable-like* if (G, A', B) is linkable for some subset A' of A such that $A \backslash A'$ is countable. In [1] a sufficient condition was given for linkability in countable webs. Here it is extended to countable-like webs.

1 Introduction

A well known conjecture of Erdös is that Menger's theorem extends to infinite graphs in the following way:

Conjecture 1.1 *In any graph (directed or undirected) $G = (V, E)$, for any two subsets A and B of V there exist a family $\mathcal{P}$ of vertex disjoint $A - B$ paths and an $A - B$ separating set S, so that S consists of the choice of precisely one vertex from each path in $\mathcal{P}$.*

In [1] this conjecture was shown to be equivalent to another conjecture on a sufficient condition for linkability (for the definition see below) in graphs. These conjectures were proved in [1] for countable graphs. The aim of this paper is to prove the latter conjecture for a more general class of graphs. We begin with recapitulation of the basic definitions from [1]:

A *web* is a triple $\Gamma = (G, A, B)$ where $G = (V, E)$ is a graph and $A, B \subseteq V$. (Such a system is sometimes called a *Gammoid*, see, e.g., [8]).

The graph may be directed or undirected, and the web is called directed or undirected in accord. Here we shall be considering only directed webs, and thus "web" in this paper is to be understood as directed. The letter Γ will be always associated with the web $\Gamma = (G = (V, E), A, B)$.

A directed graph T is called a *rooted tree* if it is acyclic and there exists in it a vertex r, called the *root* of T, such that for every $v \in V(T)$ there exists a unique directed path from r to v. (Omitting the direction from the edges, T is then a tree in the ordinary, undirected sense.) The path from r to v is denoted by Tv.

A set $\{T_i : i \in I\}$ of trees rooted at a common vertex is called *compatible* if $\cup\{E(T_i) : i \in I\}$ is the edge set of a rooted tree T, which is then denoted by $\frown\{T_i : i \in I\}$. When $|I| = 2$ we write $\frown\{T_1, T_2\} = T_1 \frown T_2$. We also say in this case that T_2 is T_1-compatible.

1

G. Hahn et al. (eds.), Cycles and Rays, 1–8.
© *1990 by Kluwer Academic Publishers. Printed in the Netherlands.*

Directed paths are a special case of rooted trees. All paths in this paper have a first vertex, but not necessarily a last vertex (i.e., they may be one-way infinite). The first vertex of a path P is denoted by $\text{in}(P)$, and the last vertex, if such exists, by $\text{ter}(P)$. If $x \in V(P)$ then xP denotes the part of P following (and including) x. A path consisting of a single vertex x is denoted by (x). Given a path P we write $\underline{P}$ for $P - \{\text{in}(P)\}$ and if P is finite we write $\overline{P} = P - \{\text{ter}(P)\}$. If P, Q are paths and $V(Q) \cap V(P) = \text{in}(Q)\} = \{\text{ter}(P)\}$ then $P*Q$ denotes the concatenation of P and Q.

If T is a rooted tree, P is a path, $x \in V(T) \cap V(P)$ and Tx^*xP is defined, we abbreviate and write $Tx^*xP = TxP$. A *warp* (a term taken from weaving) is a family of disjoint paths.

If $\mathcal{P}$ is a set of paths we write $\mathcal{P}^f = \{P \in \mathcal{P} : P \text{ is finite}\}$, $V[\mathcal{P}] = \cup\{V(P) : P \in \mathcal{P}\}$, $E[\mathcal{P}] = \cup\{E(P) : P \in \mathcal{P}\}$, $\text{in}[\mathcal{P}] = \{\text{in}(P) : P \in \mathcal{P}\}$, $\text{ter}[\mathcal{P}] = \{\text{ter}(P) : P \in \mathcal{P}^f\}$. If $x \in V[\mathcal{P}]$ we write $\mathcal{P}_x$ for the path from $\mathcal{P}$ which contains x. If $\mathcal{P}$ and $\mathcal{Q}$ are two warps and $V[\mathcal{Q}] \cap V[\mathcal{P}] = \text{in}[\mathcal{Q}] \subseteq \text{ter}[\mathcal{P}]$ we denote by $\mathcal{P}*\mathcal{Q}$ the family $\{P*Q : P \in \mathcal{P}, Q \in \mathcal{Q}$ and $\text{ter}(P) = \text{in}(Q)\}$.

For simplicity of the arguments it will be assumed for every path P mentioned in this paper that if $a \in V(P) \cap A$ then $a = \text{in}(P)$ and if $b \in V(P) \cap B$ then $b = \text{ter}(P)$. The only exceptions are the "trails" defined below.

If $C, D \subseteq V$ we say that a path P is a $C - D$ path if $V(P) \cap C = \{\text{in}(P)\}$ and $V(P) \cap D = \{\text{ter}(P)\}$. if $C = \{x\}$ for some $x \in V$ we write "$x - D$ path" for "$C - D$ path" and similarly if $D = \{y\}$ for some $y \in V$. A $C - D$ warp is a warp whose elements are $C - D$ paths. An $A - B$ warp $\mathcal{W}$ is a *linkage* if $\text{in}[\mathcal{W}] = A$. If Γ contains a linkage it is called linkable.

For a subset X of V we denote by $G[X]$ the subgraph of G spanned by X and write $G - X = G[V \backslash X]$. Then $\Gamma[X]$ is the web $(G[X], A \cap X, B \cap X)$ and $\Gamma - X = \Gamma[V \backslash X]$. If H is a subgraph of G we write $\Gamma - H$ for $\Gamma - V(H)$ (in our uses H will be a path or tree). We say that X separates C from D (or "is $C - D$ separating"; here $C, D \subseteq V$) every path from C to D contains a vertex from X.

A warp $\mathcal{W}$ is called a *wave* if $\text{in}[\mathcal{W}] \subseteq A$ and $\text{ter}[\mathcal{W}]$ separates A from B. Clearly $\mathcal{W}^f$ is then also a wave. A relation $<$ ($\leq$ if equality is allowed) is defined between waves as follows: $\mathcal{W} \leq \mathcal{U}$ if for each path $P \in \mathcal{U}$ there exists $Q \in \mathcal{W}$ such that

(a) P is an extension of Q (possibly $P = Q$), and

(b) $V(P) \cap V[\mathcal{W}] = V(Q)$.

(Note that (a) implies that $\text{in}[\mathcal{U}] \subseteq \text{in}[\mathcal{W}]$. Note also that "$\leq$" is not necessarily transitive, since (b) may fail to be transitive, but by (a) it is acyclic.)

The wave $\{(a) : a \in A\}$ is minimal in this relation, and is called the *trivial* wave.

A wave $\mathcal{W}$ is called a *hindrance* if $\text{in}[\mathcal{W}] \neq A$, and then the pair $(a, \mathcal{W})$, where a is any element of $A \backslash \text{in}[\mathcal{W}]$, is called a 1-hindrance. If Γ contains a hindrance we say that it is *hindered*. Clearly, a hindrance is a non-trivial wave.

For any two waves $\mathcal{W}$ and $\mathcal{U}$ define:
$$\mathcal{W} \to \mathcal{U} = \mathcal{W}*\{xQ : Q \in \mathcal{U} \text{ and } x \in \text{ter}[\mathcal{W}] \cap V(Q)$$
$$\text{and } V(xQ) \cap V[\mathcal{W}] = \{x\}\}$$
$$\cup\{P \in \mathcal{W} : P \text{ is infinite or}$$
$$\text{there does not exist } Q \in \mathcal{U} \text{ such that}$$
$$x = \text{ter}(P) \in V(Q) \text{ and } V(xQ) \cap V(\mathcal{W}) = \{x\}\}.$$

Let $(\mathcal{W}_\alpha : \alpha < \zeta)$ be a $(<)$-ascending chain of waves (i.e., $\mathcal{W}_\beta \leq \mathcal{W}_\alpha$ whenever $\beta < \alpha < \zeta$). We let $\mathcal{U} = \uparrow (\mathcal{W}_\alpha : \alpha < \zeta)$ be the family of paths defined by $E[\mathcal{U}] =$

$\cup\{E(P) : P \in \mathcal{W}_\alpha \text{ for some } \alpha < \zeta \text{ and in}(P) \in \cap_{\beta < \zeta}\text{in}[\mathcal{W}_\beta]\}$.

Now, let $(\mathcal{W}_\alpha : \alpha < \zeta)$ be any sequence of waves. We define $\uparrow (\mathcal{W}_\alpha : \alpha < \zeta)$ by induction on ζ, as follows. For $\zeta = 0$ we define $\uparrow \phi$ as the trivial wave. For $\zeta = 1$ we let $\uparrow (\mathcal{W}_0) = \mathcal{W}_0$. We assume inductively that $\mathcal{U}_\beta = \uparrow (\mathcal{W}_\alpha : \alpha < \beta)$ is defined for $\beta < \zeta$ and that (i) $\mathcal{U}_\beta \geq \mathcal{U}_\gamma$ and (ii) $V[\mathcal{U}_\beta] \supseteq V[\mathcal{U}_\gamma]$ whenever $\gamma \leq \beta$. If ζ is a limit ordinal we define $\uparrow (\mathcal{W}_\alpha : \alpha < \zeta) = \uparrow (\mathcal{U}_\beta : \beta < \zeta)$. If $\zeta = \beta + 1$ define: $\uparrow (\mathcal{W}_\alpha : \alpha < \zeta) = \mathcal{U}_\beta \rightarrow \mathcal{U}_\zeta$.

The formulation of some of the following definitions in [1] was erroneous, and we correct them here. The changes are in parts (a) and (b) of the definition of a 'trail'; the addition of the definition of a 'free' trail; and a change in the definition of $M(a, \mathcal{W})$.

Given a warp $\mathcal{J}$, a $\mathcal{J}$-*trail* is a sequence $T = (x_t, P_t, x_{t+1}, P_{t+1}, \ldots, P_n, x_{n+1})$, where $t = 0$ or 1, $n + 1 \geq t$, $x_i \in V$ and P_i are paths in G, and the following conditions hold:

(a) $\quad E(P_{2k+1}) \subseteq E[J] =$ for some $J \in \mathcal{J}(2k + 1 \leq n)$.

(b) $\quad V(P_{2k}) \cap V[\mathcal{J}] = \{x_{2k}, x_{2k+1}\}(2k \leq n)$.

(c) $\quad |E(P_i)| \geq 1 (i \leq n)$.

(d) $\quad x_{2k} = \text{in}(P_{2k}) = \text{in}(P_{2k-1})(0 < 2k \leq n)$

and, if $t = 0$, $x_0 = \text{in}(P_0)$;

$$x_{2k+1} = \text{ter}(P_{2k}) = \text{ter}(P_{2k+1})(2k + 1 \leq n).$$

(e) $\quad$ If $|i - j| > 1$ and $V(P_i) \cap V(P_j) \neq \phi$

then $V(P_i) \cap V(P_j) = \{x\}$, where

$$x = \text{in}(P_i) = \text{ter}(P_j) \text{ or } x = \text{in}(P_j) = \text{ter}(P_i).$$

If $t = 0$ then T is called a $\mathcal{J}$-*walk*, and if $t = 1$ (i.e., T starts with a subpath of a path from $\mathcal{J}$), then T is called $\mathcal{J}$-*track*. We say that T is a $\mathcal{J}$-*y*-*trail* (walk, track) if n is odd (i.e., T ends in a subpath of a path from $\mathcal{J}$) and that it is a $\mathcal{J}$-*n*-*trail* (walk, track) otherwise. The vertices x_i are called the *joints* of T, and we write $x_i = j_i(T)$. We write

$$FR(T) = \cup\{V(P_{2k}) : 2k \leq n\}$$

and

$$BK(T) = \cup\{V(\overline{P}_{2k+1}) : 2k + 1 \leq n\}.$$

The source of this notation is that we think of P_{2k} as going forward on edges of G and P_{2k+1} as going backwards. Indeed, we can view T as a trail (i.e., a path which may repeat some vertices), in the underlying undirected graph of G, namely $T = x_0 P_0 x_1 \overleftarrow{P}_1 \, x_2 P_2 x_3 \, \overleftarrow{P}_3, \ldots,$ where $\overleftarrow{P}_i$ denotes the path P_i taken with reversed direction. With this way of viewing $\mathcal{J}$-trails in mind, we use for them similar notation to that we use for paths. For example, $V(T)$ denotes the vertex set of T, and if $x \in V(T)$ we write Tx for the part of T up to (and including) x. We say that T is *free* if n is even and $x_{n+1} \notin V[\mathcal{J}]$.

Let now $\mathcal{W}$ be a warp and $a \in V \backslash V[\mathcal{W}]$ (resp. $a \in V[\mathcal{W}]$). For each $P \in \mathcal{W}$ let $x(P)$ be the last vertex on P (if such exists) lying on a $\mathcal{W}$-walk (resp. $\mathcal{W}$-track) starting at a, and let $x(P) = \text{in}(P)$ if no such vertex exists. We write:

$$M(a, \mathcal{W}) = M_\Gamma(a, \mathcal{W}) = \{Px : P \in \mathcal{W} \text{ and } x = x(P)\} \cup \{(x) : x \in A \backslash \{a\}\}$$

If $\mathcal{W}$ is a wave and R a free $\mathcal{W}$-walk we denote by $\mathcal{W} \triangle R$ the family $\mathcal{Y}$ of paths such that $E[\mathcal{Y}] = E(\mathcal{W}) \triangle E(R)$ (here $\triangle$ denotes symmetric difference). As is well known and easily observed, $\mathcal{W} \triangle R$ is a family of disjoint paths, and in$[\mathcal{W} \triangle R] \subseteq A$.

Here are some additional definitions to those used in [1].

A family $\mathcal{P}$ of paths is called *S-joined*, where $S \subseteq V$, if $V(P) \supseteq S$ for every $P \in \mathcal{P}$ and $V(P) \cap V(Q) = S$ whenever P and Q are distinct members of $\mathcal{P}$.

A path P is called *chordless* if it does not admit a shortcut, i.e., if $(u, v) \notin E\backslash E(P)$ whenever $u \in V(P)$ and $v \in V(uP)$. A tree T is called *chordless* if Tv is chordless for every $v \in V(T)$.

Let $\mathcal{J}$ be a warp and T a subgraph of G. We write:

$$\mathcal{J}_T = \{P \in \mathcal{J} : V(P) \cap V(T) \neq \phi\}$$

and $\mathcal{J}^T = \mathcal{J}\backslash\mathcal{J}_T$. If T consists of a single vertex x we write $\mathcal{J}_x$ and $\mathcal{J}^x$, respectively, and by a common abuse of notation, we denote by $\mathcal{J}_x$ also the single element of the set $\mathcal{J}_x$.

For a path Q and a vertex x in $V(Q)\backslash\{\text{ter}(Q)\}$ we write $Q^+(x)$ for the vertex following x on Q. If $x \in V(Q)\backslash\{\text{in}(Q)\}$ then $Q^-(x)$ denotes the vertex preceding x on Q.

For a path $P = (x_1, x_2, \ldots, x_n)$ we write $v_i(P) = x_i (1 \leq i \leq n)$.

If $\mathcal{F}$ is a family of paths and $x \in V$ we write $d_\mathcal{F}^+(x)$ (resp. $d_\mathcal{F}^-(x)$) for the cardinality of the set of paths from $\mathcal{F}$ going out of x (resp. entering x).

Let $\mathcal{J}$ be a warp and $\mathcal{W}$ a wave. A $\mathcal{J}$-walk (or track) K is said to be $\mathcal{W}$-*evading* if no vertex in $FR(K)$ is separated from B by $\text{ter}[\mathcal{W}]$ (in particular, in view of [1, Lemma 2.1], this implies that $FR(K) \cap V[\mathcal{W}] = \phi$).

If F is a set of edges and $z \in V$ we write $F\langle z\rangle = \{x : (z, x) \in F\}$. For $Z \subseteq V$ we write $F[Z] = \cup\{F\langle z\rangle : z \in Z\}$. A web $\Gamma = (G, A, B)$ is called *countable-like* if there exists an $A - B$ warp $\mathcal{J}$ such that $A\backslash\text{in}[\mathcal{J}]$ is countable.

2   A sufficient condition for linkability in countable-like webs

In this section we prove Conjecture 3.1 of [1] for countable-like webs, namely:

Theorem 2.1 *An unhindered countable-like web is linkable.*

Remark: The theorem would follow easily if we could prove for general webs Theorem 3.4 of [1], namely that if Γ is unhindered and $a \in A$ then there exists an $a - B$ path P such that $\Gamma - P$ is unhindered. This was proved in [1] only for countable webs, and although we believe it holds for general webs we do not know how to prove it.

Proof: Let $\mathcal{J}$ be an $A - B$ warp such that $A' = A\backslash\text{in}[\mathcal{J}]$ is countable. Call a vertex x in $V\backslash V[\mathcal{J}]$ $(> \aleph_0)$-*popular* if there exists an uncountable x-joined family of $\mathcal{J}$-walks from x to $B\backslash\text{ter}[\mathcal{J}]$. Let S be the set of $(< \aleph_0)$-popular vertices, and write $\hat{B} = B \cup S$. Let F be the set of ordered pairs (u, v) of vertices for which there exists an uncountable $\{u, v\}$-joined family of $\mathcal{J}$-n-walks from u to v. Let $\hat{E} = E \cup F$, $\hat{G} = (V, \hat{E})$ and $\hat{\Gamma} = \hat{\Gamma}(\mathcal{J}) = (\hat{G}, A, \hat{B})$. $\square$

The main step toward the proof of the theorem is:

Lemma 2.2 *For any $a \in A$ there exists in $\hat{\Gamma}$ an $a - \hat{B}$ path P such that $\Gamma - V(P)$ is unhindered.*

Before proving the lemma, let us explain how it implies the theorem. Enumerate the elements of A' as $(a_{2i} : i < \beta)$, where $\beta \leq \omega$. By the lemma there exists an $a_0 - \hat{B}$ path P_0 in $\hat{\Gamma}$ such that $\Gamma - V(P_0)$ is unhindered. Add $\text{in}[\mathcal{J}_{P_0}]$ to A', and put its elements in the first vacant places in the sequence $(a_k : k < \omega)$, which was partly filled by the elements of

A'. Apply now the lemma to $\Gamma - V(P_0)$ where $\mathcal{J}^{P_0}$ replaces $\mathcal{J}$ and the role of a is taken by the first a_i, $i > 0$, appearing the list (a_k). By the lemma there exists a path P_1 from this a_i to $\hat{B}$, such that $\Gamma - V(P_0) - V(P_1)$ is unhindered. We then add in$[\mathcal{J}_{P_1} \backslash \mathcal{J}_{P_0}]$ to the list (a_k), in the first vacant places. Continuing this γ steps, where $\gamma \leq \omega$, the list (a_k) is exhausted, i.e., each a_k equals in(P_j) for some path P_j chosen in the process. Let A'' be the set of vertices a_k defined in this process. Then $A'' \supseteq A'$, and whenever $Q \in \mathcal{J}$ and $V(Q) \cap V(P_i) \neq \phi$ for some i there holds in $(Q) \in A''$. Hence

$$\mathcal{J}' = \{P_i : i < \gamma\} \smile (\mathcal{J} \backslash \underset{i \, < \, \gamma}{\overset{\cup}{}} \mathcal{J}_{P_i})$$

is a linkage of A into $\hat{B}$ in $\hat{\Gamma}$.

From $\mathcal{J}'$ we construct a linkage of Γ as follows: let $(e_i : i < \delta)$ be an enumeration of $E[\mathcal{J}'] \cap (F \backslash E)$, where $\delta \leq \omega$. For each $e_i = (u_i, v_i)$ there exists an uncountable set of $\{u_i, v_i\}$-joined $\mathcal{J}$-walks. Since $|\mathcal{J}' \Delta \mathcal{J}| \leq \aleph_0$ there exists also an uncountable set of $\{u_i, v_i\}$-joined $\mathcal{J}'$-walks. Since $\delta \leq \omega$, for each $i < \delta$ we can choose a $\mathcal{J}'$-walk T_i from u_i to v_i so that $\mathcal{J}'_{T_i} \cap \mathcal{J}'_{T_j} = \phi$ whenever $i \neq j$. Then $(E[\mathcal{J}'] \backslash (F \backslash E)) \Delta E[\{T_i : i < \delta\}]$ is the edge set of an $A - \hat{B}$ linkage $\mathcal{K}$ in Γ. Clearly $|\mathcal{K} \Delta \mathcal{J}| \leq \aleph_0$, and hence $U = \text{ter}[\mathcal{K}] \backslash B$ is countable. Also, by our construction $U \subseteq S$. Hence we can choose a $\mathcal{J}$-walk $T(u)$ from u to $B \backslash \text{ter}[\mathcal{K}]$ for each $u \in U$, in such a way that

$$\mathcal{K}_{T(u)} \cap \mathcal{K}_{T(v)} = \phi$$

whenever $u \neq v$. Then $(E[\mathcal{K}] \backslash (F \backslash E)) \Delta E[\{T(u) : u \in U\}]$ is the edge set of an $A - B$ linkage in Γ.

Proof of Lemma 2.2: For $n < \omega$ we construct inductively finite trees T_n in $\hat{\Gamma}$ rooted at a, countable families $\mathcal{G}_n$ of finite paths in $\hat{\Gamma}$, enumerations e_n of $\mathcal{G}_n$ and waves $\mathcal{W}_n$ in $\Gamma - T_n$. Let $h : \omega \to \omega \times \omega$ be a bijection such that if $h(n) = (k, \ell)$ then $k \leq n$. Let $\mathcal{W}_0$ be the trivial wave, and $T_0 = \mathcal{G}_0 = \ell_0 = \phi$. Assume that T_k, $\mathcal{W}_k$, $\mathcal{G}_k$ and e_k are defined for all $k \leq n$, and that $\Gamma - V(T_n)$ is unhindered. Let $\mathcal{J}^n = \mathcal{J}^{T_n}$, and let $\mathcal{G}_{n+1}$ be the set of finite paths Q in $\hat{\Gamma}$ satisfying the following conditions:

(A) Q is T_n-compatible (*implying in $(Q) = a$*), and $V(Q) \not\subseteq V(T_n)$.
(B) ter$(Q) \in \hat{B} \cup V[\mathcal{J}^n]$.
(C) $V(Q) \cap \hat{B} \subseteq \{\text{ter}(Q)\}$.
(D) Q is chordless.
(E) If $x \in V(Q) \cap V[\mathcal{J}^n]$, $y = Q^-(x)$ and $\mathcal{J}_x \notin \mathcal{J}_{Q_y}$ (that is, x is the first vertex on Q lying on $\mathcal{J}_x$) then there exists a $\mathcal{J}^{T_n \frown Q_y} - n$-track $T = T(Q)$ from x to $B \backslash \text{ter}[\mathcal{J}^n]$ in $\Gamma - Qy$, such that

$$FR(T) \cap F[V(Qy)] = \phi.$$

Assertion a: $\mathcal{G}_{n+1}$ is countable.

Proof of Assertion a: Assume that $\mathcal{G}_{n+1}$ is uncountable. Then there exist an uncountable subfamily $\mathcal{G}'$ of $\mathcal{G}_{n+1}$, a path R and a vertex z such that $z = \text{ter}(R)$, $z \in V(Q) \backslash \{\text{ter}(Q)\}$ and $Qz = R$ for every $Q \in \mathcal{G}'$, and $Q_1^+(z) \neq Q_2^+(z)$ whenever Q_1, Q_2 are distinct paths in $\mathcal{G}'$. Let $W = T_n \frown R$. By (C), $z \notin \hat{B}$. We claim the following:

(H) There exists an uncountable family $\mathcal{G}''$ of $\mathcal{G}'$ such that each path in $\mathcal{G}''$ meets $\mathcal{J}^W$, and denoting by $x(Q)$ the first vertex on Q belonging to $V[\mathcal{J}^W]$ $(Q \in \mathcal{G}'')$, there holds

$$V(Q_1 x(Q_1)) \cap V(Q_2 x(Q_2)) = V(R)$$

whenever Q_1, Q_2 are distinct paths in $\mathcal{G}''$.

Suppose that (H) fails. Since every path zQ for Q in $\mathcal{G}'$ is finite, we may assume that all such paths in $\mathcal{G}'$ have the same length, m. For each $1 \le i \le m$ let $V_i = \{v_i(zQ) : Q \in \mathcal{G}'\}$ (thus $V_1 = \{z\}$). By the assumption on $\mathcal{G}'$ the set V_2 is uncountable. By the negation assumption on (H), $V_2 \cap V[\mathcal{J}^W]$ is countable, and hence $V_2' = V_2 \backslash V[\mathcal{J}^W]$ is uncountable. Suppose, if possible, that V_3 is countable. Then there exists a vertex $u \in V_3$ such that $u = v_3(zQ)$ for uncountably many paths Q in $\mathcal{G}'$. But then $(z, u) \in F$, contradicting the assumption that the paths in $\mathcal{G}_{n+1}$ are chordless. Thus V_3 is uncountable, and hence we can choose an uncountable subfamily $\mathcal{G}_3'$ of $\mathcal{G}'$ such that $V(Q_1 v_3(Q_1)) \cap V(Q_2 v_3(Q_2)) = V(R)$ for every pair Q_1, Q_2 of distinct paths in $\mathcal{G}_3'$. By the assumption that (H) fails $V_3 \cap V[\mathcal{J}^W]$ is countable, and hence $V_3' = \{v_3(zQ) : Q \in \mathcal{G}_3'\}$ is uncountable. Repeating this argument we finally obtain an uncountable subfamily $\mathcal{G}_m'$ of $\mathcal{G}'$, such that $V(Q) \cap V[\mathcal{J}^W] = \phi$ for any $Q \in \mathcal{G}_m'$, and $V(Q_1) \cap V(Q_2) = V(R)$ for any pair Q_1, Q_2 of distinct elements in $\mathcal{G}_m'$. Since $\mathcal{J}_R$ is finite, $\{Q \in \mathcal{G}_m' : V(zQ) \cap V[\mathcal{J}] \subseteq \{z\}\}$ is uncountable. Thus $\{zQ : Q \in \mathcal{G}_m', V(zQ) \cap V[\mathcal{J}] \subseteq \{z\}\}$ is an uncountable z-joined family of paths in $V \backslash (V[\mathcal{J}] \backslash \{z\})$ (which are therefore also $\mathcal{J}$-walks) from z to $\hat{B}$. But this clearly implies that $z \in \hat{B}$, a contradiction.

Let $\mathcal{G}''$ be as in (H), and for every $Q \in \mathcal{G}''$ let $T(Q)$ be the $\mathcal{J}^{T_n^\frown Q}y$-$n$-track $(y = Q^-(x))$ from $x(Q)$ to B, as in condition (E). We may assume that all tracks $T(Q)$ have the same number of joints, k. For each $Q \in \mathcal{G}''$ and odd i, $3 \le i \le k$, write:

$$I_i(Q) = zQx(Q)T(Q)j_i(T(Q)) \text{ and } H_i(Q) = j_{i-1}(T(Q))I_i(Q).$$

Let $\mathcal{K}_3 = \{H_3(Q) : Q \in \mathcal{G}''\}$. Since $FR(T(Q)) \cap F\langle z\rangle = \phi$ for any $Q \in \mathcal{G}''$ (see (E)), there holds $d_{\mathcal{K}_3}^-(v) \le \aleph_0$ for each vertex v.

Hence there exists an uncountable subfamily $\mathcal{G}_3''$ of $\mathcal{G}''$ such that $V(I_3(Q_1)) \cap V(I_3(Q_2)) = \{z\}$ for any $Q_1 \ne Q_2$ in $\mathcal{G}_3''$. Repeating this argument we eventually reach an uncountable subfamily $\mathcal{G}_k''$ of $\mathcal{G}''$ such that $V(I_k(Q_1)) \cap V(I_k(Q_2)) = \{z\}$ for any $Q_1 \ne Q_2$ in $\mathcal{G}_k''$. But this implies that $z \in S$, a contradiction. This proves the assertion.

We now return to the proof of the lemma. Let e_{n+1} be an enumeration of $\mathcal{G}_{n+1}$. Write $h(n + 1) = (k, \ell)$. Suppose first that $e_k(\ell)$ is undefined (i.e., $|\mathcal{G}_k| < \ell + 1$) or, writing $e_k(\ell) = Q$, there holds $V(Q) \cap V[W_n] \ne \phi$. Then let $T_{n+1} = T_n$ and $W_{n+1} = W_n$. Next consider the case that $Q = e_k(\ell)$ satisfies $V(Q) \cap V[W_n] = \phi$. If $\Gamma - V(T_n^\frown Q)$ is unhindered let $T_{n+1} = T_n^\frown Q$ and $W_{n+1} = W_n$. If, on the other hand, $\Gamma - V(T_n^\frown Q)$ is hindered, let x be the first vertex on Q such that $\Gamma - V(T_n^\frown Qx)$ is hindered. Let $T_{n+1} = T_n^\frown \overline{Qx}$. By [1, Lemma 2.8] there exists a wave $\mathcal{U}$ in $\Gamma - T_{n+1}$ such that $x \in \text{ter}[\mathcal{U}]$. Let $W_{n+1} = W_n \to \mathcal{U}$. By [1, Lemmas 2.3 and 2.4] W_{n+1} is a wave in $\Gamma - T_{n+1}$. By [1, Lemma 2.4] $\text{ter}[W_{n+1}]$ separates x from B in $\Gamma - T_n$.

Having constructed the trees T_n and waves W_n for all $n < \omega$, we define $T = \cup\{T_n : n < \omega\}$ and $W = \uparrow (W_n : n < \omega)$. By [1, Lemma 2.6] W is a wave in $\Gamma - T$.

Note that by our construction T is chordless.

Assertion b. There does not exist in $\hat{\Gamma}$ a W-evading $\mathcal{J}^T$-walk from $V(T)$ to $B \backslash V[\mathcal{J}^T]$.

Proof of Assertion b: Suppose that there exists a walk as in the assertion. Let K be such a walk having a minimal number of joints. Let $z = \text{in}(K)$ and $v = j_1(K)$. We may clearly assume that $V(K) \cap V(T) = \{z\}$. Write $Q = TzKv$, and let $\tilde{Q}$ be a chordless path in $\hat{\Gamma}$ from a to v such that $V(\tilde{Q}) \subseteq V(Q)$ and the vertices of $\tilde{Q}$ appear in the same order they have in Q. By replacing z, if necessary, we may assume that z is the last vertex on $\tilde{Q}$ belonging to $V(T)$. Since T is chordless, $\tilde{Q}z = Tz$. Since K is finite there exists $n < \omega$ for which $z \in V(T_n)$ and $\mathcal{J}_K \cap \mathcal{J}^{T_n} = \mathcal{J}_K \cap \mathcal{J}^T$. Then K is a $\mathcal{W}_n$-evading $\mathcal{J}^{T_n}$-walk from $V(T)$ to $B \backslash V[\mathcal{J}^{T_n}]$ (note that being $\mathcal{W}$-evading implies being $\mathcal{W}_n$-evading). Suppose that there exists a vertex $t \in FR(vK) \cap F\langle s \rangle$ for some $s \in V(\overline{\tilde{Q}v})$. Then $K' = Ts^*(s,t)^* tK$ is a walk satisfying the requirements in the assertion and having fewer joints than K, contradicting the assumption on K. Thus

$$FR(vK) \cap F[V(\overline{\tilde{Q}v})] = \phi.$$

Hence the track vK shows that $\tilde{Q}$ satisfies condition (E) at the $(n+1)$-st stage of the inductive construction, and thus $\tilde{Q} \in \mathcal{G}_{n+1}$. Thus $\tilde{Q} = e_{n+1}(\ell)$ for some ℓ. Let $m \geq n+1$ be such that $h(m) = (n+1, \ell)$. The fact that $x = \tilde{Q}^+(z) \notin V(T)$ implies that $T_{m+1} = T_m$. Since K is $\mathcal{W}$-evading, it is $\mathcal{W}_m$-evading and so

$$FR[K] \cap V[\mathcal{W}_m] = \phi.$$

Also, since $\mathcal{W}_m$ is a wave in $\Gamma - V(T_m)$, it follows that $\Gamma - V(T_m^\frown \tilde{Q}x)$ is hindered and $\mathcal{W}_{m+1} = (\mathcal{W}_m \rightarrow U)$, where U is a wave in $\Gamma - V(T_{m+1})$ such that $x \in \text{ter}(U)$. By [1, Lemma 2.4] $\text{ter}(\mathcal{W}_{m+1})$ separates x from B in $\Gamma - T$ (since $V(\mathcal{W}_{m+1}) \subseteq V(\mathcal{W})$). This contradicts the assumption on K, and the assertion is proved.

For each $J \in \mathcal{J}^T$ let $M(J)$ be the set of points reachable from T by a $\mathcal{W}$-evading $\mathcal{J}^T$-walk. For any path $Q \in \mathcal{W}$ such that $z = \text{ter}(Q) \in V(J)$ let $H(Q) = QzJw$, where w is the last vertex on zJ such that $w \in M(J)$ and $V(zJw) \cap V[\mathcal{W}] = \{z\}$, if such a vertex exists, and let $H(Q) = Q$ if no such vertex exists. Also, for any $Q \in \mathcal{W}$ such that $\text{ter}(Q) \notin V[\mathcal{J}^T]$ let $H(Q) = Q$. Let $\mathcal{K} = \{H(Q) : Q \in \mathcal{W}\}$.

Assertion c. $\text{ter}[\mathcal{K}]$ separates $V[\mathcal{W}]$ from B in $\Gamma - T$.

Proof of Assertion c: Let R be a path in $\Gamma - T$ from some vertex in $V[\mathcal{W}]$ to B. Since $\mathcal{W}$ is a wave in $\Gamma - T$, the last vertex u on R belonging to $V[\mathcal{W}]$ is in $\text{ter}[\mathcal{W}]$. Suppose that $u \notin \text{ter}[\mathcal{K}]$. Then $u \in V(J)$ for some $J \in \mathcal{J}^T$, and there exists a $\mathcal{W}$-evading $\mathcal{J}^T$-n-walk K from $V(T)$ to some vertex w on uJ, such that $V(uJw) \cap V[\mathcal{W}] = \{u\}$. Let t be the last vertex on R lying on some $\mathcal{W}$-evading $\mathcal{J}^T$-walk L, such that $\text{in}(L) \in V(T)$. Since K shows that u is such a vertex, t follows u on R, and hence there holds $V(tR) \cap V[\mathcal{W}] = \phi$. Consider the following cases:

1) $t \in BK(L)$ or $t \notin V[\mathcal{J}^T]$. Let $s = R^+(t)$ (note that by Assertion b $t \neq \text{ter}(R)$). Then s lies on the $\mathcal{W}$-evading $\mathcal{J}^T$-walk $LtRs$, contradicting the definition of t.

2) $t = j_{2k+1}(L)$ for some k. Let $I = \mathcal{J}_t$. Then, since $\mathcal{W}$ is a wave in $\Gamma - T$, there holds $V(ItR) \cap V[\mathcal{W}] \neq \phi$, and since $V(tR) \cap V[\mathcal{W}] = \phi$ it follows that $V(It) \cap \text{ter}[\mathcal{W}] \neq \phi$.

By the definition of $\mathcal{K}$ it follows that unless $t \in \text{ter}[\mathcal{K}]$ (in which case the assertion is proved), there exists a $\mathcal{W}$-evading $\mathcal{J}^T$-n-walk M from $V(T)$ to some vertex on tI. Let y be the first vertex on tI belonging to $FR(M)$. Then either $t \in BK(M)$ or $t \in BK(M')$, where $M' = My \overleftarrow{I} t$. In both possibilities we are back in case 1.

From Assertion c there immediately follows:

Assertion d. $\mathcal{K}$ is a wave in $\Gamma - T$.

Assertion e. $\mathcal{K}$ is a hindrance in Γ.

Proof of Assertion e: Let R be an $A - B$ path. If $V(R) \cap V(T) = \phi$ then $V(R) \cap \text{ter}[\mathcal{W}] \neq \phi$, since $\mathcal{W}$ is a wave in $\Gamma - T$. By Assertion c it then follows that $V(R) \cap \text{ter}[\mathcal{K}] \neq \phi$. So, assume that $V(R) \cap V(T) \neq \phi$, and let t be the last vertex on R belonging to $V(T)$. If $V(tR) \cap V[\mathcal{W}] \neq \phi$ then $V(tR) \cap \text{ter}[\mathcal{K}] \neq \phi$ by Assertion c. Hence we may assume that $V(tR) \cap V[\mathcal{W}] = \phi$. If $V(tR) \cap V[\mathcal{J}^T] = \phi$ then tR is a $\mathcal{W}$-evading $\mathcal{J}^T$-n-walk from T to $B \backslash V[\mathcal{J}^T]$, contradicting Assertion b. Hence we may assume that tR meets $\mathcal{J}^T$. Let u be the first vertex on tR such that $u \in V(I)$ for some $I \in \mathcal{J}^T$. Since $\mathcal{W}$ is a wave in $\Gamma - T$, the path IuR contains a vertex from ter$[\mathcal{W}]$, which must belong to $V(\overline{Iu})$. Hence, by the definition of $\mathcal{K}$, there holds $u \in V[\mathcal{K}]$. By Assertion d and [1, Lemma 2.1] it follows that uR contains a vertex from ter$[\mathcal{K}]$. We have thus shown that $\mathcal{K}$ is a wave in Γ. But $a \notin$ in$[\mathcal{K}]$, so $\mathcal{K}$ is, in fact, a hindrance.

Assertion e provides a contradiction to the assumption that Γ is unhindered, and thus the lemma is proved. As already explained, this completes the proof of the theorem.

Reference

[1]   R. Aharoni, Menger's theorem for countable graphs, *J. Combinatorial Theory Ser. B* **43**(1987), 303–313.

DECOMPOSITION INTO CYCLES I: HAMILTON DECOMPOSITIONS

B. ALSPACH

Department of Mathematics and Statistics
Simon Fraser University
Burnaby, B.C., V5A 1S6
Canada

J.-C. BERMOND, D. SOTTEAU
C.N.R.S., UA410 Informatique
Bât. 490
Université Paris-Sud
91405 Orsay
France

Abstract

In this part we survey the results concerning the partitions of the edge-set of a graph into Hamilton cycles or into Hamilton cycles and a single perfect matching.

By and large the terminology of [15] will be followed. The topic under discussion is the partitioning of the edge-set of a graph into Hamilton cycles. An obvious necessary condition for such a partition to exist is that the graph be regular of even degree. However, if a graph is regular of odd degree, then the closest one can come to such a partition is a partition of the edge-set into a single perfect matching (also called a 1-factor) and Hamilton cycles. The following definition is used in order to incorporate odd degree and even degree into a single definition.

Definition 1 Let G be a regular graph with edge-set $E(G)$. It is said to have a *Hamilton decomposition* (or be *Hamilton decomposable*) if either

(i) $\deg(G) = 2d$ and $E(G)$ can be partitioned into d Hamilton cycles, or

(ii) $\deg(G) = 2d + 1$ and $E(G)$ can be partitioned into d Hamilton cycles and a perfect matching.

---

*This article was written while the second author was visiting Simon Fraser University; support of Advanced Systems Foundations and Simon Fraser University is gratefully acknowledged.

G. Hahn et al. (eds.), Cycles and Rays, 9–18.
© *1990 by Kluwer Academic Publishers. Printed in the Netherlands.*

The earliest results about Hamilton decompositions deal with the complete graph K_n and are attributed to Walecki in [32].

Theorem 1 *The complete graph K_n has a Hamilton decomposition for all n.*

Proof: The following functional notation makes Walecki's construction easy to describe. If H is a subgraph of the graph K_n and $f : V(H) \to V(G)$ is a 1-1 function, then $f(H)$ denotes the subgraph of K_n whose vertex-set is $f(V(H))$ where $f(u)$ is adjacent to $f(v)$ in $f(H)$ if and only if u is adjacent to v in H. The result is trivially true for $n = 1$ and $n = 2$. Let $n = 2m + 1 \geq 3$ be odd. Let the vertices of K_n be labelled $u_0, u_1, u_2, \ldots, u_{2m}$, let C be the Hamilton cycle $u_0 u_1 u_2 u_{2m} u_3 u_{2m-1} u_4 u_{2m-2} \ldots u_{m+3} u_m u_{m+2} u_{m+1} u_0$ and let σ be the permutation $(u_0)(u_1 u_2 u_3 u_4 \ldots u_{2m-1} u_{2m})$. Then $C, \sigma(C), \sigma^2(C), \ldots, \sigma^{m-1}(C)$ is a Hamilton decomposition of K_n.

When $n = 2m \geq 4$ is even, let C be the Hamilton cycle

$$u_0 u_1 u_2 u_{2m-1} u_3 u_{2m-2} \ldots u_{m-1} u_{m+2} u_m u_{m+1} u_0$$

and let σ be the permutation $(u_0)(u_1 u_2 u_3 \ldots u_{2m-2} u_{2m-1})$. Then, $C, \sigma(C), \ldots, \sigma^{m-2}(C)$ are $m - 1$ edge disjoint Hamilton cycles. The remaining edges $u_0 u_m, u_{m-1} u_{m+1}, u_{m-2} u_{m+2}, \ldots, u_1 u_{2m-1}$ form a perfect matching. $\square$

Corollary 1 *The complete graph K_{2m} has a decomposition into Hamilton paths.*

Proof: Simply take a Hamilton decomposition of K_{2m+1} and remove a vertex and its incident edges from K_{2m+1}. $\square$

The above constructions of Walecki have been used in a wide variety of settings. One of them leads to the following problem. Consider the Walecki construction for the Hamilton decomposition of K_{2m+1}, $m \geq 1$. There are m Hamilton cycles each of which may be oriented in two ways to produce a Hamilton directed cycle. If each of the m Hamilton cycles is given an orientation, the resulting regular tournament is called a *Walecki tournament*. There are 2^m different orientation schemes but different orientations may produce isomorphic tournaments.

Problem 1 (Alspach) *Enumerate the Walecki tournaments of order $2m + 1$.*

The results about the complete graph lead one to consider just how large the degree of a regular graph must be before it is guaranteed to have a Hamilton decomposition.

Kotzig (unpublished) conjectured that any self-complementary regular graph of order $4k + 1$ has a Hamilton decomposition. This is a particular case of:

Conjecture 1 (Nash-Williams [35]) *Every $2k$-regular graph with at most $4k+1$ vertices has a Hamilton decomposition.*

A very strong result has been obtained recently by Häggkvist [20] who has proved that for any $p < 4$ every sufficiently large $2k$-regular graph with at most pk vertices has a Hamilton decomposition.

Most of the remaining results that deal with Hamilton decompositions of graphs involve various products of graphs. Since terminology for products is not uniform, the definitions are included.

The *cartesian product* $G_1 \times G_2$ of G_1 and G_2 (also called the cartesian sum) has vertex-set $V(G_1) \times V(G_2)$ with (u_1, u_2) adjacent to (v_1, v_2) if and only if either $u_1 = v_1$ and u_2 is adjacent to v_2 in G_2 or $u_2 = v_2$ and u_1 is adjacent to v_1 in G_1.

The *lexicographic product* $G_1 \otimes G_2$ of G_1 and G_2 (also called the composition, or tensor product, or wreath product) is the graph with vertex-set $V(G_1) \times V(G_2)$ and an edge joining (u_1, u_2) to (v_1, v_2) if and only if either u_1 is adjacent to v_1 in G_1 or $u_1 = v_1$ and u_2 is adjacent to v_2 in G_2.

The *conjunction* $G_1 \cdot G_2$ of G_1 and G_2 (also called the cartesian product) is the graph with vertex-set $V(G_1) \times V(G_2)$ and an edge joining (u_1, u_2) to (v_1, v_2) if and only if both u_1 and v_1 are adjacent in G_1 and u_2 and v_2 are adjacent in G_2.

The cartesian product has probably attracted the most attention and the following conjecture is the prominent unsolved problem.

Conjecture 2 (Bermond [10]) *If G_1 and G_2 have decompositions into Hamilton cycles, then so does $G_1 \times G_2$.*

The above conjecture is a general version of earlier conjectures and results. Kotzig [29] proved that $C_r \times C_n$ can be decomposed into two Hamilton cycles and conjectured that $C_r \times C_s \times C_n$ could be decomposed into three Hamilton cycles. The latter conjecture was solved by Foregger [19]. Kotzig also conjectured that if $G_1, G_2, \ldots, G_p$ can all be decomposed into n Hamilton cycles then $G_1 \times G_2 \times \ldots \times G_p$ can be decomposed into pn Hamilton cycles. Independent of any knowledge of Kotzig's conjecture, Myers [34] proved that $C_r \times C_r$ and $K_r \times K_r$ have Hamilton decompositions.

Ringel [36] earlier proved that the 2^n-cube, $n \geq 1$, has a Hamilton decomposition and posed the problem of whether or not the $2n$-cube has a Hamilton decomposition for $n \geq 1$ [36, Problem 2]. Some recent work disposes of Ringel's problem and unifies the above work.

Theorem 2 (Aubert and Schneider [5]) *If C is a cycle and G is a 4-regular graph which is decomposable into two Hamilton cycles, then $C \times G$ can be decomposed into three Hamilton cycles.*

The proof of Theorem 2 is constructive and has a handful of cases to consider. It has several corollaries, including the results below, which we now state.

Corollary 2 *If G_1 has a decomposition into n_1 Hamilton cycles and G_2 has a decomposition into n_2 Hamilton cycles, where $n_1 \leq n_2 \leq 2n_1$, then $G_1 \times G_2$ has a Hamilton decomposition.*

Proof: Let $\oplus$ denote edge disjoint union of spanning subgraphs. Then $G_1 = H_1 \oplus H_2 \oplus \ldots \oplus H_{n_1}$ and $G_2 = H_1' \oplus H_2' \oplus \ldots \oplus H_{n_2}'$ where the H_i's and H_i''s are Hamilton cycles. Then letting $n_2 = n_1 + d$,

$$G_1 \times G_2 = (H_1 \times (H_1' \oplus H_{n_1+1}')) \oplus (H_2 \times (H_2' \oplus H_{n_1+2}')) \oplus \ldots$$
$$\oplus (H_d \times (H_d' + H_{n_1+d}')) \oplus (H_{d+1} \times H_{d+1}') \oplus \ldots \oplus (H_{n_1} \times H_{n_1}').$$

Each of the terms in the preceding expression has a Hamilton decomposition by either Theorem 2 or Kotzig's result. The result then follows. □

Corollary 3 *The cartesian product* $C_{i_1} \times C_{i_2} \times \ldots \times C_{i_n}$ *has a Hamilton decomposition.*

Proof: This follows easily by induction on n. The initial values are handled by Kotzig's result and Theorem 2 or Foregger's result. For larger n write

$$C_{i_1} \times C_{i_2} \times \ldots \times C_{i_n} = (C_{i_1} \times \ldots \times C_{i_m}) \times (C_{i_{m+1}} \times \ldots \times C_{i_n})$$

where $m = \lfloor \frac{n}{2} \rfloor$. □

Proposition 1 *The n-cube has a Hamilton decomposition for all n.*

Proof: For even n, the n-cube is nothing more than $C_4 \times C_4 \times \ldots \times C_4$ so that the result follows from Corollary 3. For odd n, view the n-cube as two $(n-1)$-cubes joined by a perfect matching. Remove one edge from each Hamilton cycle of a Hamilton decomposition of the $(n-1)$-cube so that the removed eges have no vertices in common. Do the same for the other $(n-1)$-cube. Use the perfect matching to form Hamilton cycles in the n-cube. This can always be done [3] so that the n-cube has a Hamilton decomposition. □
From 9 it follows also that Kotzig's conjecture above is true.

Corollary 4 *If* $G_1, G_2, \ldots, G_p$ *can all be decomposed into n Hamilton cycles, then* $G_1 \times G_2 \times \ldots \times G_p$ *can be decomposed into pn Hamilton cycles.*

There has been some recent progress on Conjecture 2 by Stong [38]. His results are summarized below.

Theorem 3 *If* G_1 *has a decomposition into* n_1 *Hamilton cycles and* G_2 *has a decomposition into* n_2 *Hamilton cycles,* $n_1 \leq n_2$, *then* $G_1 \times G_2$ *has a Hamilton decomposition if any one of the following is true:*
  (i) $n_2 \leq 3n_1$,
  (ii) $n_1 \geq 3$,
  (iii) G_1 *has an even number of vertices, or*
  (iv) $|V(G_2)| \geq 6 \lceil \frac{n_2}{n_1} \rceil - 3$.

Aubert and Schneider also proved the following result [6].

Theorem 4 *The cartesian product* $K_n \times K_m$ *has a Hamilton decomposition for all n and m.*

Alspach and Rosenfeld [4] studied the cartesian product of a cubic graph and K_2 for Hamilton decomposability. They give a necessary and sufficient condition which implies that $G \times K_2$ has a Hamilton decomposition if G is cubic and has a perfect 1-factorization, that is, G has a 1-factorization such that the union of any pair of 1-factors is a Hamilton cycle. They asked for other conditions on a cubic graph G that guarantee that $G \times K_2$ has a Hamilton decomposition. The following example is one condition that one might suggest.

Question 1 (Alspach and Rosenfeld [4]) *If G is a 3-connected cubic graph having a Hamilton cycle, does* $G \times K_2$ *have a Hamilton decomposition?*

A first result about the lexicographic product is the following.

Theorem 5 (Auerbach and Laskar [8] and Hetyei [21]) *The complete multipartite graph $K_{r;s}$ has a Hamilton decomposition.*

A general study of Hamilton decomposition of complete regular s-partite graphs has been done by Hilton and Roger [22].

The analogue of Conjecture 2 for lexicographic products asked in [10] has been solved.

Theorem 6 (Baranyai and Szasz [9]) *If G_1 can be decomposed into m_1 Hamilton cycles, G_2 can be decomposed into m_2 Hamilton cycles and G_2 has n vertices, then $G_1 \otimes G_2$ can be decomposed into $m_2 + m_1 n$ Hamilton cycles.*

The preceding result supercedes several special cases that had been proved earlier. For example, the statement of the theorem does not indicate what happens in the degenerate case that $G_2 = \overline{K}_n$ (the complement of K_n). The result is still true in this case and follows easily from the following result.

Theorem 7 (Laskar [31] and Hetyei [21]) *The graph $C_r \otimes \overline{K}_n$ has a Hamilton decomposition.*

However, the generalization of Theorem 7 to arbitrary Hamilton decomposable graphs is unsettled. That is, if G_1 is connected and has a decomposition into Hamilton cycles and a perfect matching, it is not known whether or not $G_1 \otimes G_2$ has a Hamilton decomposition. On the other hand, if G_1 has a decomposition into one or more Hamilton cycles and G_2 is Hamilton decomposable, then $G_1 \otimes G_2$ has a Hamilton decomposition.

In [10] Bermond has shown that the following two results hold.

Theorem 8 *The conjunction of two cycles $C_r \cdot C_s$ can be decomposed into two Hamilton cycles if and only if at least one of r and s is odd.*

Corollary 5 *If both G_1 and G_2 can be decomposed into Hamilton cycles and one of G_1 or G_2 has odd order, then the conjunction $G_1 \cdot G_2$ has a Hamilton decomposition.*

Many of the special cases discussed above are, in fact, Cayley graphs on abelian groups. This led Alspach [2] to ask the following question.

Question 2 (Alspach [2]) *Does every connected Cayley graph on an abelian group have a Hamilton decomposition?*

If the degree of the graph is 2, then the answer is obviously yes. If the degree is 3, the answer is again yes since such a graph has a Hamilton cycle as proved by many authors (see the survey of Witte and Galian [42]). The case of degree 4 recently has been solved by Bermond, Favaron and Maheo [13] and the answer is again yes. The answer is also yes for degree 5 [3]. In the case of degree 4, when the group is the additive group of integers modulo n, the Cayley graphs are known as double loop graphs and the problem was asked by Bermond, Illiades and Peyrat [14]. More generally the construction of graphs admitting a decomposition into k Hamilton cycles and having the smallest diameter possible is an open problem arising in the context of computer loop networks (see the survey by Bermond, Comellas and Hsu [12]).

There is one positive result about the more general class of vertex-transitive graphs.

Theorem 9 (Alspach [1]) *Every connected vertex-transitive graph of order $2p$, p prime, $p \equiv 3(mod\ 4)$ has a Hamilton decomposition.*

The graph O_k (called the *odd graph*) is a k-regular graph whose vertices are indexed by the $(k-1)$-subsets of a $(2k-1)$-set. Two vertices are adjacent if and only if their indexing subsets are disjoint. For example, O_3 is the Petersen graph. Meredith and Lloyd [33] proved that O_4, O_5 and O_6 admit a Hamilton decomposition and more generally they conjectured.

Conjecture 3 [33] *The odd graph O_m has a Hamilton decomposition, for $m \geq 4$.*

The graph B_k (called boolean graph) is a k-regular bipartite graph with one part consisting of vertices by the $(k-1)$-subsets of a $(2k-1)$-set, and the other part of vertices indexed by its k-subsets. Adjacency is given by containment. Notice that in other words B_k is the subgraph induced by the middle two levels of the boolean lattice representation of a $(2k-1)$-cube.

The boolean graphs are conjectured to be hamiltonian. However it may be interesting to notice that, for $k = 2^m$, $m \geq 1$, if O_k has a Hamilton decomposition, then so does B_k (see [17] for more details).

There are some more general decomposition problems involving Hamilton cycles that should be mentioned at this time.

Conjecture 4 (Kotzig [28]) *The complete graph K_{2m} has a perfect 1-factorization for all $m \geq 2$.*

This conjecture for some time has been known to be true for $2m = p+1$ and $2p$, p a prime, and for $2m = 16, 28, 244$ and 344. Recently, Seah and Stinson [37] have shown it is true for 36, Ihrig, Seah and Stinson [24] have shown it is true for 50, and Kobayashi and Kiyasu-Zen'iti [27] have shown it is true for 1,332 and 6,860. The following result is proved in [18] for other reasons but it is related to Conjecture 4.

Theorem 10 *There is a 1-factorization $\{I_1, I_2, \ldots, I_{2m-1}\}$ of K_{2m} such that $I_1 \cup I_j$ is a Hamilton cycle for all $j \geq 2$.*

There are two conjectures related to Theorem 10 given in [18]. One of them conjectures that if $I'_1, I'_2, \ldots, I'_{2m-2}$ are any $2m-2$ 1-factors of K_{2m}, then there is a 1-factorization $\{I_1, I_2, \ldots, I_{2m-1}\}$ of K_{2m} such that $I_i \cup I'_i$ is a Hamilton cycle for $i = 1, \ldots, 2m-2$. The second, and weaker, conjecture is the same except that the conclusion is that $I_i \cap I'_i = \emptyset$ for all $i = 1, \ldots, 2m-2$.

Definition 2 A *Kotzig factorization* of K_{2m+1} consists of a decomposition into Hamilton cycles $H_1, H_2, \ldots, H_m$ and a decomposition into m-matchings $M_1, M_2, \ldots, M_{2m+1}$ such that $|E(H_i) \cap E(M_j)| = 1$ for all i and j.

Theorem 11 (Horton [23]) *The complete graph K_{2m+1} has a Kotzig factorization for all $m \geq 1$.*

A Kotzig factorization is a special case of a more general factorization considered in [3] and discussed by Alspach at the "Cycles in Graphs 1982" workshop held at Simon Fraser University in July 1982. Let $\mathcal{F} = \{F_1, F_2, \ldots, F_r\}$ be any 2-factorization of a $2r$-regular graph G. An r-matching M is said to be *orthogonal* to $\mathcal{F}$ if $|M \cap F_i| = 1$ for $i = 1, 2, \ldots, r$. Alspach asked the following two questions at the workshop. (Also see Discrete Math. 69 (1988), p.106).

Question 3 *If $F_1, F_2, \ldots, F_r$ is any 2-factorization of a $2r$-regular graph G, does there always exist an orthogonal r-matching?*

Question 4 *What conditions on a 2-factorization $\mathcal{F}$ of a $2r$-regular graph G guarantee that the edge-set of G can be partitioned into r-matchings each of which is orthogonal to $\mathcal{F}$?*

Recently Kouider and Sotteau [30] gave a positive answer to Question 3 for $2r$-regular graphs G of order $m \geq 3.23r$.

Jaeger [25] proved that if G can be decomposed into an even number of Hamilton cycles, then its line graph $L(G)$ is 1-factorable. The key of the proof lies in the following result: If G can be decomposed into two Hamilton cycles, then its line graph $L(G)$ can be decomposed into three Hamilton cycles. Kotzig proved that if G is a cubic Hamilton graph, then its line graph can be decomposed into two Hamilton cycles. That leads to the following conjecture.

Conjecture 5 (Bermond [11]) *If G has a Hamilton decomposition, then its line graph can be decomposed into Hamilton cycles.*

In a personal letter, Jackson has asked a question for which an affirmative answer implies Conjecture 5. A *separating transition* in a graph G is a set of two adjacent edges e_1 and e_2 such that $G - \{e_1, e_2\}$ is disconnected. If G is a $2r$-regular graph, a necessary condition for $L(G)$ to have a Hamilton decomposition is that G have no separating transitions. Jackson has asked whether or not this condition is also sufficient. He has also conjectured that a 3-connected, $2r$-regular graph has a line-graph which is Hamilton decomposable. He has verified the latter conjecture when $r = 2$.

Also let us mention that Thomason [39] has proved that if a $2k$-regular graph has a Hamilton decomposition, then it has at least $(3k - 2)(3k - 5) \ldots 7 \cdot 4$ Hamilton decompositions. Colbourn [16] gave all the non-isomorphic Hamilton decompositions of K_n for $n = 7, 9$ and those which have a non-trivial automorphism group for $n = 11$.

Finally let us conclude with some results concerning Hamilton decompositions of directed graphs. Some of these decompositions are obtained from the decompositions of the underlying undirected graph in particular when the directed graph is symmetric by associating to each Hamilton cycle of the underlying graph two opposite directed Hamilton cycles. For example from Theorem 1, we have that the complete symmetric directed graph K_{2n+1}^* has a decomposition into $2n$ directed Hamilton cycles. The following strong result cannot be obtained in that way. (For earlier results see the survey [10]).

Theorem 12 (Tillson [41]) *If $2n \geq 8$, K_{2n}^* can be decomposed into $2n - 1$ directed Hamilton cycles.*

Aubert and Schneider [7] studied the decompositions into directed Hamilton cycles of the symmetric digraph G^* associated to a cubic graph G consisting of a cycle plus a perfect matching. In particular they proved that if the number of vertices of G (which must be even) is a multiple of 4 then G^* has no Hamilton decomposition.

Recent results have been obtained concerning products of directed graphs and tournaments.

A characterization of the Cayley digraphs of finite abelian groups which can be decomposed into two directed Hamilton cycles has been given by Keating [26]. That implies that the cartesian product of two directed cycles C_r and C_s can be decomposed into directed Hamilton cycles if and only if there exist positive integers u and v such that $u+v = gcd(r,s)$ and $gcd(uv, rs) = 1$ (result first obtained by Lindgren). In [26] can also be found results on the conjunction of two digraphs. The conjecture that if the two digraphs G_1 and G_2 can be decomposed into directed Hamilton cycles then their lexicographic product can also be decomposed into directed Hamilton cycles is still open (see [9] for partial results). Even the analogous version of Theorem 5 for complete multipartite digraphs is still open.

Substantial progress has been made on Kelly's conjecture that every regular tournament can be decomposed into directed Hamilton cycles by Thomassen [40] who proved that every regular tournament of order n can be covered by a collection of $12n$ directed Hamilton cycles and by Häggkvist [20] who proved that the conjecture is true for sufficiently large regular tournaments. Furthermore he proved that every sufficiently large k-regular digraph with at most $2k$ vertices has a Hamilton decomposition.

References

[1]   B. Alspach, Hamiltonian partitions of vertex-transitive graphs of order $2p$, *Congressus Numerantium* **28**(1980) 217–221.

[2]   B. Alspach, Research Problem 59, *Discrete Math.* **50**(1984), 115.

[3]   B. Alspach, K. Heinrich and G. Liu, Orthogonal factorizations of graphs, preprint.

[4]   B. Alspach and M. Rosenfeld, On Hamilton decompositions of prisms over simple 3-polytopes, *Graphs and Combinatorics* **2**(1986), 1–8.

[5]   J. Aubert and B. Schneider, Décompositions de la somme cartésienne d'un cycle et de l'union de 2 cycles Hamiltoniens, *Discrete Math.* **38**(1982), 7–16.

[6]   J. Aubert and B. Schneider, Décomposition de $K_m + K_n$ en cycles Hamiltoniens, *Discrete Math.* **37**(1981), 19–27.

[7]   J. Aubert and B. Schneider, Graphes orientés indécomposables en circuits Hamiltoniens, *J. Combin. Theory Ser. B* **32**(1982), 347–349.

[8]   B. Auerbach and R. Laskar, On decompositions of r-partite graphs into edge-disjoint Hamiltonian circuits, *Discrete Math.* **14**(1976) 265–268.

[9]   Z. Baranyai and Gy. R. Szasz, Hamiltonian decomposition of lexicographic product, *J. Combin. Theory Ser. B* **31**(1981), 253–261.

[10] J.-C. Bermond, Hamilton decomposition of graphs, directed graphs and hypergraphs, in *Advances in Graph Theory, Annals of Discrete Mathematics* **3**(1978), 21–28.

[11] J.-C. Bermond, Problem 97, *Discrete Math.* **71**(1988), 275–276.

[12] J.-C. Bermond, F. Comellas and F. Hsu, Distributed loop computer networks, a survey, submitted IEEE Trans. on Computers.

[13] J.-C. Bermond, O. Favaron and M. Maheo, Hamiltonian decomposition of Cayley graphs of degree four, *J. Combin. Theory Ser. B*, to appear.

[14] J.-C. Bermond, G. Illiades and C. Peyrat, An optimization problem in distributed loop computer networks, in *Proc. Coll. 3rd Int. Conf. on Combinatorial Mathematics*, New York, June 1985, Annals New York Acad. Sci., to appear.

[15] J. A. Bondy and U. S. R. Murty, *Graph Theory with Applications*, MacMillan Press, 1976.

[16] C. J. Colbourn, Hamiltonian decompositions of complete graphs, *Ars Combinatoria* **14**(1982) 261–270.

[17] D. Duffus, P. Hanlon and R. Roth, Matching and Hamiltonian Cycles in Some Families of Symmetric Graphs, preprint.

[18] H. Fleischner, A. J. W. Hilton and B. Jackson, On the maximum number of pairwise compatible Euler cycles, preprint.

[19] M. F. Foregger, Hamiltonian decompositions of product of cycles, *Discrete Math.* **24**(1978) 251–260.

[20] R. Häggkvist, Seminar Orsay 1986.

[21] G. Hetyei, On Hamilton circuits and 1-factors of the regular complete n-partite graphs, *Acta Acad. Pedagog., Civitate Press Ser.* 6, **19**(1975), 5–10.

[22] A. J. W. Hilton and C. A. Rodger, Hamiltonian decompositions of complete regular r-partite graphs, *Discrete Math.* **58**(1986), 63–78.

[23] J. D. Horton, The construction of Kotzig factorizations, *Discrete Math.* **43**(1983), 199–206.

[24] E. Ihrig, E. Seah and D. Stinson, A perfect one-factorization of K_{50}, *J. Combinatorial Math. and Combinatorial Computing* **1**(1987), 217–219.

[25] F. Jaeger, The 1-factorization of some line graphs, *Discrete Math.* **46**(1983), 89–92.

[26] K. Keating, Multiple-ply Hamiltonian graphs and digraphs, in *Cycles in graphs, Annals of Discrete Mathematics* **27**(1985), 81–88.

[27] M. Kobayashi and Kiyasu-Zen'iti, Perfect 1-factorization of K_{1332} and K_{6860}, preprint.

[28] A. Kotzig, Problem in Theory of Graphs and its Applications, Academic Press 1963, 162 and 63–82.

[29] A. Kotzig, Every cartesian product of two circuits is decomposable into two Hamiltonian circuits, Rapport 233, Centre de Recherches Mathématiques, Montréal 1973.

[30] M. Kouider and D. Sotteau, On the existence of a matching orthogonal to a 2-factorization, to appear in *Discrete Math.*

[31] R. Laskar, Decomposition of some composite graphs into Hamiltonian cycles, in *Proc. 5th Hungarian Coll. Keszthely 1976*, North Holland, 1978, 705–716.

[32] D. E. Lucas, Recréations Mathématiques, Vol. II. Gauthier Villars, Paris, 1892.

[33] G. H. J. Meredith and E. K. Lloyd, The footballers of Croam, *J. Combin. Theory Ser. B* 15(1973), 161–166.

[34] B. R. Myers, Hamiltonian factorization of the product of a complete graph with itself, *Networks* 2(1972) 1–9.

[35] C. St. J. A. Nash-Williams, Hamiltonian arcs and circuits, in *Recent Trends in Graph Theory, Lecture Notes in Math.* 185, Springer Verlag (1971) 197–210.

[36] G. Ringel, Über drei kombinatorische Probleme am n-dimensionalen Würfel and Würfelgitter, *Abh. Math. Sem. Univ. Hamburg* 20(1956), 10–19.

[37] E. Seah and D. Stinson, A perfect one factorization of K_{36}.

[38] F. Stong, On Kotzig's conjecture, preprint.

[39] A. G. Thomason, Hamiltonian cycles and uniquely edge colourable graphs, in *Annals of Discrete Mathematics* 3(1978), 259–268.

[40] C. Thomassen, Hamilton circuits in regular tournaments, in *Cycles in Graphs, Annals of Discrete Mathematics* 27(1985), 159–162.

[41] T. Tillson, A Hamiltonian decomposition of K_{2n}, $2n \geq 8$, *J. Combin. Theory Ser. B* 29(1980), 68–74.

[42] D. Witte and J. A. Gallian, A survey: Hamiltonian cycles in Cayley graphs, *Discrete Math.* 51(1984), 293–304.

AN ORDER- AND GRAPH-THEORETICAL CHARACTERISATION OF WEAKLY COMPACT CARDINALS

G.D. BADENHORST, T. STURM
Department of Mathematics
University of Natal
King George V Avenue
Durban 4001
South Africa

Abstract

An uncountable cardinal m is weakly compact if every ordered set P satisfies the condition: If every chain and antichain of P is of cardinality $< m$, then $|P| < m$.

We assume throughout that m is a cardinal. We recall that a *chain* in an ordered set $A = (A', \leq_A)$ is a subset of A' such that any two elements are comparable, while an *antichain* in A is a set of pairwise incomparable elements. We introduce two cardinal predicates P and Q.

$P(m)$: Every ordered set $A = (A', \leq_A)$ of cardinality $|A'| = m$ contains a chain of size m or an antichain of size m.

$Q(m)$: Every graph $G = (V(G), E(G))$ of cardinality $|V(G)| = m$ contains a clique of size m or an independent set of size m.

It is obvious that $P(m)$ and $Q(m)$ for $m < 3$ and that non $P(m)$ and non $Q(m)$ for $3 \leq m < \omega_0$ or for infinite, singular m's. It was shown by Dushnik and Miller [2, Theorem 5.24] that $P(\omega_0)$ and non $P(\omega_1)$. We have $Q(\omega_0)$ by the classical Ramsey Theorem (see e.g., [1, Corollary 8.8]).

We will give a full characterisation of P and Q. Recall that m is said to be *weakly compact* if $m > \omega_0$ and every edge bicoloration of the complete graph K_m has a monochromatic m-clique. An equivalent formulation (see e.g., [1, Theorem 8.23]): m is weakly compact iff m is uncountable, and for every linear order $\leq$ on m, there is an $A \subseteq m$ such that $|A| = m$ and either the chain $(A, \leq \cap A^2)$ or its dual $(A, \leq^{-1} \cap A^2)$ is a well ordered set.

Theorem 1 *The following conditions are equivalent:*
- *(1)* m *is a weakly compact cardinal;*
- *(2)* $m > \omega_0$ *and* $P(m)$*;*
- *(3)* $m > \omega_0$ *and* $Q(m)$*.*

Proof: $(3 \Rightarrow 2)$: Let A be an ordered set. Take the graph $G = (A', \leq \cup \leq^{-1})$. An $X \subseteq A'$ is a chain (or an antichain) of A iff X is a clique (or, respectively, an independent subset) of G. Now, the implication is clear.

G. Hahn et al. (eds.), Cycles and Rays, 19–20.

$(2 \Rightarrow 1)$: We will employ Dushnik and Miller's idea [2, pp. 608–609], mutatis mutandis. Assume $m > \omega_0$ and m not weakly compact. Then there exists a chain $(m, \leq)$ such that we have, for every $X \subseteq m$, if $|X| = m$ then neither $(X, \leq \cap X^2)$ not $(X, \leq^{-1} \cap X^2)$ is a well ordered set.

Take $\prec \overset{\mathrm{Def}}{=} \subseteq \cap \leq$, it is an order on m. Let C be a chain of $(m, \prec)$, then C is a subchain of the well ordered set $(m, \subseteq)$, hence it is a well ordered subset of $(m, \leq)$. Thus $|C| < m$.

Since $x, y \in m$ are $\prec$-incomparable iff

$$\text{either } x < y \text{ and } y \subset x \ , \text{ or } y < x \text{ and } x \subset y,$$

every antichain A of $(m, \prec)$ is a well ordered subset of $(m, \leq^{-1})$; especially $|A| < m$. Thus we have non $P(m)$.

$(1 \Rightarrow 3)$ follows immediately from the definition of weak compactness. $\square$

Corollary 2 $P = Q$.

Corollary 3 $P(m)$ *if and only if* $m \in \{0, 1, 2, \omega_0\}$ *or* m *is weakly compact.*

Acknowledgement

The predicate Q was introduced thanks to a remark of Professor H.A. Jung. The second author was partially supported by NSERC Canada Grant A-5047, FCAR Québec EQ-0539, and South African FRD Grant 883-474-10.

References

[1]   W. W. Comfort and S. Negrepontis: The Theory of Ultrafilters. Springer-Verlag, Berlin 1974.

[2]   B. Dushnik and E. W. Miller: Partially Ordered Sets. *Amer. J. Math.* **63**(1941), 600–610.

SMALL CYCLE DOUBLE COVERS OF GRAPHS

J.A BONDY
Department of Combinatorics and Optimization
University of Waterloo
Waterloo
Ontario, N2L 3G1
Canada

Abstract

A *cycle double cover* (CDC) of a graph G is a collection $\mathcal{C}$ of cycles of G such that each edge of G belongs to exactly two members of $\mathcal{C}$. P.D. Seymour has conjectured that every 2-edge-connected graph admits a CDC. The purpose of this article is to introduce a refinement of Seymour's conjecture in which the number of cycles in the CDC plays a significant role. We define a *small cycle double cover* (SCDC) of a graph G to be a CDC $\mathcal{C}$ of G such that $|\mathcal{C}| \leq n - 1$, where n is the number of vertices in G, and conjecture that every simple 2-edge-connected graph admits an SCDC. This conjecture and its many ramifications are discussed.

1 The Small Cycle Double Cover Conjecture

We begin by recalling a theorem on the decomposition of the edge set of an even graph into cycles, implicit in the work of Euler [12].

Definition 1.1 A graph G is *even* if each vertex of G is of even degree.

Definition 1.2 A *cycle decomposition* (CD) of a graph G is a set $\mathcal{C}$ of cycles of G such that each edge of G belongs to exactly one member of $\mathcal{C}$.

Theorem 1.1 (L. Euler) *Every even graph admits a CD.*

How can this theorem be extended to graphs with vertices of odd degree? A natural way to overcome the discrepancy in parity is to consider cycle double covers instead of cycle decompositions.

Definition 1.3 A *cycle double cover* (CDC) of a graph G is a collection $\mathcal{C}$ of cycles of G such that each edge of G belongs to exactly two members of $\mathcal{C}$.

G. Hahn et al. (eds.), Cycles and Rays, 21–40.
© *1990 by Kluwer Academic Publishers. Printed in the Netherlands.*

Example 1.1 Two CDC's of K_4:

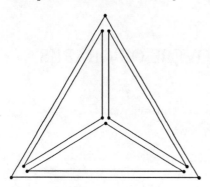 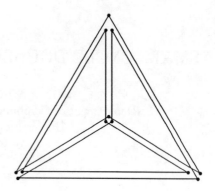

Remark 1.1 Let G be a connected graph.

(i) If G admits a CDC, G is 2-edge-connected (because each edge of G lies in a cycle).

(ii) If G is 2-edge-connected *and planar*, G admits a CDC (consisting of the facial cycles in some plane representation of G).

Szekeres [36] and Seymour [34], motivated by different considerations, each proposed the following conjecture.

Conjecture CDC (The Cycle Double Cover Conjecture) *Every 2-edge-connected graph admits a CDC.*

Remark 1.2 This conjecture has been studied extensively; an excellent reference is the survey article by Jaeger [26]. Classes of graphs for which the conjecture has been verified include planar graphs (by Remark 1.1(ii)), 3-edge-colourable 3-regular graphs (because each pair of colours induces a 2-factor, and the cycles of the resulting three 2-factors constitute a CDC), 4-edge-connected graphs (by a theorem of Jaeger [25]), and graphs which have Hamilton paths (by a theorem of Tarsi [37]; see also Goddyn [18]). Also, it has been shown that every 2-edge-connected graph admits both a cycle quadruple cover (Bermond, Jackson and Jaeger [1]) and a cycle sextuple cover (Fan [14]).

Before introducing our refinement of Conjecture CDC, we recall a conjectured refinement of Theorem 1.1 due to Hajós (see Lovász [28]).

Definition 1.4 A *small cycle decomposition* (SCD) of a graph G is a CD $\mathcal{C}$ of G such that $|\mathcal{C}| \leq (n-1)/2$, where $n = |V(G)|$.

Conjecture SCD (G. Hajós) *Every simple even graph admits an SCD.*

Remark 1.3 The statement in [28] of Conjecture SCD is marginally weaker than the one given here, the bound on $|\mathcal{C}|$ being $n/2$ instead of $(n-1)/2$. However, Dean [7] has observed that the two are, in fact, equivalent.

Our conjecture bears the same relationship to Conjecture CDC as does Conjecture SCD to Theorem 1.1.

Definition 1.5 A *small cycle double cover* (SCDC) of a graph G on n vertices is a CDC $\mathcal{C}$ of G such that $|\mathcal{C}| \leq n - 1$.

Conjecture SCDC (The Small Cycle Double Cover Conjecture) *Every simple 2-edge-connected graph admits an SCDC.*

Remark 1.4 The complete graph K_n shows that this conjecture, if true, is sharp: to double cover the $n(n-1)/2$ edges of K_n requires at least $n-1$ cycles, since each is of length at most n. A much more general class of extremal graphs for Conjecture SCDC is the class of trigraphs, which we now define.

Definition 1.6 A *tritree* of a graph G is a spanning tree T of G such that every fundamental cycle of G with respect to T is a triangle.

Definition 1.7 A *trigraph* is a graph which admits a tritree.

Example 1.2 A spanning star of a simple graph is a tritree, called a *star tritree*.

Example 1.3 The square of a tree is a trigraph.

Definition 1.8 Let G be a trigraph and T a tritree of G. A cycle C of G is a *T-star cycle* of G if there is a vertex $u \in V(G)\backslash V(C)$ such that $uv \in E(T)$ for all $v \in V(C)$.

Trigraphs are discussed in detail in [4]. The following results are proved there.

Proposition 1.1 *Let G be a trigraph, C a cycle of G and T a tritree of G. Then*

$$|E(C) \cap E(T)| = \begin{cases} 0 & \text{if } C \text{ is a } T\text{-star cycle of } G, \\ 2 & \text{otherwise} \end{cases}$$

Corollary 1.1 *Let G be a trigraph on n vertices and $\mathcal{C}$ a CDC of G. Then*

$$|\mathcal{C}| \geq n - 1,$$

with equality if and only if

$$|E(C) \cap E(T)| = 2 \quad \text{for all } C \in \mathcal{C} \text{ and all } T \in \mathcal{T},$$

where $\mathcal{T}$ is the set of tritrees of G.

Corollary 1.1 shows that every trigraph is an extremal graph for Conjecture SCDC. We conjecture that, conversely, every extremal graph for Conjecture SCDC is a trigraph.

Conjecture 1.1 *Let G be a simple 2-edge-connected graph on n vertices. If*

$$|\mathcal{C}| \geq n - 1 \quad \text{for every } CDC \ \mathcal{C} \text{ of } G,$$

then G is a trigraph.

In Section 4, we shall see some evidence in support of this conjecture.

Before considering several special cases of Conjecture SCDC, we make a few elementary observations.

Proposition 1.2 *The complete graph K_n, $n \geq 3$, and the complete bipartite graph $K_{p,q}$, $q \geq p \geq 2$, admit SCDCs.*

Proof: For $n = 2m + 1 \geq 3$, the cycles

$$v v_i v_{i+1} v_{i-1} v_{i+2} v_{i-2} \cdots v_{i+m-1} v_{i-m+1} v_{i+m} v, \ 0 \leq i \leq 2m - 1,$$

where subscripts are reduced modulo $2m$, form an SCDC of K_n.
For $n = 2m \geq 4$, the cycles

$$v v_i v_{i+1} v_{i-1} v_{i+2} v_{i-2} \cdots v_{i+m-1} v_{i-m+1} v, \ 0 \leq i \leq 2m - 2,$$

where subscripts are reduced modulo $2m - 1$, form an SCDC of K_n.
For $q \geq p \geq 2$, the cycles

$$u_0 v_i u_1 v_{i+1} \cdots u_{p-1} v_{i+p-1} u_0, \ 0 \leq i \leq q - 1,$$

where subscripts are reduced modulo q, form an SCDC of $K_{p,q}$.

As is the case with Conjecture CDC, Conjecture SCDC can easily be reduced to 3-connected cyclically 4-edge-connected graphs.

Proposition 1.3 *If every simple 3-connected cyclically 4-edge-connected graph admits an SCDC, then every simple 2-edge-connected graph admits an SCDC.*

Proof: Suppose that every simple 3-connected cyclically 4-edge-connected graph admits an SCDC. We prove, by induction on the number n of vertices, that every simple 2-edge-connected graph admits an SCDC. This is clearly so if $n = 3$. Let G be a simple 2-edge-connected graph on n vertices, where $n \geq 4$. We may assume that G is not both 3-connected and cyclically 4-edge-connected.

If G has a 1-vertex cut $\{v\}$, then $G = G_1 \cup G_2$, where $G_1 \cap G_2 = \{v\}$ and $|V(G_i)| = n_i < n$, $i = 1, 2$. By the induction hypothesis, G_i admits an SCDC C_i, $i = 1, 2$. Thus G admits the CDC $C = C_1 \cup C_2$, where

$$|C| = |C_1| + |C_2| \leq (n_1 - 1) + (n_2 - 1) = n - 1.$$

If G has a 2-vertex cut $\{u, v\}$, then $G = G_1 \cup G_2$, where $V(G_1) \cap V(G_2) = \{u, v\}$ and $|V(G_i)| = n_i < n$, $i = 1, 2$. By the induction hypothesis, $G_i + uv$ admits an SCDC C_i, $i = 1, 2$. Let C_i^j, $j = 1, 2$, be the two cycles in C_i which include the edge uv, $i = 1, 2$. If $uv \in E(G)$, define

$$C = C_1 \cup C_2 \cup \{C_1^1 \triangle C_2^1\} \setminus \{C_1^1, C_2^1\}.$$

If $uv \notin E(G)$, define

$$C = C_1 \cup C_2 \cup \{C_1^1 \triangle C_2^1, \ C_1^2 \triangle C_2^2\} \setminus \{C_1^1, C_2^1, C_1^2, C_2^2\}.$$

In both cases, C is an SCDC of G.

Thus we may assume that G is 3-connected but not cyclically 4-edge-connected. It follows that

$$G = G_1 \cup G_2 \cup \{x_1 y_1, \ x_2 y_2, \ x_3 y_3\},$$

where $G_1 \cap G_2 = \emptyset$, the x_i are distinct vertices of G_1, and the y_i are distinct vertices of G_2. Denote by H_1 the graph obtained from G by contracting the subgraph G_2 to a single vertex z_1, and by H_2 the graph obtained from G by contracting the subgraph G_1 to a single vertex z_2. By the induction hypothesis, H_i admits an SCDC C_i, $i = 1, 2$. Let C_i^j, $j = 1, 2, 3$, be the three cycles in C_i which include z_i, $i = 1, 2$, where we assume that C_1^j does not include $z_1 x_j$ and C_2^j does not include $z_2 y_j$, and let $P_i^j = C_i^j - z_i$, $i = 1, 2$, $j = 1, 2, 3$. Define

$$C' = \{P_1^j \cup P_2^j \cup \{x_i y_i : i \neq j\} : j = 1, 2, 3\}$$

and

$$C'' = \{C_i^j : i = 1, 2, j = 1, 2, 3\}.$$

Then

$$C = (C_1 \cup C_2 \cup C') \setminus C''$$

is an SCDC of G.

Remark 1.5 Conjecture SCDC can not be reduced to 3-regular graphs (as can Conjecture CDC). Indeed, in Section 5, we discuss a strengthening of Conjecture SCDC for 3-regular graphs.

2 Planar Graphs

We have observed that the facial cycles in a planar embedding of a 2-edge-connected graph constitute a CDC of G. If G has not too many edges, Euler's Formula implies that this CDC will be small. More precisely:

Proposition 2.1 *Let G be a 2-edge-connected planar graph on n vertices and at most $2n - 3$ edges. Then the facial cycles in a planar embedding of G constitute an SCDC of G.*

Simple planar graphs on n vertices may, of course, have as many as $3n - 6$ edges, and it is not evident that every such graph admits an SCDC. Thus the restriction of Conjecture SCDC to planar graphs is worthy of special mention.

Conjecture 2.1 *Every simple 2-edge-connected planar graph admits an SCDC.*

The case of simple triangulations has, however, been settled.

Theorem 2.1 (J.A. Bondy and K.Seyffarth) *Let G be a simple triangulation (of any surface). Then G admits an SCDC.*

Proof: Let v be any vertex of G, and let C_v denote the cyclic sequence of neighbours of v determined by the embedding of G. Since each face of G is a triangle, C_v is a cycle of G. We call C_v the *vertex cycle* of G corresponding to the vertex v. Note that the collection

$$C = \{C_v : v \in V\}$$

is a CDC of G, because each edge of G belongs to precisely two facial triangles of G and thus to precisely two vertex cycles of G. However, C is not an SCDC of G because

$$|C| = |V| = n.$$

To obtain an SCDC of G, we modify $\mathcal{C}$, eliminating one vertex cycle C_u, as follows.

For each of the vertex cycles C_v, where v is a neighbour of u, we replace the segment $v^- u v^+$ of C_v (where v^- and v^+ are the vertices immediately preceding and following v on C_u) by the path $v^- u v v^+$, and denote the resulting cycle by C_v'. Then the cycle collection

$$\mathcal{C}' = \{C_v : \ v \in V \setminus (N(u) \cup \{u\})\} \cup \{C_v' : \ v \in N(u)\}$$

is an SCDC of G.

In order to eliminate two vertex cycles, C_u, C_w, in the above manner, the distance between vertices u and w must be at least three. Seyffarth [32] has proved that a simple plane triangulation on n vertices whose maximum degree Δ is no more than $(2n/3) - 1$ must have a diameter of at least three, and that this bound on Δ is best possible, provided that $\Delta \geq 8$. A corollary is the following result.

Theorem 2.2 (K. Seyffarth) *Let G be a simple plane triangulation on n vertices with maximum degree $\Delta \leq (2n/3) - 1$. Then G admits a CDC $\mathcal{C}$ such that $|\mathcal{C}| = n - 2$.*

3   The Perfect Path Double Cover Conjecture

Definition 3.1 A *path double cover* (PDC) of a graph G is a collection $\mathcal{P}$ of paths of G such that each edge of G belongs to exactly two paths of $\mathcal{P}$. A *small path double cover* (SPDC) of a graph G on n vertices is a PDC $\mathcal{P}$ of G such that $|\mathcal{P}| \leq n$. A *perfect path double cover* (PPDC) of a graph G is a PDC $\mathcal{P}$ of G such that each vertex of G occurs exactly twice as an end of a path of $\mathcal{P}$; a path of length zero is considered to have two (identical) ends.

Remark 3.1 If a PPDC $\mathcal{P}$ of a graph G contains a path $P = v$ of length zero, and if $Q = u \ldots v \ldots w$ is a path of $\mathcal{P}$ which contains v as an internal vertex, then the path collection

$$\mathcal{P}' = \mathcal{P} \cup \{Q_1, Q_2\} \setminus \{P, Q\},$$

where $Q_1 = u \ldots v$ and $Q_2 = v \ldots w$, is also a PPDC of G. Thus, if a graph G admits a PPDC, it admits a PPDC in which paths of length zero are present only at isolated vertices of G.

Example 3.1 Lovász [28] has proved that a simple graph G on n vertices in which every vertex is of odd degree admits a decomposition $\mathcal{P}$ into $n/2$ paths. Since each vertex of G must clearly be an end of at least one path of $\mathcal{P}$, and since there are exactly n ends of paths of $\mathcal{P}$, each vertex of G is an end of exactly one path of $\mathcal{P}$. Therefore, the path collection consisting of two copies of each path of $\mathcal{P}$ is a PPDC of G.

Conjecture PPDC (The Perfect Path Double Cover Conjecture) *Every simple graph admits a PPDC.*

Since a PPDC of a graph on n vertices consists of exactly n paths, a weaker conjecture is:

Conjecture SPDC (The Small Path Double Cover Conjecture) *Every simple graph* *admits an SPDC.*

The following theorem is proved in [4].

Theorem 3.1 *Conjecture PPDC, and Conjecture SCDC restricted to trigraphs, are equivalent.*

We conclude this section with a refinement of Conjecture PPDC.

Definition 3.2 Let $\mathcal{P}$ be a collection of paths in a graph G. To each path P of $\mathcal{P}$, we associate an edge e_P with the same ends as P. The *associated graph* of $\mathcal{P}$ is the graph $H = G(\mathcal{P})$ defined by

$$V(H) = V(G), \quad E(H) = \{e_P : P \in \mathcal{P}\}.$$

Definition 3.3 An *eulerian perfect path double cover* (EPPDC) of a graph G is a PPDC $\mathcal{P}$ of G whose associated graph is a cycle.

Conjecture EPPDC (The Eulerian Perfect Path Double Cover Conjecture) *Every simple graph admits an EPPDC.*

Path double covers are discussed in detail in [3].

4 Weighted Graphs

Definition 4.1 A *weighted graph* G is one in which each edge e is assigned a nonnegative number $w(e)$, called the *weight* of e. The *weight* of a subgraph H of G is defined by

$$w(H) = \sum\{w(e) : e \in E(H)\}.$$

An *optimal cycle* (or *optimal path*) of G is a cycle (or path) of maximum weight.

Suppose that Conjecture SCDC holds. Let G be a simple 2-edge-connected weighted graph on n vertices, and let $\mathcal{C}$ be an SCDC of G. Then

$$\sum\{w(C) : C \in \mathcal{C}\} = 2w(G)$$

and so the average weight of the cycles in $\mathcal{C}$ is

$$\frac{\sum\{w(C) : C \in \mathcal{C}\}}{|\mathcal{C}|} = \frac{2w(G)}{|\mathcal{C}|} \geq \frac{2w(G)}{n-1}.$$

Conjecture SCDC thus implies that an optimal cycle of G has weight at least $2w(G)/(n-1)$. This implication of Conjecture SCDC has, indeed, been verified [6].

Theorem 4.1 (J.A. Bondy and G. Fan) *Let G be a simple 2-edge-connected weighted graph on n vertices. Then G contains a cycle of weight at least $2w(G)/(n-1)$.*

A straightforward consequence of Theorem 4.1 is the following implication of Conjecture PPDC, proposed in [5] and proved in [17].

Corollary 4.1 (A. Frieze, C. McDiarmid and B. Reed) *Let G be a simple weighted graph on n vertices. Then G contains a path of weight at least $2w(G)/n$.*

Remark 4.1 Corollary 4.1 is valid for *arbitrary* weights, as can be seen by deleting all edges of negative weight and applying Corollary 4.1; Theorem 4.1, on the other hand, can not be so extended.

Remark 4.2 An unweighted graph can be regarded as a weighted graph in which each edge is assigned weight one; the weight of a graph is then its number of edges and the weight of a path or cycle its length. Corollary 4.1 and Theorem 4.1 thus generalize to weighted graphs two extremal theorems of Erdős and Gallai [10] on longest paths and cycles in unweighted graphs.

The extremal graphs for Theorem 4.1 and Corollary 4.1 have been completely determined [6], and are of interest because they lend support to Conjecture 1.1. In order to describe them, we need two more definitions.

Definition 4.2 A simple 2-edge-connected weighted graph G on n vertices is *cycle-extremal* if its optimal cycles are of weight precisely $2w(G)/(n-1)$. A simple weighted graph G on n vertices is *path-extremal* if its optimal paths are of weight precisely $2w(G)/n$.

Definition 4.3 Let G be a weighted graph, and let $\mathcal{H}$ be a collection of subgraphs of G. If there is an assignment of a positive real number α_H to each $H \in \mathcal{H}$ such that, for every $e \in E(G)$,

$$w(e) = \sum \{\alpha_H : e \in H \in \mathcal{H}\},$$

we say that G is a *weighted union* of the members of $\mathcal{H}$, and write

$$G = \sum \{\alpha_H H : H \in \mathcal{H}\}.$$

Theorem 4.2 (J.A. Bondy and G. Fan)

1. *Let G be a simple 2-edge-connected cycle-extremal weighted graph on n vertices. Then either $w(G) = 0$ or G is a weighted union of tritrees.*

2. *Let G be a simple path-extremal weighted graph on n vertices. Then either $w(G) = 0$ or G is a uniformly-weighted complete graph.*

The following conjecture, put forward in [6], is implied by Conjecture SCDC and implies Theorem 4.1.

Conjecture 4.1 *Every simple 2-edge-connected graph admits a collection $\mathcal{C}$ of cycles such that*

$$G = \sum \{\alpha_C C : C \in \mathcal{C}\},$$

where

$$\sum \{\alpha_C : C \in \mathcal{C}\} \leq \frac{n-1}{2}.$$

Remark 4.3 Seymour [34] proves that every 2-edge-connected graph can, indeed, be expressed as a weighted sum of cycles. He does not, however, give an upper bound on the sum of the coefficients α_C.

Remark 4.4 Seymour [35] has pointed out that the analogous assertion for paths follows from Corollary 4.1 by duality.

5 3-Regular Graphs

It was noted, in Remark 1.5, that Conjecture SCDC can not be reduced to 3-regular graphs. In fact, we propose a stronger conjecture for these graphs.

Conjecture 5.1 *Let G be a simple 2-connected 3-regular graph on n vertices, where $n \geq 6$. Then G admits a CDC $\mathcal{C}$ such that $|\mathcal{C}| \leq n/2$.*

Remark 5.1 Conjecture 5.1 can not be extended to 2-connected 3-regular graphs in general. But it is easily seen that, if Conjecture 5.1 holds, all counterexamples to its extension may be constructed from the 3-connected 3-regular graphs on two and four vertices by the recursive operation of subdividing an edge twice and joining the resulting two new vertices by a second edge.

Remark 5.2 If G is a simple 3-regular 3-edge-colourable graph on n vertices, with edge-colouring $\{F_1, F_2, F_3\}$, then the cycles in the three 2-factors $G[F_i \cup F_j]$, $1 \leq i < j \leq 3$, form a CDC $\mathcal{C}$ of G such that $|\mathcal{C}| \leq 3n/4$, because each 2-factor has at most $n/4$ components; moreover, $|\mathcal{C}| \leq n/2$ if G has no 4-cycles, and $|\mathcal{C}| \leq (n/2) + 1$ if one of the above 2-factors is a Hamilton cycle of G.

Example 5.1 (L.A. Goddyn) The following graphs show that Conjecture 5.1, if true, is sharp:

(a) three sporadic examples on eight and ten vertices:

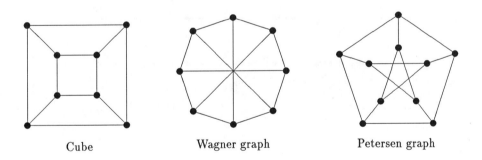

Cube   Wagner graph   Petersen graph

(b) for $n \geq 8$, the infinite family of *cubic ladders*:

Remark 5.3 *Ad hoc* methods are needed to verify that the sporadic graphs of Example 5.1(a) do not admit CDCs with fewer than $n/2$ cycles. To see that the cubic ladders

of Example 5.1(b) do not admit such CDCs, on the other hand, it suffices to observe that no cycle uses more than two of the $n/2$ bold edges. Aside from small examples, the cubic ladders are, in fact, the only simple 2-connected 3-regular graphs on n vertices containing $n/2$ edges with this property.

Theorem 5.1 *Let G be a simple 2-connected 3-regular graph on n vertices, and let S be a set of edges of G, no three of which lie on a common cycle of G. Then*

(a) *if G has connectivity three,*
$$|S| \leq 3,$$
with equality if and only if S is a 3-edge-cut of G.

(b) *if G has connectivity two,*
$$|S| \leq n/2,$$
with equality if and only if G is a cubic ladder.

Proof: Assertion (a) is due to Lovász [29]. We prove (b) by induction on n, the theorem being vacuously satisfied for $n = 4$ and $n = 6$.

A simple 3-regular graph of connectivity two has the following structure: there exist two disjoint subgraphs, G_1 and G_2, and nonadjacent vertices x_i, y_i in G_i, $i = 1, 2$, such that G is the edge-disjoint union of G_1, G_2, and a *ladder* G_3 whose ends are x_1, y_1, x_2 and y_2:

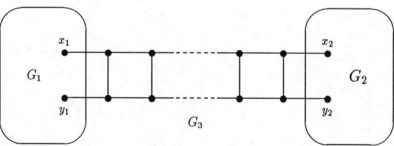

Let
$$m = n/2,$$
$$m_i = |V(G_i)|/2 \quad \text{and} \quad G_i' = G_i + x_i y_i, \ i = 1, 2.$$
Then $m_i \geq 2$ and G_i' is a simple 2-connected 3-regular graph, $i = 1, 2$. Let
$$S_i = S \cap E(G_i), \ i = 1, 2,$$
and denote by S_3 and S_4 the sets of 'vertical' and 'horizontal' edges of G_3 which belong to S, respectively. Thus
$$S = S_1 \cup S_2 \cup S_3 \cup S_4.$$
Moreover,
$$|S_3| \leq m_3,$$

where
$$m_3 = m - m_1 - m_2,$$

and, since there is a cycle of G which includes every horizontal edge,

$$|S_4| \leq 2.$$

Suppose $S_i = S$, $i = 1$ or 2. Since no three edges of S_i lie on a cycle of G, no three edges of S_i lie on a cycle of G_i'. Therefore, by the induction hypothesis, applied to G_i',

$$|S| = |S_i| \leq \max\{3, m_i\} < m.$$

Thus, we may assume that $S_i \neq S$, $i = 1, 2$. Define

$$S_i' = S_i \cup \{x_i y_i\}, \ i = 1, 2.$$

Let P_i be an (x_i, y_i)-path in G_i, and let Q_i be an (x_i, y_i)-path in $G - E(G_i)$ which includes an edge of S, $i = 1, 2$. The cycle $P_i \cup Q_i$ includes at most two edges of S, and so P_i includes at most one edge of S_i. It follows that every cycle of G_i' includes at most two edges of S_i'. By the induction hypothesis, applied to G_i',

$$|S_i'| \leq \max\{3, m_i\}, \ i = 1, 2,$$

whence
$$|S_i| \leq \max\{2, m_i - 1\} \leq m_i, \ i = 1, 2.$$

Suppose $S_i = \emptyset$, $i = 1$ or 2; say $S_1 = \emptyset$. Then

$$|S| = |S_2| + |S_3| + |S_4| \leq \max\{m_3 + 4, m - m_1 + 1\} \leq m$$

with equality only if $|S_2| = 2$ and $|S_4| = 2$. But if $|S_2| \geq 1$, then $|S_4| \leq 1$, since G contains a cycle through one edge of S_2 and every edge of S_4. Thus equality cannot hold in this case.

We may therefore assume that $S_i \neq \emptyset$, $i = 1, 2$. Consequently, $S_4 = \emptyset$, and

$$|S| = |S_1| + |S_2| + |S_3| \leq m_1 + m_2 + m_3 = m$$

with equality if and only if
$$|S_i| = 2 = m_i, \ i = 1, 2,$$

and
$$|S_3| = m_3.$$

Thus $G_i' \cong K_4$, $i = 1, 2$, and G is a cubic ladder.

Conjecture 5.1 can be reduced to simple 3-connected 3-regular graphs.

Proposition 5.1 *If Conjecture 5.1 holds for all simple 3-connected 3-regular graphs, then it holds for all simple 2-connected 3-regular graphs.*

Proof: Suppose that Conjecture 5.1 holds for all simple 3-connected 3-regular graphs, and let G be a 3-regular simple graph of connectivity two on n vertices. We prove, by induction on n, that G admits a CDC $\mathcal{C}$ such that $|\mathcal{C}| \leq n/2$. Define m, G_i, x_i, y_i, G_i', m_i, $i = 1, 2$, and G_3, m_3, as in the proof of Theorem 5.1. Then $m_i \geq 2$ and G_i' is a simple 2-connected 3-regular graph, $i = 1, 2$. By the induction hypothesis, G_i' has a CDC $\mathcal{C}_i$ such that

$$|\mathcal{C}_i| \leq \begin{cases} m_i + 1 & \text{if } G_i' \cong K_4, \\ m_i & \text{otherwise,} \end{cases} \quad i = 1, 2.$$

Also, the graph G_3' defined by

$$G_3' = G_3 + \{x_1 y_1, x_2 y_2\}$$

has a CDC $\mathcal{C}_3$ such that

$$|\mathcal{C}_3| = m_3 + 2,$$

consisting of $(m_3 + 1)$ 4-cycles and one $(2m_3 + 4)$-cycle. Let

$$\mathcal{C}' = \mathcal{C}_1 \cup \mathcal{C}_2 \cup \mathcal{C}_3.$$

By pairing up the two cycles of $\mathcal{C}_i$ through $x_i y_i$ with the two cycles of $\mathcal{C}_3$ through $x_i y_i$, $i = 1, 2$, and taking symmetric differences, we obtain a CDC $\mathcal{C}$ of G such that

$$|\mathcal{C}| = |\mathcal{C}_1| + |\mathcal{C}_2| + |\mathcal{C}_3| - 4 \leq m.$$

(In fact, $|\mathcal{C}| < m$ unless $G_i' \cong K_4$, $i = 1, 2$, in which case G is a cubic ladder.)

Remark 5.4 A stronger conjecture than Conjecture 5.1 appears to be valid for simple 3-connected 3-regular graphs. We know of no such graph on n vertices which fails to admit a CDC $\mathcal{C}$ such that $|\mathcal{C}| \leq (n + 10)/4$. A still stronger conjecture may well be valid for 3-connected cyclically 4-edge-connected 3-regular graphs.

Conjecture 5.1 has the following implication for weighted graphs.

Conjecture 5.2 *Let G be a simple 2-connected 3-regular weighted graph on n vertices, where $n \geq 6$. Then G contains a cycle of weight at least $4w(G)/n$.*

Remark 5.5 The sharpness of Conjecture 5.2, if true, is demonstrated by the cubic ladder of Example 5.1(b), weighted so that the bold edges have weight one and the remaining edges have weight zero. One might expect the three sporadic graphs of Example 5.1(a) also to admit such extremal weightings. However, this is not the case.

Each edge of a cube lies in four of its six Hamilton cycles. Therefore, the average weight of these Hamilton cycles, in a weighted cube G, is $2w(G)/3$; moreover, the uniform weighting shows that this bound is best possible. A similar argument, based on the 9-cycles of the Petersen graph, shows that any weighted Petersen graph G must contain a cycle of weight at least $3w(G)/5$; again, the uniform weighting demonstrates that this bound can not be improved. For the Wagner graph, the eight 7-cycles and eight 8-cycles obtained by rotating those depicted below together form a 10-cover of the graph. Thus, in a weighted Wagner graph G, one of these cycles must be of weight at least $5w(G)/8$; the weighting in which the edges of the outer cycle are assigned weight one and the chords of this cycle each receive weight two shows that this bound is sharp.

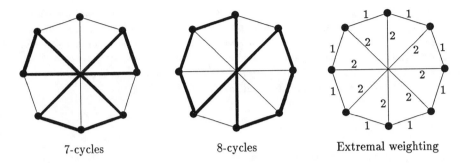

| 7-cycles | 8-cycles | Extremal weighting |

Not surprisingly, Conjecture 5.2 can be reduced to simple 3-connected 3-regular graphs.

Proposition 5.2 *If Conjecture 5.2 holds for all simple 3-connected 3-regular weighted graphs, then it holds for all simple 2-connected 3-regular weighted graphs.*

Proof: Suppose that Conjecture 5.2 holds for all simple 3-connected 3-regular weighted graphs, and let G be a simple 3-regular weighted graph of connectivity two on n vertices. We prove, by induction on n, that G contains a cycle of weight at least $4w(G)/n$. Define x_i, y_i, $i = 1, 2$, G_i, G'_i, m_i, $i = 1, 2, 3$, and m, as in the proof of Proposition 5.1, where the weights $w(x_i y_i)$ of the edges $x_i y_i \in G'_i$, $i = 1, 2$, are set at zero. Then $m_i \geq 2$ and G'_i is a simple 2-connected 3-regular weighted graph, $i = 1, 2$. Let

$$w_i = w(G'_i),\ i = 1, 2, \quad \text{and} \quad w_3 = w(G) - w_1 - w_2.$$

By the induction hypothesis, if C_i is an optimal cycle in G'_i, $i = 1, 2$, and C is an optimal cycle in G, then

$$w(C) \geq w(C_i) \geq 2w_i/m_i,\ i = 1, 2,$$

so

$$(m_i - 1)w(C) \geq \frac{2(m_i - 1)}{m_i} w_i = 2w_i - 2w_i/m_i,\ i = 1, 2. \tag{1}$$

Since G'_i is 2-connected, x_i and y_i are connected by disjoint paths to C_i, $i = 1, 2$. It follows that x_i and y_i are connected by a path P_i in G'_i of weight

$$w(P_i) \geq w(C_i)/2 \geq w_i/m_i,\ i = 1, 2. \tag{2}$$

Consider the CDC $\mathcal{C}_3$ of G'_3 described in the proof of Proposition 5.1. We modify the cycles of $\mathcal{C}_3$ which include either or both of the edges $x_i y_i$, $i = 1, 2$, replacing each occurrence of $x_i y_i$ by the path P_i, and call the resulting collection of $(m_3 + 2)$ cycles $\mathcal{C}'_3$. Using inequality (2), we have

$$\begin{aligned}
(m_3 + 2)w(C) &\geq \sum\{w(C): C \in \mathcal{C}'_3\} \\
&\geq 2w_3 + 2w(P_1) + 2w(P_2) \\
&\geq 2w_3 + 2w_1/m_1 + 2w_2/m_2.
\end{aligned} \tag{3}$$

Adding inequalities (1) and (3), we obtain

$$mw(C) \geq 2w(G),$$

whence

$$w(C) \geq 2w(G)/m = 4w(G)/n.$$

Remark 5.6 A stronger conjecture than Conjecture 5.2 should be valid for simple 3-connected 3-regular weighted graphs; we know of no such graph G on n vertices which fails to contain a cycle of weight at least $w(G)/n^c$, where c is an absolute constant, $c < 1$. A still stronger conjecture may well be valid for 3-connected cyclically 4-edge-connected 3-regular weighted graphs; we know of no such graph G which fails to contain a cycle of weight at least $cw(G)$, where c is an absolute constant, $c > 0$.

6   Related Questions

Certain questions about decompositions and covers of graphs by paths and cycles are, of course, closely related to the topic of small cycle double covers. Here, we review some of them, and discuss their interconnections.

6.1   Decompositions

We first recall the conjecture of G. Hajós on small cycle decompositions which was mentioned in Section 1.

Conjecture SCD (G. Hajós) Every simple even graph admits an SCD.

An analogous conjecture for paths, also cited in [28], was made by T. Gallai.

Definition 6.1 A *path decomposition* (PD) of a graph G is a collection $\mathcal{P}$ of paths of G such that each edge of G belongs to exactly one path of $\mathcal{P}$. A *small path decomposition* (SPD) of a graph G on n vertices is a PD $\mathcal{P}$ of G such that $|\mathcal{P}| \leq (n+1)/2$.

Conjecture SPD (T. Gallai) Every simple connected graph admits an SPD.

Remark 6.1 Conjecture SPD implies that every simple connected graph on n vertices admits a PDC $\mathcal{P}$ such that $|\mathcal{P}| \leq n+1$, and Conjecture SCD implies that every simple even graph admits an SCDC. Lovász [28] has proved that every simple connected graph on n vertices admits a decomposition into at most $n/2$ paths and cycles. If each vertex is of odd degree, such a decomposition must necessarily consist of $n/2$ paths; thus Lovász' theorem implies Conjecture SPD in this case. Donald [8] has refined Lovász' argument and established that every simple connected graph on n vertices admits a decomposition into at most $(3n_0/4) + (n_1/2)$ paths, where n_0 and n_1 denote the numbers of vertices of even and odd degree, respectively. Lovász [28] asked whether a simple connected graph on n_0 even and n_1 odd vertices admits a decomposition into at most $(n_0/2)$ cycles and $(n_1/2)$ paths. At first sight, this seems to be a generalization of Conjecture SCD, but it is easily seen that the two are, in fact, equivalent. Favaron and Kouider [15] and Granville and Moisiadis [20] have verified Conjecture SCD for graphs of maximum degree four; Favaron and Kouider [15] have also verified Conjecture SPD for such graphs.

A refinement of Conjecture SCD for regular graphs of high degree was proposed by Nash-Williams [30].

Conjecture HD (C.St.J.A. Nash-Williams) Every simple $2k$-regular graph on at most $4k + 1$ vertices admits a decomposition into Hamilton cycles.

Remark 6.2 Häggkvist [21] has announced a solution of a slightly weaker form of Conjecture HD: *for any constant $c < 4$, there is an integer k_c such that, if $k > k_c$, every $2k$-regular simple graph on at most ck vertices admits a decomposition into Hamilton cycles.* Jackson [24] has proved that every simple k-regular graph on n vertices, where $n \leq 2k + 1$, contains $\lfloor (3k - n + 1)/6 \rfloor$ edge-disjoint Hamilton cycles.

Decompositions into cycles and edges were considered by Erdős and Gallai (see [9]).

Conjecture CED (P. Erdős and T. Gallai) Every simple graph on n vertices admits a decomposition into at most cn cycles and edges, where c is an absolute constant.

Dean [7] discusses versions of Conjecture SCD for eulerian digraphs of various types.

6.2 Covers

Definition 6.2 A *path cover* (PC) of a graph G is a collection $\mathcal{P}$ of paths of G such that each edge of G belongs to at least one path of $\mathcal{P}$. A *cycle cover* (CC) of a graph G is a collection $\mathcal{C}$ of cycles of G such that each edge of G belongs to at least one cycle of $\mathcal{C}$.

We propose the following conjecture.

Conjecture SCC Every simple 2-connected graph on n vertices admits a cycle cover $\mathcal{C}$ such that $|\mathcal{C}| \leq (2n - 1)/3$.

Example 6.1 Let G be the graph formed from a complete bipartite graph $K_{3,n-3}$, where $n \geq 5$, by joining all three pairs of vertices in the part of cardinality three. Then G can not be covered by fewer than $(2n - 3)/3$ cycles. To see this, assign weight one to the edges of $K_{3,n-3}$ and weight two to the three additional edges. Then G has weight $3(n - 1)$, and no cycle of G has weight more than six. Moreover, the total weight of the cycles in any cycle cover $\mathcal{C}$ of G is at least $3(n - 1) + (n - 3)$, because, at each vertex of degree three, some edge must belong to at least two cycles of $\mathcal{C}$. Therefore,

$$|\mathcal{C}| \geq (4n - 6)/6 = (2n - 3)/3.$$

Remark 6.3 Conjecture SCC implies the conjecture of Erdős, Goodman and Pósa [11], proved by Pyber [31], that every simple graph on n vertices can be covered by at most $n - 1$ cycles and edges.

The cardinality $|\mathcal{C}|$ of a cycle cover $\mathcal{C}$ is a measure of the efficiency of the cover. Another measure that has been studied is the length $\|\mathcal{C}\|$ of the cover.

Definition 6.3 The *length* $\|\mathcal{C}\|$ of a cycle cover $\mathcal{C}$ is defined by

$$\|\mathcal{C}\| = \sum \{|E(C)| : C \in \mathcal{C}\}.$$

The minimum length of a cycle cover is bounded below by the number of edges in the graph. The following conjecture, concerning an upper bound for this minimum length, is stated in [1].

Conjecture 6.1 *Every simple 2-edge-connected graph on e edges admits a cycle cover $\mathcal{C}$ such that $\|\mathcal{C}\| \leq 8e/5$.*

Remark 6.4 No counterexample to the sharper bound of $7e/5$ is known (this bound being attained by the Petersen graph) and Bermond, Jackson and Jaeger [2] have proposed such a strengthening of Conjecture 6.1. They have proved [1] that a 2-edge-connected graph on e edges admits a cycle cover $\mathcal{C}$ such that $\|\mathcal{C}\| \leq 5e/3$. An alternative proof of this result, and a generalization to weighted graphs, have been obtained by Fan [13]. Itai and Rodeh [23] ask whether a 2-connected graph on n vertices and e edges admits a cycle cover $\mathcal{C}$ such that $\|\mathcal{C}\| \leq e + n - 1$. Fraisse [16] has established an upper bound of $e + 5(n - 1)/4$ and a generalization to weighted graphs. Itai, Lipton, Papadimitriou and Rodeh [22] conjecture that the problem of determining whether a simple graph admits a cycle cover $\mathcal{C}$ such that $\|\mathcal{C}\| \leq k$ is NP-complete.

6.3   Path and Cycle Lengths

Most of the questions discussed so far have concerned the *number* of paths or cycles in a decomposition, cover, or double cover of a graph. Here, we examine briefly what might be said about their *length*. A rather general problem is the following:

Problem 6.1 *Let k be a positive integer. Which {graphs, oriented graphs, digraphs} admit {decompositions, covers, double covers, multicovers} by {paths, cycles} of length {exactly, at least} k?*

We mention several instances of this problem which appear to be interesting.

Question 6.1 *Does every simple k-regular graph, except K_{k+1}, for k even, admit a decomposition into paths of length at least k?*

Question 6.2 *Does every simple 2-connected $2k$-regular graph admit a decomposition into cycles of length at least $2k + 1$?*

Remark 6.5 Affirmative answers to these questions would imply the truth of Conjectures SPD and SCD, respectively, for regular graphs. Also, a proof of Conjecture HD would provide an affirmative answer to the latter question for graphs on at most $4k + 1$ vertices.

Remark 6.6 The hypothesis of regularity in Questions 6.1 and 6.2 could, of course, be replaced by a constraint on the minimum degree. We know of no graph which provides a negative answer to either of these more general questions.

Question 6.3 *Which graphs admit decompositions into paths of length k?*

Question 6.4 *Which graphs admit double covers by paths of length k?*

Definition 6.4 A *k-perfect path double cover* (k-PPDC) is a PPDC in which each path is of length k.

Question 6.5 *Which graphs admit k-PPDCs?*

Remark 6.7 For $k = 2$, these questions have been settled, in [27] and [3]. For $k = 3$, however, they remain open. It is known that a simple 3-regular graph admits a decomposition into paths of length three if and only if it has a perfect matching, and that every simple 3-regular graph admits a 3-PPDC. We conjecture [3] that every simple k-regular graph admits a k-PPDC.

Many of the theorems and conjectures mentioned in this article are interrelated. The diagram overleaf summarizes these relationships.

Acknowledgements
The research for this paper was funded by the Natural Sciences and Engineering Research Council of Canada. The work was completed at the Technical University of Denmark, where the author was a Visiting Professor in the Department of Mathematics. I thank Carsten Thomassen and other members of the department for their kind hospitality during my visit.

Thanks are also due to the referees for a thorough reading of the first draft and for their helpful remarks.

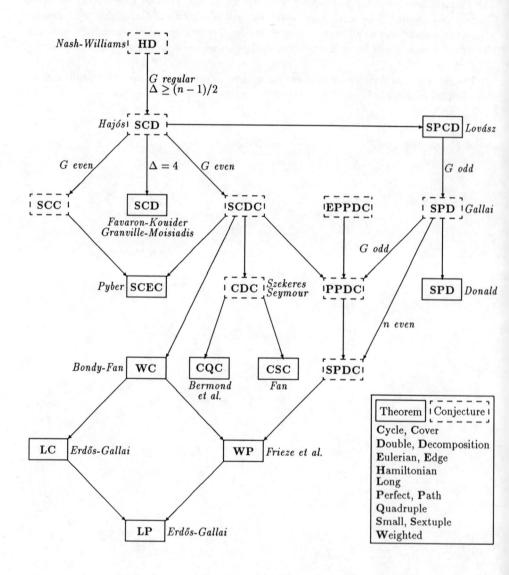

References

[1] J.C. Bermond, B. Jackson and F. Jaeger, Shortest coverings of graphs with cycles, *J. Combinatorial Theory Ser. B* **35**(1983), 297–308.

[2] J.C. Bermond, B. Jackson and F. Jaeger, personal communication.

[3] J.A. Bondy, Perfect path double covers of graphs, *J. Graph Theory*. To appear.

[4] J.A. Bondy, Trigraphs, *Discrete Math.*. To appear.

[5] J.A. Bondy and G. Fan, Optimal paths and cycles in weighted graphs, in *Graph Theory in Memory of G.A. Dirac* (L.D. Andersen, C. Thomassen and B. Toft, eds.), Ann. Discrete Math. **41** (1989), 53–70.

[6] J.A. Bondy and G. Fan, Cycles in weighted graphs, *Combinatorica*. To appear.

[7] N. Dean, What is the smallest number of dicycles in a dicycle decomposition of an eulerian digraph?, *J. Graph Theory* **10**(1986), 299–308.

[8] A. Donald, An upper bound for the path number of a graph, *J. Graph Theory* **4**(1980), 189–201.

[9] P. Erdős, On the combinatorial problems which I would most like to see solved, *Combinatorica* **1**(1981), 25–42.

[10] P. Erdős and T. Gallai, On maximal paths and circuits of graphs, *Acta Math. Acad. Sci. Hungar.* **10**(1959), 337–356.

[11] P. Erdős, A. Goodman and L. Pósa, The representation of a graph by set intersections, *Canad. J. Math.* **18**(1966), 106–112.

[12] L. Euler, Solutio problematis ad geometriam situs pertinentis, *Comm. Acad. Sci. Imp. Petropol.* **8**(1736), 128–140.

[13] G. Fan, Covering weighted graphs by even subgraphs, *J. Combinatorial Theory, Ser. B.* To appear.

[14] G. Fan, 6-flows and cycle 6-covers, preprint.

[15] O. Favaron and M. Kouider, Path partitions and cycle partitions of eulerian graphs of maximum degree 4, *Studia Sci. Math. Hungar.* **23**(1988), 237–244.

[16] P. Fraisse, Cycle covering in bridgeless graphs, *J. Combinatorial Theory, Ser B.* **39**(1985), 146–152.

[17] A. Frieze, C.D. McDiarmid and B. Reed, On a conjecture of Bondy and Fan, Research Report CORR 89-16, University of Waterloo, 1989.

[18] L.A. Goddyn, Cycle double covers of graphs with Hamilton paths, *J. Combinatorial Theory, Ser. B* **46**(1989), 253–254.

[19] L.A. Goddyn, *Cycle Covers of Graphs*, Ph.D. Thesis, University of Waterloo, 1988.

[20] A. Granville and A. Moisiadis, On Hajós' conjecture, in *Proceedings of Sixteenth Manitoba Conference on Numerical Mathematics and Computing* (D.S. Meek, R.G.Stanton and G.H.J. van Rees, eds.), Congressus Numerantium **56**, Utilitas Mathematica Publishing Inc., Winnipeg, 1987, pp. 183–187.

[21] R. Häggkvist, Third British Combinatorial Conference, Southampton, 1983.

[22] A. Itai, R.J. Lipton, C.H. Papadimitriou and M. Rodeh, Covering graphs by simple circuits, *SIAM J. Comput.* **10**(1981), 746–754.

[23] A. Itai and M. Rodeh, Covering a graph by circuits, in *Automata, Languages and Programming*, Notes in Computer Science No. 62, Springer, Berlin, 1978, pp. 289–299.

[24] B. Jackson, Edge-disjoint Hamilton cycles in regular graphs of large degree, *J. London Math. Soc.* **19**(1979), 13–16.

[25] F. Jaeger, Flows and generalized coloring theorems in graphs, *J. Combinatorial Theory, Ser B.* **26**(1979), 205–216.

[26] F. Jaeger, A survey of the cycle double cover conjecture, in *Cycles in Graphs* (B.R. Alspach and C.D. Godsil, eds.), Ann. Discrete Math. **27**, North-Holland, Amsterdam, 1985, pp. 1–12.

[27] A. Kotzig, From the theory of finite regular graphs of degree three and four, *Časopis Pěst. Mat.* **82**(1957), 76–92 (in Czech).

[28] L. Lovász, On covering of graphs, in *Theory of Graphs* (P. Erdős and G.O.H. Katona, eds.), Academic Press, New York, 1968, pp. 231–236.

[29] L. Lovász, Problem 5, *Period. Math. Hungar.* **4** (1974), 82.

[30] C.St.J.A. Nash-Williams, Hamiltonian lines in graphs whose vertices have sufficiently large valencies, in *Combinatorial Theory and its Applications III* (P. Erdős, A. Rényi and V.T. Sós, eds.), Colloq. Math. Soc. Janos Bolyai, **4**(1970), pp. 813–819.

[31] L. Pyber, An Erdős-Gallai conjecture, *Combinatorica* **5**(1985), 67–79.

[32] K. Seyffarth, Maximal planar graphs of diameter two, *J. Graph Theory*. To appear.

[33] K. Seyffarth, personal communication.

[34] P.D. Seymour, Sums of circuits, in *Graph Theory and Related Topics* (J.A. Bondy and U.S.R. Murty, eds.), Academic Press, New York, 1979, pp. 341–355.

[35] P.D. Seymour, personal communication.

[36] G. Szekeres, Polyhedral decomposition of cubic graphs, *Bull. Austral. Math. Soc.* **8**(1973), 367–387.

[37] M. Tarsi, Semi-duality and the cycle double cover conjecture, *J. Combinatorial Theory, Ser B.* **41**(1986), 332–340.

κ-TRANSFORMATIONS, LOCAL COMPLEMENTATIONS AND SWITCHING

A. BOUCHET*

Centre de Recherche en Informatique et Combinatoire
Université du Maine
72017 Le Mans Cedex
France

Abstract

The κ-transformation of an Euler tour T of a 4-regular graph H at a vertex v is the Euler tour which has the same transitions as T except at v. The local complementation of a simple graph G at a vertex v is the simple graph obtained by replacing the subgraph induced by G on $\{w : vw \in E(G)\}$ by its complement.

We shall state some problems and conjectures concerning κ-transformations, local complementations and switchings. One of these problems is to determine the maximum of the minimum number of κ-transformations between two Euler tours of H. We prove that it is $\leq 3/2|V(H)|$.

The κ-transformation of an Euler tour T of a 4-regular graph H at a vertex v is the Euler tour which has the same transitions as T except at v. This transformation has been introduced by A. Kotzig who proved that any Euler tour is accessible from another one by successive κ-transformations.

The local complementation of a simple graph G at a vertex v is the simple graph obtained by replacing the subgraph induced by G on $\{w : vw \in E(G)\}$ by its complement. This transformation was also introduced by Kotzig in relation with κ-transformations.

Local complementations are related to κ-transformations only when they are performed on a special class of graphs, the circle graphs or alternance graphs. In the general case there is also a natural frame to deal with local complementations. It is provided by isotropic systems, a combinatorial and algebraic structure which unifies some properties common to 4-regular graphs and pairs of dual binary matroids. We shall insist on switchings, a transformation which generalizes κ-transformations to isotropic systems.

We shall state some problems and conjectures concerning κ-transformations, local complementations and switchings. One of these problems is to determine the maximum of the minimum number of κ-transformations between two Euler tours of H. We prove that it is $\leq 3/2|V(H)|$.

---

*Partially supported by PRC Mathématiques et informatique.

G. Hahn et al. (eds.), Cycles and Rays, 41–50.

1 κ-transformations and local complementations

Let H be a connected 4-regular graph, and let $T = (v_0, e_0, v_1, e_1, \ldots, v_{2n-1}, e_{2n-1})$ be an Euler tour of H which meets successively the vertices $v_0, v_1, \ldots, v_{2n-1}$ and the edges $e_0, e_1, \ldots, e_{2n-1}$. We point out that the sequence defining T is considered up to a rotation and up to a reversion. We notice that each vertex of H occurs precisely twice, and we say that the vertex-sequence $\mu = \mathrm{VS}(T) = (v_0, v_1, \ldots, v_{2n-1})$ is a double occurence word on $V(H)$.

Let v be a vertex of H, and suppose that $v = v_0 = v_i, i \neq 0$. The κ-transformation of T at v is the Euler tour $T * v$ obtained by replacing the subsequence $e_0, v_1, e_1, \ldots, e_{i-1}$ by its reverse $e_{i-1}, \ldots, e_1, v_1, e_0$ (we could as well replace $e_i, v_{i+1}, \ldots, e_{2n-1}$ by its reverse). For any word $m = w_1 w_2 \ldots w_q$ on $V(G)$, we define inductively $T * m = (T * w_1 w_2 \ldots w_{q-1}) * w_q$.

Property 1.1 (Accessibility Property A. Kotzig [15]) *If T' and T'' are any two Euler tours of a 4-regular graph H, then there exists a word m on $V(H)$ such that $T'' = T' * m$.*

An alternance of T is a nonordered pair xy such that we meet alternatively $\ldots x \ldots y \ldots x \ldots y \ldots$ when reading $\mathrm{VS}(T)$. The alternance graph of T is the simple graph $\mathrm{Alt}(T)$ defined on the vertex-set $V(H)$ whose edges are the alternances of T. Thus a family of simple graphs, $\ell_H = \{\mathrm{Alt}(T) : T \text{ is an Euler tour of } H\}$ is attached to the 4-regular graph H. We describe now the relations among the members of ℓ_H.

Let G be a simple graph. The local complementation of G at one of its vertices v is the simple graph $G * v$ obtained by replacing the subgraph induced by G on $\{w : vw \in E(G)\}$ by the complementary subgraph. As above we define inductively $G * m$ for any word on $V(G)$, and we say that two simple graphs G' and G'' are locally equivalent if there exists m such that $G'' = G' * m$. This is actually an equivalence relation because $(G * v) * v = G$ for every vertex v of G, which implies $(G * m) * \tilde{m} = G$ for every word m on $V(G)$ if $\tilde{m}$ denotes the reverse of m. It is easy to verify the following property.

Property 1.2 *If T is an Euler tour of a 4-regular graph H and v is a vertex of H, then $\mathrm{Alt}(T * v) = \mathrm{Alt}(T) * v$.*

Properties (1.1) and (1.2) imply that ℓ_H is a class of local equivalence, and by definition, each member of ℓ_H is an alternance graph. More generally, the isotropic systems which are defined in the next section are combinatorial and algebraic structures which can be associated to any class of local equivalence.

κ-transformations, local complementations and their relations were introduced by A. Kotzig [15, 16]. Alternance graphs are also known as stack sorting graphs, overlapping graphs, circle graphs. The reader will refer to M. C. Golumbic [14] for a survey on these graphs.

2   Switchings

Isotropic systems are introduced in [1] and [2]. All the properties stated in this section are proved in these two papers.

We consider a 2-dimensional vector space K over $\mathrm{GF}(2)$ and we let $(x, y) \to xy$ denote the bilinear form over K such that $xy = 1$ if and only if $0 \neq x \neq y \neq 0$. For each finite set V, we consider K^V as a $2|V|$ -dimensional vector space over GF(2), and we define over it the bilinear form $(A, B) \rightarrow AB := \sum(A(v)B(v) : v \in V)$. This bilinear form is antisymmetric and nondegenerate. Therefore every subspace L of K^V which is totally isotropic (i.e. $A, B \in L \Rightarrow AB = 0$) is such that $\dim(L) \leq |V|$, and the equality holds if and only if L is maximal to be totally isotropic. An isotropic system is a pair $S = (L, V)$ with a finite set V and a totally isotropic subspace L of K^V which is maximal.

A vector $A \in K^V$ is said to be complete if $A(v) \neq 0$ for every $v \in V$. For any complete vector A and any $P \subseteq V$, we let AP be the vector of K^V which is defined by $AP(v) = A(v)$ if $v \in P$ and $AP(v) = 0$ if $v \notin P$. We let $\hat{A} = \{AP : P \subseteq V\}$. Notice that $\hat{A}$ is a subspace of K^V because the ground field is GF(2), and that $\dim(\hat{A}) = |V|$.

Property 2.1 *If $S = (L, V)$ is an isotropic system, then there exists a complete vector A such that $L \cap \hat{A} = \{0\}$.*

Any A satysfying (2.1) is called an Eulerian vector of S.

Property 2.2 *For any Eulerian vector A of an isotropic system $S = (L, V)$ and any $v \in V$, there exists precisely one Eulerian vector $A * v$ of S such that $A * v(v) \neq A(v)$ and $A * v(w) = A(w)$ for every $w \in V - v$.*

The transformation $A \rightarrow A * v$ is called a switching.

Property 2.3 (Accessibility Property) *For any two Eulerian vectors A' and A'' of an isotropic system $S = (L, V)$, there exists a word m on V such that $A'' = A' * m$.*

The relation between switchings and κ-transformations is the following one. Let H be a 4-regular graph over the vertex-set V. A transition of H at a vertex v is a pair of edges of H which have v as a common end. A bitransition at v is a pair of disjoint transitions at v. If $T = (v_0, e_0, v_1, e_1, \ldots, v_{2n-1}, e_{2n-1})$ is an Euler tour of G, and $v = v_i = v_j$, then $\{\{e_{j-1}, e_j\}, \{e_{i-1}, e_i\}\}$ is the bitransition of T at v. Let us associate to each bitransition at v a nonnull value of K, called its code, in such a way that the correspondance is bijective, which is possible because there are three bitransitions at v and also three nonnull values in K. We say that H is a coded graph if the coding is defined for each vertex of H. Then each Eulerian tour T of H is coded by a complete vector $\Gamma(T)$, where $\Gamma(T)(v)$ is the code of the bitransition of T at v for every $v \in V$. Clearly T is uniquely determined by $\Gamma(T)$, but not every complete vector of K^V is the code of an Euler tour.

Property 2.4 *For every coded graph H, there exists precisely one isotropic system $S = (L, V)$, whose Eulerian vectors are the codes of the Eulerian tours of H. Moreover, if an Eulerian vector A of S and an Eulerian tour T of H satisfy $A = \Gamma(T)$, then $A * v = \Gamma(T) * v$ for every $v \in V$.*

An isotropic system S which can be associated with a coded graph as in (2.4) is called a graphic system.

The relation between switchings and local complementations is the following one. Two vectors $A, B \in K^V$ are said to be supplementary if $0 \neq A(v) \neq B(v) \neq 0$ for every $v \in V$.

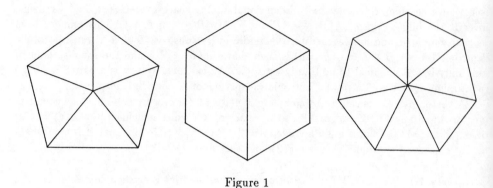

Figure 1

Property 2.5 *Let $P = (G, A, B)$ be a triple with a simple graph G and two supplementary vectors A and B of K^V, $V = V(G)$. For each $v \in V$ let $n(v) = \{w : vw \in E(G)\}$. If L is the subspace of K^V generated by $(An(v) + Bv : v \in V)$, then $S = (L, V)$ is an isotropic system. Moreover A is an Eulerian vector of S.*

The triple P is called a graphic presentation of S, and the graph G is called a fundamental graph of S.

Property 2.6 *For every Eulerian vector A of an isotropic system $S = (L, V)$ there exists precisely one graphic presentation $P = (G, A, B)$ of S. If $v \in V$ and $A' = A * v$ and (G', A', B') is the graphic presentation associated to the Eulerian vector A', then $G' = G * v$.*

The preceding property implies that any class of local equivalence is the set of the fundamental graphs of some isotropic system. Finally we have the following relation between alternance graphs and fundamental graphs of isotropic systems.

Property 2.7 *Let $S = (L, V)$ be the graphic system associated to a coded graph H, and let A be the code of an Eulerian tour T of H. If (G, A, B) is the graphic presentation of S associated to A, then $G = Alt(T)$.*

3   Problems and conjectures

Until recently no good characterization was known for alternance graphs except for those which are bipartite found by H. de Fraysseix [12]. Polynomial recognition algorithms have been announced in 1985 by W. Naji [19], C. P. Gabor, W. Hsu and K. J. Supowit [13], A. Bouchet [3] (proofs in [4]).

Conjecture 3.1 *A graph G is an alternance graph if and only if no graph locally equivalent to G contains an induced subgraph isomorphic to one of the graphs depicted in Fig. (1).*

The conjecture can be interpreted as follows in the theory of isotropic systems. Alternance graphs are the fundamental graphs of the graphic systems which play the same role as graphic matroids in the theory of matroids. If G is a fundamental graph of an isotropic system S, then taking an induced subgraph of a graph locally equivalent to G yields a fundamental graph of an isotropic system S', and S is related to S' like a matroid is related to one of its minors (see [1] and [2]). Then the conjecture amounts to characterize graphic systems by three excluded minors, which is similar to a classical theorem of Tutte for graphic matroids [22].

Let us consider a simple graph G provided with an orientation, which means that an initial end and a final end have been distinguished for each edge. The adjacency matrix of this oriented graph is the antisymmetric $(0, \pm 1)$-matrix $A = (A_{vw} : v, w \in V(G))$ such that $A_{vw} = +1$ if and only if vw is an edge oriented from v to w. For every $W \subseteq V(G)$ let $A[W] = (A_{vw} : v, w \in W)$. The orientation of G is unimodular if $det(A[W]) \in \{-1, 0, +1\}$ for every $W \subseteq V(G)$. It is proved in [9] that every alternance graph can be provided with an unimodular orientation. Thus if G is an alternance graph, every graph locally equivalent to G can be provided with an unimodular orientation.

Conjecture 3.2 *A simple graph G is an alternance graph if and only if every graph locally equivalent to G can be provided with an unimodular orientation.*

We shall get another good characterization of alternance graphs if the following also holds.

Conjecture 3.3 *The class of the simple graphs which can be provided with an unimodular orientation is in NP and in co-NP.*

For a simple graph G, a subset $V' \subseteq V(G)$, and $V'' = V(G) \backslash V'$, the cut-matrix of V' is the matrix $\Pi = (\Pi_{v'v''} : v' \in V', v'' \in V'')$ with coefficients in GF(2) and such that $\Pi_{v'v''} = 1$ if and only if $v'v'' \in E(G)$. The function $c : V' \rightarrow rank(\Pi)$ is the connectivity function of G. For example $c(V') = 0$ if and only if V' is an union of components of G. We have $c(V') = 1$ if and only if there exist nonempty sets $W' \subseteq V'$ and $W'' \subseteq V''$ such that the cocyle $\delta(V')$ is equal to the edge-set of the complete bipartite graph on the color-classes W' and W'', which means that $\{V', V''\}$ is a split of G in the sense of W. Cunningham [10] if $|V'|, |V''| \geq 2$. It is easy to verify that locally equivalent graphs have the same connectivity function (see [5] for a proof and an interpretation in terms of isotropic systems).

Conjecture 3.4 *Two graphs are locally equivalent if and only if they have the same connectivity function.*

If vw is an edge of a simple graph G, then we can verify that $G * vwv = G * wvw$ (see [2] for details). The transformation $G \rightarrow G * vwv$ is the complementation along the edge vw. If G is bipartite then we easily verify that $G * vwv$ also is bipartite. The reader will refer to [6] for an interpretation of edge complementation in terms of a coded graph.

Conjecture 3.5 *If G' and G'' are two locally equivalent bipartite graphs, then G'' can be obtained from G' through successive edge complementations.*

Let M be a binary matroid on the set V, and let X be a base of M. The fundamental graph of M w.r.t. X is the simple graph $F = \text{Fd}(X)$ defined by $V(F) = V$ and $E(F) = \{vw : v \in X, w \notin X, X - v + w$ is a base of $M \}$. In other terms, for each $w \in V \backslash X$, the subset $\{w\} \cup \{v : v \in X, vw \in E(F)\}$ is the fundamental circuit of M contained in $X \cup \{w\}$. Since any binary matroid is determined by its fundamental circuits w.r.t. to a base, the fundamental graph F determines M completely. It is easy to verify that $Fd(X) * vwv = Fd(X - v + w)$ holds for every edge vw of F, $v \in X$. If r is the rank function of M, the connectivity function of M is $\gamma : V' \to r(V') + r(V \backslash V') - r(V), V' \subseteq V$. If c is the connectivity function of F, then it is easy to verify that $\gamma = c$ (refer to K. Truemper [21] whose partial representations are precisely our fundamental graphs).

Theorem 3.1 *If Conjectures (3.4) and (3.5) are true, then any binary matroid is uniquely determined by its connectivity function.*

Proof: Let M' and M'' be two binary matroids with the same connectivity function. These matroids are defined on the same set. Let X' and X'' be bases of M' and M'' respectively. The fundamental graphs $F' = Fd(X')$ and $F'' = Fd(X'')$ have the same connectivity function, and they are bipartite. Conjectures (3.4) and (3.5) imply that F'' can be obtained from F' through successive edge complementations. Since a binary matroid is determined by any of its fundamental graphs, it follows that $M' = M''$.

W. Cunningham asked whether a matroid is determined by its connectivity function. P. Seymour gave a negative answer by constructing a matroid which has the same connectivity function as the uniform matroid U_4^8 and is not isomorphic to U_4^8. If Conjectures (3.4) and (3.5) are true, the answer will be positive for binary matroids.

The switching graph of an isotropic system $S = (L, V)$ is the graph SW whose vertices are the Eulerian vectors of S, and whose edges are the unordered pairs $A'A''$ of Eulerian vectors such that one can find $v \in V$ satisfying $A'' = A' * v$ (or equivalently $A' = A'' * v$). If S is a graphic system associated with a coded graph H, then following (2.4), SW could as well be defined on the set of the Eulerian tours of H with each edge $T'T''$ corresponding to a κ-transformation $T'' = T' * v$. We search for an upper bound of diam(SW), the diameter of SW (a lower bound is clearly $|V|$). In [8] we proved that diam(SW) $\leq 2|V| - 1$, but in this paper we considered a generalization of isotropic systems, the Eulerian systems, and the upper bound is sharp for these Eulerian systems. We prove in the next section that diam(SW) $\leq \frac{3}{2|V|}$ holds for isotropic systems. But this new upper bound is still not sharp.

Conjecture 3.6 *If SW is the switching graph of an isotropic system $S = (L, V)$, then $diam(SW) \leq |V| + 1$.*

We also can define the graph SWF whose vertices are the fundamental graphs of S and whose edges are the unordered pairs $F'F''$ of fundamental graphs such that one can find $v \in V$ satisfying $F'' = F' * v$. Then if we let $\phi(A)$ be the fundamental graph of S which is associated with an Eulerian vector A, (2.6) implies $\phi(A * m) = \phi(A) * m$ for every word m on V. Thus any upper bound on diam(SW) is also an upper bound for the number of local complementations to transform a simple graph F' into a locally equivalent graph F''. The mapping ϕ is a covering of SW above SWF (see [8]), and so there exists an integer $k \geq 1$, called the index of S, such that $|\phi^{-1}(F)| = k$ for every fundamental graph F of S. The determination of the index is important because it allows to enumerate the number of graphs which are locally equivalent to a given one (work in progress).

Conjecture 3.7 *The index k of an isotropic system S is such that either $k = 2^S$ or $k = 2^S + 2$ for some integer $s \geq 0$.*

M. Mulder makes the two following conjectures [18].

Conjecture 3.8 *A connected graph G satisfies $G * v = G$ for every vertex v if and only if either $G = K_1$ or $G = K_2$.*

Conjecture 3.9 *A connected graph G and its complement $\overline{G}$ are in the same local equivalence class if and only if G is selfcomplementary.*

Finally we recall two conjectures which are now solved. The first one also has been made by M. Mulder [17].

Property 3.1 *Any two locally equivalent trees are isomorphic.*

Property 3.2 *Any graph locally equivalent to a cycle is hamiltonian.*

The first conjecture is proved in [7]. D. Fon-der-Flass proves both conjectures in [11].

4 An upper bound for diam(SW)

Let $S = (L, V)$ be an isotropic system, and let (G, A, B) be a graphic presentation of S. The two following properties are proved in [2].

Property 4.1 $A * u = A + B\{u\}, u \in V$.

Property 4.2 $A * uvu = A + A\{u, v\} + B\{u, v\}, uv \in E(G)$.

The neighborhood function of G is the linear function $n : 2^V \to 2^V$ such that $n(v) = \{w : vw \in E(G)\}$ for every $v \in V$. Thus following (2.5), any $X \in L$ can be expressed like $X = \mathrm{An}(P) + BP$ for some $P \subseteq V$.

Theorem 4.1 *The diameter of the switching graph of S is at most equal to $3/2|V|$.*

Proof: For any two Eulerian vectors A' and A'', let $D(A', A'') = \{v \in V : A'(v) \neq A''(v)\}$, and let $s(A', A'')$ be the minimal length of a word m on V satisfying $A'' = A' * m$. We prove by induction on $d = |D(A', A'')|$ that $s(A', A'') \leq 3/2|D(A', A'')|$, which implies $s(A', A'') \leq 3/2|V|$. If $d = 0$ we have obviously $s(A', A'') = 0$, so that the inequality actually holds when $d < 1$. We suppose $d \geq 1$, we let $U = D(A', A'')$, we let (G', A', B') and (G'', A'', B'') be the graphic presentations associated to A' and A'' respectively, and we let n' and n'' be the neighborhood functions of G' and G'' respectively.

If there exists $u \in U$ such that $A' * u(u) = A''(u)$, we have $|D(A' * u, A'')| = d - 1$. Using induction we get $s(A', A'') \leq s(A' * u, A'') + 1 \leq 3/2|D(A' * u, A'')| + 1 < 3/2|D(A', A'')|$.

Thus we suppose $A' * u(u) \neq A''(u)$ for every $u \in U$ and exchanging the roles of A' and A'', we also can assume that $A'' * u(u) \neq A'(u)$. Following (4.1), we have $A' * u(u) = A'(u) + B'(u)$, so that $A'(u) + B'(u) \neq A''(u)$. The three values $A'(u), A''(u), B'(u)$ are nonnull and such that $B'(u) \neq A'(u) \neq A''(u)$. We have either $B'(u) \neq A''(u)$ or $B'(u) = A''(u)$. The first case would imply that $B'(u), A'(u)$ and $A''(u)$ are the three nonnull elements of K, so that $A'(u) + B'(u) + A''(u) = 0$, a contradiction with $A'(u) + B'(u) \neq A''(u)$. Therefore $B'(u) = A''(u)$. Exchanging the roles of A' and A'', we also have $B''(u) = A'(u)$.

Now let us choose some $u \in U$, which is possible because $U \neq \emptyset$. The vector $A'n'(u) + B'\{u\}$ belongs to L, so that it can be expressed in terms of the graphic presentation (G'', A'', B''), say

$$A'n'(u) + B'\{u\} = A''n''(P) + B''P \text{ for some } P \subseteq V. \tag{1}$$

We claim that $P \subseteq U$. Otherwise we could find an element $x \in P\backslash U$. Since $x \notin U$ we have $A'(x) = A''(x)$, so that $(A'n'(u)+A''N''(P))(x)$ is either equal to 0 or $A''(x)$. We have also $B'\{u\}(x) = 0$ because $x \notin U$, and $B''P(x) = B''(x)$ because $x \in P$. Thus (1) implies either $B''(x) = 0$ or $B''(x) = A''(x)$, a contradiction because A'' and B'' are supplementary.

Now we prove that U either contains an edge of G' or an edge of G''. Suppose the contrary. Projecting each member of (1) over K^U, we get

$$A'[n'(u) \cap U] + B'\{u\} = A'[n''(P) \cap U] + B''[P \cap U]. \tag{2}$$

We have $n'(u) \cap U = \emptyset$ because $u \in U$ and U is independent in G', $n''(P) \cap U = \emptyset$ because $P \subseteq U$ and U is independent in G'', and $P \cap U = P$ because $P \subseteq U$. Thus (2) yields $B''P = B'\{u\}$, which implies $P = \{u\}$ and $B'(u) = B''(u)$, a contradiction with $B'(u) = A''(u) \neq A'(u) = B''(u)$.

If uv is an edge of G'' contained in U, let $A = A'' * uvu$. Following (4.2) we have $A(u) = B''(u) = A'(u), A(v) = B''(v) = A'(v)$ and $A(w) = A''(w)$ for every $w \in V\backslash\{u,v\}$. Therefore $|D(A, A')| = d - 2$. By induction we get $s(A', A'') \leq s(A' * uvu, A'') + 3 \leq 3/2(d - 2) + 3 = 3/2|(A', A'')|$. If uv is an edge of G' contained in U, we let $A = A' * uvu$, and we conclude similarly.

5   Hamiltonicity of SW

To check some of the preceding conjectures with a computer, it is necessary to explore the whole set of the graphic presentations (or equivalently of the Eulerian vectors) of an isotropic system. This can be done efficiently by means of a recursive procedure PATH(S, A, v). The parameters are an isotropic system $S = (L, V)$, an Eulerian vector A of S and some $v \in V$. PATH generates a hamiltonian path of SW, the switching graph of S, ending at A and $A * v$ (so that every edge of SW belongs to a hamiltonian cycle). The following material introduced in [1, 2] is needed. For $v \in V$ and $x \in K\backslash\{0\}$, let $L|_x^v$ be the canonical projection over K^{V-v} of $\{A \in L : A(v) = 0 \text{ or } A(v) = x\}$. Then $S|_x^v = (V - v, L|_x^v)$ is an isotropic system called an elementary minor of S, and a graphic presentation of $S|_x^v$ can be derived from a graphic presentation of S. For $v, w \in V, v \neq w$, every Eulerian vector B of S satisfies either $B * (vw)^2 = B$ or $B * (vw)^3 = B$, and the actual case can be checked from a graphic presentation of S. The following version of PATH(S, A, v) generates a succession $m = (v_1, v_2, \dots, v_k)$ of elements of V, and the hamiltonian path is $A_0, A_1, \dots, A_k$ defined by $A_0 = A$ and $A_i = A_{i-1} * v_i$ for $1 \leq i \leq k$. The validiy of the algorithm easily follows from the properties stated in [1, 2] for the elementary minors of an isotropic system. The notation $A\backslash v$ stands for the canonical projection of A over K^{V-v}.

```
begin
    if V = {v} then output v
    else begin
        let w ∈ V − v;
        let p be such that A * (vw)^p = A;
        for i := 1 to p do begin
            PATHS(S|^v_{A(v)}, A\v, v);
            A := A * vw;
            if i < p then output v
        end
    end
end.
```

The sequence m produced by the algorithm is reminiscent of the Gray code (see [20], p.10) for generating a hamiltonian path of the hypercube, and actually if the graph with no edge is a fundamental graph of S, then SW is equal to a hypercube κ, and m is a Gray code for κ. Fu-ji Zhang and Xiao-Fong Guo [23] studied an analogue of SW for an arbitrary Euler graph H, and proved that every edge of SW belongs to an hamiltonian cycle.

Added in proof. Some conjectures have been proved or disproved since the presentation of this paper. D.G. Fon-der-Flaass disproved Conjectures 3.4, 3.6 and 3.9. He proved Conjecture 3.5 and improved Theorem 4.1 with the inequality diam(SW) $\leq \frac{10|V|}{9}$, which is sharp. The author proved Conjecture 3.7. P. Seymour proved Cunningham's conjecture for binary matroids (see Theorem 3.1).

References

[1]   A. Bouchet, Isotropic systems, *European Journal of Combinatorics* **8**(1987), 231-244.

[2]   A. Bouchet, Graphic presentations of isotropic systems, *J. Combinatorial Theory Ser. B* **45**(1988), 58-76.

[3]   A. Bouchet, Un algorithme polynômial pour reconnaître les graphes d'alternance, *C. R. Acad. Sc. Paris*, t. 300, Série I, **16**(1985), 569-572.

[4]   A. Bouchet, Reducing prime graphs and recognizing circle graphs, *Combinatorica* **7**(1987), 243-254.

[5]   A. Bouchet, Connectivity of isotropic systems, to appear in *Proceedings of the 3rd International Conference on Combinatorial Mathematics*, New York, 1985.

[6]   A. Bouchet, Tutte-Martin polynomials and orienting vectors of isotropic systems, submitted.

[7]   A. Bouchet, Transforming trees by successive local complementations, *J. Graph Theory* **12**(1988), 195-207.

[8]   A. Bouchet, Digraph decompositions and Eulerian systems, *SIAM Journal on Algebraic and Discrete Methods* **8**(1987), 323-337.

[9] A. Bouchet, Unimodularity and circle graphs, *Discrete Mathematics* **66**(1987), 203-208.

[10] W. Cunningham, Decomposition of directed graphs, *SIAM Journal on Algebraic and Discrete Methods* **32**(1982), 214-228.

[11] G. D. Fon-der-Flaass, letter, december 1966.

[12] H. de Fraysseix, Local complementations and interlacement graphs, *Discrete Mathematics* **33**(1981), 29-35.

[13] C. P. Gabor, W Hsu, K.J. Supowit, Recognizing circle graphs in polynomial time, *Proc. 26 IEEE Annual Symposium*, 1985.

[14] M. C. Golumbic,*Algorithmic Graph Theory and Perfect Graphs*, Academic Press, New-York, 1980.

[15] A. Kotzig, Eulerian lines in finite 4-valent graphs and their transformations, in *Theory of Graphs* (Erdös and Katona, eds.), Proceedings of the Colloqium held at Tihany (Hungary), Sept. 1966, 219-230, North-Holland, Amsterdam, 1968.

[16] A. Kotzig, Quelques remarques sur les transformations κ, séminaire Paris, 1977.

[17] H. Mulder, in Problem Session, Meeting on Graph Theory, Oberwolfach, 1986.

[18] M. Mulder, letter, July 1987.

[19] W. Naji, Reconnaissance des graphes de cordes, *Discrete Mathematics* **54**(1985), 329-337.

[20] A. Nijenhuis, H. Wilf, *Combinatorial Algorithms*, Academic Press, New York, 1975.

[21] K. Truemper, Elements of a decomposition theory for matroids, in *Progress in Graph Theory* (Bondy and Murty eds.), Proceedings of the Silver Jubilee Conference on Combinatorics held in Waterloo, 1982, Academic Press, Toronto, 1984.

[22] W. T. Tutte, Lectures on matroids, *Journal of Research of the National Bureau of Standards* **69B**(1965), 1-47.

[23] Fu-ji Zhang, Xiao-Fong Guo, Hamilton cycles in Euler tour graph, *J. Combinatorial Theory Ser. B* **40**(1986), 1-8.

TWO EXTREMAL PROBLEMS IN INFINITE ORDERED SETS AND GRAPHS: INFINITE VERSIONS OF MENGER AND GALLAI-MILGRAM THEOREMS FOR ORDERED SETS AND GRAPHS

J.M. BROCHET
Department of Mathematics and Statistics
University of Calgary
Calgary, Alberta, T2N 1N4
Canada

M. POUZET*
Institut de Mathématiques-Informatique
Laboratoire d'Algèbre Ordinale
Université Claude Bernard (Lyon I)
69622 Villeurbanne Cedex
France

Abstract

We survey two infinite extremal problems: 1) Versions of Menger's theorem for infinite orders; 2) extensions of the Gallai-Milgram theorem to infinite graphs. We mention for example the two following results: 1) If an ordered set P contains at most k pairwise disjoint maximal chains, where k is finite, then for any finite family of maximal chains of P there is a set of size k which intersects each of these chains.
2) We define a path to be a directed graph whose transitive closure is a linear ordering. If each finite set of vertices of an undirected graph G with no infinite independent set is contained in the union of k pairwise disjoint paths, where k is finite, then G is covered by k pairwise disjoint paths.

1 Introduction

We survey two extremal problems from the area of infinitary combinatorics: Menger properties for infinite ordered sets and Gallai-Milgram properties for infinite graphs. The reader will find the original theorems in [15] and [13]. The first question tries to relate the maximum number of pairwise disjoint maximal chains and the minimum size of a set intersecting every maximal chain ("cutset"), whereas the second question seeks to relate the minimum number of paths necessary to cover a graph with the maximum size of its

---

*The second author has been partially supported by the C.N.R.S., P.R.C. Math.-info.

G. Hahn et al. (eds.), Cycles and Rays, 51–74.
© *1990 by Kluwer Academic Publishers. Printed in the Netherlands.*

independent sets, assumed to be finite. The introduction of a new notion of path is necessary in the second problem: We define a path to be a graph whose transitive closure is a
linear ordering.

In each problem, we try to extend a classical result about finite structures to infinite
structures. While we restrict here our attention to the field of combinatorics, similar
questions occur in different areas like, for example, the conjecture about the existence of
invariant non trivial closed subspaces for continuous linear operators in Hilbert spaces.
Such questions naturally suggest to ask what general techniques are available for bridging
the gap between the finite and the infinite.

The compactness theorem, whether from logic or topology, is now a classical tool used
for going from the finite to the infinite, as illustrated by Dilworth's theorem [12]. Sometimes
compactness can also be used to deduce a result about a finite structure from the infinite
version, as in Ramsey's theorem [20]. There are some variants of compactness which can be
more convenient, for example Rado's selection principle, the coherence lemma of Fraïssé,
and König's tree argument for the denumerable case.

Slightly more sophisticated than the compactness theorem are tools like Baire's theorem
and its variants, such as the omitting-type theorem, which can often be used for countable
structures, and Martin's axiom which may apply to uncountable structures of size less than
the continuum.

Of course Zorn's lemma or equivalently transfinite inductions are also heavily employed
to go from the finite to the infinite, even if the application is not direct. We can define on
a well-founded ordered set a function with ordinal numbers as values (the height-function),
and then do an induction on these ordinals. The well-founded ordered set to be considered
is not always itself obvious. For example, in a kind of extension of Dilworth's theorem
for some ordered sets with no infinite antichains, Abraham uses in [1] the height of the
sets of the (finite) antichains ordered by the reverse-inclusion (this height-function was also
used by Milner and Prikry in [16] §4 to give a direct proof by induction of a result due to
Pouzet). Another chain decomposition theorem for some other ordered sets is proved by
induction by Todorcevic in [21], by considering a height-function on the set of the chains
of size 2.

Recently a new method, involving stationary sets to define obstructions, was used by
Aharoni, Nash-Williams, and Shelah in [5] to extend the marriage theorem of Hall, and by
Aharoni in [2] for the extension of the duality theorem of König for bipartite graphs.

In spite of these results, there are still very few non-standard methods and evidently
some fascinating open problems for which the known methods fail. Let us mention, for
example the question of unfriendly partitions of a graph: Is it possible to split the vertex-
set of an infinite countable (undirected) graph into two parts such that each vertex has no
more neighbours in its cell of the partition than it has in the other one?

We discuss Menger properties for infinite ordered sets in §2. The problem originates in
a question of Zaguia and the second author who asked in [22] if the minimum size of a set
intersecting every maximal chain was equal to the maximum number of pairwise disjoint
maximal chains in a chain-complete ordered set. We mention a more general result (modulo
a finiteness assumption) that we conjectured with Woodrow and proved with Aharoni in
[6]. *Let k be a finite integer and $\mathcal{P}$ be an arbitrary ordered set in which there is no family
of $k + 1$ pairwise disjoint maximal chains. Then for any finite family of maximal chains
of $\mathcal{P}$ there is a set of size $\leq k$ which intersects each of the chains in the finite family.* We

also cite some results from [9], due to the first author, about ordered sets which are the transitive closure of finitely many chains: *There is a function h such that if x is an arbitrary element of an ordered set which is the transitive closure of at most k chains, then there is a set of at most $h(k) - 1$ elements incomparable with x and which meets each maximal chain avoiding x). A fortiori, there are at most $f(k) \le h(k)$ pairwise disjoint maximal chains. The minimal function f grows faster than any polynomial function.*

In §3 we survey partial answers for Gallai-Milgram properties for infinite graphs. Those were obtained in [11]. As already emphasized, an infinite path is any graph whose transitive closure is a linear ordering. We deduce a Gallai-Milgram property for undirected graphs from a stronger statement: *An undirected graph with no infinite independent set is covered by $\le k$ pairwise disjoint paths provided that each finite set is contained in the union of $\le k$ pairwise disjoint finite paths.* Hence, an undirected graph without $(k + 1)$ independent vertices is covered by $\le k$ paths. We also discuss the following result. *If $\mathcal{G} = (V, E)$ is a countable path, $\mathcal{H} = (H, \cdot)$ a finite group, θ a function from E into H, then there is a set $E' \subseteq E$ of edges such that $\mathcal{G}' = (V, E')$ is still a path and $\tilde{\theta}(p) = \theta(x_0, x_1) \cdot \theta(x_1, x_2) \cdot \ldots \cdot \theta(x_{n-1}, x_n) \in H$ depends only upon the extremities x_0, x_n for any finite path $p = (x_0, x_1, \ldots, x_n)$ of $\mathcal{G}'$.*

In §4, we mention some results, due to Aharoni, Milner, Prikry [7] and also Milner, Shelah [17], on the problem of unfriendly partitions of a graph, previously mentioned. We also cite results and questions concerning the profile function of a relational structure (i.e. the number of non-isomorphic substructures of a given cardinality; cf [18] and [19]).

Finally we would like to express our thanks to R. Aharoni, E. C. Milner and R. E. Woodrow for useful conversations.

2   The Menger theorem for ordered sets

A *cutset* of a family of non-empty sets $\mathcal{F}$ is a subset $S \subseteq \cup \mathcal{F}$ which has a non-empty intersection with each member F of $\mathcal{F}$, and we define

$$\mathrm{cut}(\mathcal{F}) = \min\{|S| : S \text{ is a cutset of } \mathcal{F}\}.$$

$\mathcal{F}$ is a *disjoint family* if the members of $\mathcal{F}$ are pairwise disjoint, and for any family $\mathcal{F}$ of non-empty sets we define

$$\mathrm{disj}(\mathcal{F}) = \sup\{|\mathcal{F}'| : \mathcal{F}' \subseteq \mathcal{F} \text{ and } \mathcal{F}' \text{ is disjoint}\}.$$

Clearly $\mathrm{disj}(\mathcal{F}) \le \mathrm{cut}(\mathcal{F})$ and we say that $\mathcal{F}$ is a *Menger family* if the equality holds, i.e. if $\mathrm{disj}(\mathcal{F}) = \mathrm{cut}(\mathcal{F})$. We say that $\mathcal{F}$ is *finitely Menger* if the common value is finite.

2.1   The Menger theorem for finite ordered sets

For a *finite directed graph* $\mathcal{G} = (V, E)$ and any sets $A, B \subseteq V$, the *Menger theorem*, [15], says that *the family of the (A, B)-paths* (finite paths starting in A and ending in B) *is Menger*. For extensions of this theorem to infinite graphs, see [3] and [4].

For a *finite* ordered set $\mathcal{P} = (P, \le)$, a maximal chain is exactly an (A, B)-path (i.e. a finite path starting in A and ending in B) in the cover graph, where A is the set of the minimal elements of $\mathcal{P}$ and B is the set of the maximal elements of $\mathcal{P}$ (and where (x, y) is an edge in the cover graph if $x < y$ and x and y are consecutive). Thus the *family* $M(\mathcal{P})$ *of the maximal chains of a finite ordered set $\mathcal{P}$ is Menger*.

2.2   Extension to the infinite: Counter-examples

2.2.1   First counter-example

In the ordered set $\mathcal{P}$ depicted in figure 1, $\mathrm{disj}(M(\mathcal{P})) = 1$ but $\mathrm{cut}(M(\mathcal{P})) = \omega$. Thus the family of the maximal chains of an infinite ordered set is not necessarily Menger. Therefore two questions can be asked:

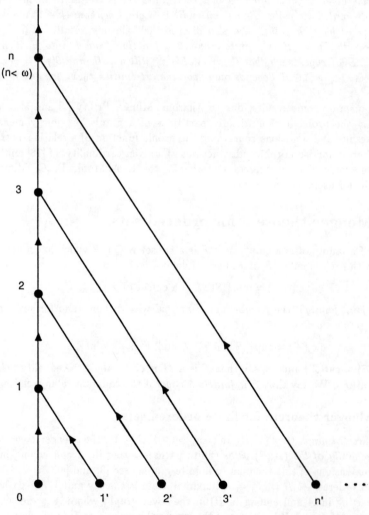

Figure 1

1. What additional conditions need to be imposed on a partially ordered set $\mathcal{P}$ in order to ensure that $M(\mathcal{P})$ is Menger?

2. Is there some weakening of the Menger condition that is satisfied by an arbitrary partial order?

2.2.2 Second counter-example

In the first example one may note that the maximal chains have no supremum and so we can ask if $M(\mathcal{P})$ is Menger for *chain-complete* ordered sets $\mathcal{P}$ (i.e. any non-empty chain has an infimum and a supremum in $\mathcal{P}$). In fact, this was conjectured to be the case by Zaguia in his thesis, [22]. But even in this case some additional conditions are needed. For the ordered set $\mathcal{P}$ shown in figure 2 provides a counter-example since it is chain-complete (and even well-founded), disj$(M(\mathcal{P})) = \omega$ (but this supremum is not attained) and cut$(M(\mathcal{P})) = \kappa$ where κ is any infinite cardinal (cf [6]).

2.3 A general result

Most of the results that we obtain have some finiteness assumption in them. For example, *if all the chains in $\mathcal{P}$ are finite*, we deduce as in (2.1) from a result on graphs due to Aharoni, [3], that *there is a disjoint family of maximal chains and a cutset consisting of a choice of exactly one point of each element of this disjoint family* (and so $M(\mathcal{P})$ is Menger).

The main result (valid for an arbitrary ordered set $\mathcal{P}$) obtained with Aharoni and conjectured with Woodrow is the following:

Proposition 2.1 [6] *Let $\mathcal{P}$ be an ordered set. If disj$(M(\mathcal{P})) = k$ is finite, then cut$(C) \leq k$ for any finite subset C of $M(\mathcal{P})$.*

Actually we can replace $M(\mathcal{P})$ by $M_A(\mathcal{P})$, the family of those maximal chains disjoint from a fixed set $A \subseteq P$.

The result was previously obtained for $k = 1$ by a direct method in [10] (induction and introduction of a "special" chain). From this result for $k = 1$ can be deduced a slightly stronger result:

Proposition 2.2 [10] *Let $\mathcal{P}$ be an ordered set and a finite family $C \subseteq M(\mathcal{P})$. There are $C_1, C_2 \in M(\mathcal{P})$ such that $C_1 \cup C_2 \subseteq \cup C$ and $C_1 \cap C_2 = \cap C$.*

With Dilworth's theorem, [12], we get another corollary:

Corollary 2.1 [10] *If the maximal chains of a finite width ordered set $\mathcal{P}$ pairwise intersect, then they have a common point.*

More precisely *for any family $C \subseteq M(\mathcal{P})$, there are $C_1, C_2 \in M(\mathcal{P})$ such that $C_1 \cap C_2 = \cap C$*; (this last property also holds for ordered sets $\mathcal{P}$ such that $M(\mathcal{P})$ is compact, but not for chain-complete ordered sets $\mathcal{P}$; moreover if $M(\mathcal{P})$ is compact, then we can ensure that $C_1 \cup C_2 \subseteq \cup C$, but not if we assume only that $\mathcal{P}$ has finite width.)
This corollary cannot be extended to $k \geq 2$, as shown by the following example of Kierstead [13].

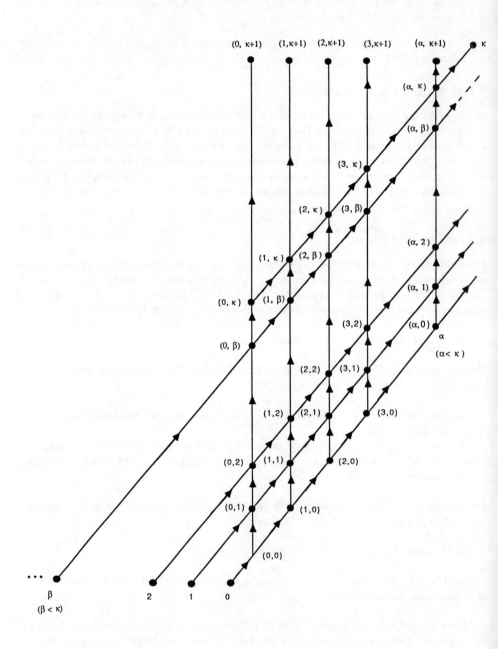

Figure 2

The ordered set $\mathcal{P} = (P, \leq)$, pictured in figure 3, is of width 2, but $\operatorname{cut}(M(\mathcal{P})) = 4$ while $\operatorname{disj}(M(\mathcal{P})) = 2$.

Let us mention the following question.

Problem. [10] Does the conclusion of the corollary still hold under the weaker assumption that the antichains are finite?

Besides a direct proof using the technique of (infinite) alternating paths, we can reduce the main theorem, by another approach, to the following proposition and the two next lemmas.

Proposition 2.3 [10] *Let $\mathcal{P}$ be an ordered set. If $M(\mathcal{P})$ is closed in 2^P (for the product topology) (i.e. $M(\mathcal{P})$ is compact), then $M(\mathcal{P})$ is finitely Menger.*

We can replace $M(\mathcal{P})$ by $M_A(\mathcal{P})$ for any subset A of P.

Definition 2.1 [6] *If $\mathcal{P} = (P, \leq)$ is an ordered set and if $\mathcal{C} \subseteq M(\mathcal{P})$, we define the ordered set $\chi(\mathcal{C}) = (\cup\mathcal{C}, \leq)$, where $\leq'$ is the transitive closure of $\cup\{\leq \mid C : C \in \mathcal{C}\}$ and $\leq \mid C = \{(a, b) \in \leq : a, b \in C\}$.*

Lemma 2.1 [6] *If $\mathcal{C}$ is a finite family of maximal chains of the ordered set $\mathcal{P}$, then $M(\chi(\mathcal{C})) \subseteq M(\mathcal{P})$.*

Lemma 2.2 [6] *If the ordered set $\mathcal{P}$ is such that $\mathcal{P} = \chi(\mathcal{C})$ for some finite family $\mathcal{C} \subseteq M(\mathcal{P})$, then $M(\mathcal{P})$ is closed in 2^P.*

In the above proposition, compactness is used to reduce the problem to a finite set. The ordered sets $\mathcal{P}$ where $M(\mathcal{P})$ is compact have been studied by Bell and Ginsburg, who showed in [8] that they are characterized by the two equivalent properties:

1. *Any cutset of $M(\mathcal{P})$ contains a finite cutset.*

2. *For any $x \in P$, there is a finite set $F \subseteq \operatorname{inc}(x) = \{y \in P : x \text{ and } y \text{ are incomparable}\}$, such that $\{x\} \cup F$ is a cutset of $M(\mathcal{P})$.*

For example in the proof of Lemma 2.2, we do an induction, using (1) to deduce "only" (2) which seems apparently weaker. The idea to prove the lemmas is the following one. If $\mathcal{P} = \chi(\mathcal{C})$ where $\{C_i : 0 \leq i < k\} = \mathcal{C} \subseteq M(\mathcal{P})$ is finite, then $x < y$ means that there exists a finite sequence $(x_0, \ldots, x_n)$ such that $x = x_0 < x_1 < \, < x_n = y$ and $x_i, x_{i+1}(i < n)$ are in a same C_{j_i}. Now if n is minimal, then the j_i have to be distinct, and $n \leq k$. Therefore there are only finitely many possibilities for the sequences $(j_0, \ldots, j_{n-1})$. Thus, if for example x is above every element of a chain A, we can "homogenize" these sequences for a cofinal subset of A, and do an induction on the length of this sequence. For example, using this kind of method, to prove Lemma 2.1, we first prove the following useful fact:

Fact. *If $P = \chi(\mathcal{C})$ where $\mathcal{C} \subseteq M(\mathcal{P})$ is finite, and if A, B are two chains of $\mathcal{P}$ such that for any $a \in A$, there is $b \in B$ such that $a < b$, then*

1. *either for any $b \in B$, there is $a \in A$ such that $b < a$*

2. *or there is $b \in B$ such that $a < b$ for any $a \in A$.*

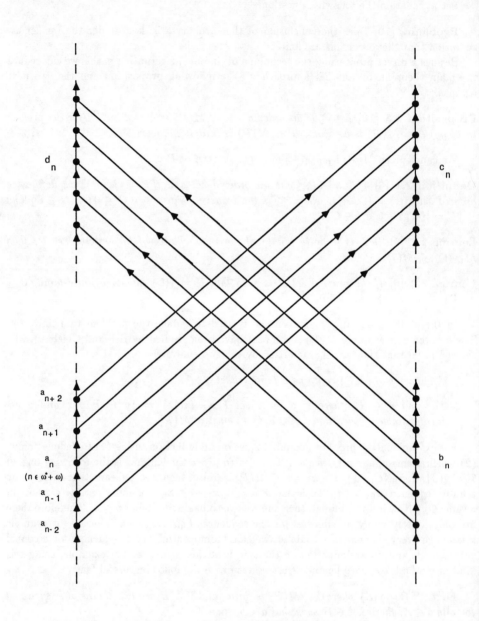

Figure 3

Finally let us mention the following problems.

Problem 1. If $\mathcal{P}$ is an ordered set such that $\mathrm{disj}(M(\mathcal{P})) = k$ is finite, must one have $\mathrm{cut}(\mathcal{C}) \leq k$ for any compact subset of $M(\mathcal{P})$?

We can remark that the lemmas corresponding to Lemma 2.1 and Lemma 2.2, with $\mathcal{C}$ compact instead of $\mathcal{C}$ finite, are both false. We can also formulate a stronger problem:

Problem 2. Let $\mathcal{P}$ be an ordered set, k an integer ≥ 1, $\mathcal{C}$ a subfamily of $M(\mathcal{P})$ such that any finite subfamily of $\mathcal{C}$ has a cutset of size k. If $\mathcal{C}$ is compact, must $\mathcal{C}$ have a cutset of size k?

Note that under these assumptions (without the compactness assumption on $\mathcal{C}$), $\mathcal{C}$ can be decomposed into k subsets $\mathcal{C}_1, \mathcal{C}_2, \ldots, \mathcal{C}_k$, such that any finite subfamily of $\mathcal{C}_i$ has a non-empty intersection. But even if the $\mathcal{C}_i$ are compact, a standard compactness argument does not imply directly that their intersection is non-empty. For example if T is the countable binary tree, then the family $\mathcal{F} = \{T \backslash C : C \text{ is a maximal chain of } T\}$ is compact and $\cap \mathcal{F} = \emptyset$ while $\cap \mathcal{F}' \neq \emptyset$ for any finite subfamily $\mathcal{F}'$ of $\mathcal{F}$.

2.4 The case of chain-complete ordered sets

From the main theorem in (2.3), we deduce easily, using the compactness of the maximal chains for the order-topology, that $M(\mathcal{P})$ is Menger for chain-complete ordered sets $\mathcal{P}$ such that $\mathrm{disj}(M(\mathcal{P}))$ is finite (this last assumption is essential, as shown by the second counter-example in (2.1)). This result was first proved by Kierstead (1986) (using another approach) in [14] and the case $\mathrm{disj}(M(\mathcal{P})) = 1$ is due to Sands (1985).

More precisely, we got with Aharoni the slightly better result:

Proposition 2.4 [6] *Let $\mathcal{P}$ be an ordered set such that $\mathrm{disj}(M(\mathcal{P})) = k$ is finite. If there are k pairwise disjoint complete maximal chains, then $M(\mathcal{P})$ is Menger.*

Again, we can replace $M(\mathcal{P})$ by $M_A(\mathcal{P})$ for $A \subseteq P$, assuming that the k complete chains are in $M_A(\mathcal{P})$.

This property can also be proved directly (see [6]) by a method which provides a way to construct a cutset of size k (via a technique of alternating paths): For the given ordered set $\mathcal{P} = (P, \leq)$, and $x, y \in P$, we let $[x \rightarrow) = \{z \in P : z \geq x\}(\leftarrow y] = \{z \in P : z \leq y\}$, $[x, y] = [x \rightarrow) \cap (\leftarrow y]$. The concatenation of two finite sequences $u = (u_0, \ldots, u_\ell)$ and $v = (v_0, \ldots, v_m)$ is the finite sequence $u * v = (u_0, \ldots, u_\ell, v_0, \ldots, v_m)$. For a disjoint family $\mathcal{C} \subseteq M(\mathcal{P})$ of size $k = \mathrm{disj}(M(\mathcal{P}))$, we define an *alternating path* as a finite sequence $s = (D_0, x_0, x_0', \ldots, D_n, x_n)$ such that:

1. x_j and x_j' are in a same $C \in \mathcal{C}$ and $x_j' < x_j (j < n)$.

2. x_n is in some $C \in \mathcal{C}$.

3. $x_j' < x_{j+1} (j < n)$.

4. D_j is a maximal chain of $[x_{j-1}', x_j]$ such that $D_j \cap (\cup \mathcal{C}) = \{x_{j-1}', x_j\}(0 < j)$, and D_0 is a maximal chain of $(\leftarrow x_0]$ such that $D_0 \cap (\cup \mathcal{C}) = \{x_0\}$.

A *complete alternating path* is a sequence $S*(x'_n, D_{n+1})$ where $s = (D_0, x_0, x'_0, ..., D_n, x_n)$ is an alternating path, x_n and x'_n are in the same $C \in \mathcal{C}, x'_n < x_n$, and D_{n+1} is a maximal chain of $[x'_n \rightarrow)$ such that $D_{n+1} \cap (\cup \mathcal{C}) = \{x'_n\}$. The complete alternating path is called *clean* if the following conditions hold:

1. $x_j \leq x'_{j'}$, if $j < j'$ and x_j and $x'_{j'}$ are in a same $C \in \mathcal{C}$.

2. $D_j \cap D'_j \subseteq \cup \mathcal{C}$ if $j \neq j'$.

Now for each $C \in \mathcal{C}$ it is possible to define $a_C \in C$ as the supremum of the points of C which are end-points of an alternating path (a_C is not necessarily itself such an end-point). The claim is that $\{a_C : C \in \mathcal{C}\}$ is a cutset of $M(\mathcal{P})$. The main fact, using the completeness of the chains $C \in \mathcal{C}$, is the following one: If D is a maximal chain avoiding $\{a_C : C \in \mathcal{C}\}$, then the quotient chain $[D \cap (\cup \mathcal{C})]/_{\mathcal{R}}$ is finite, where, on $D \cap (\cup \mathcal{C}), x \mathcal{R} y$ means $I(x, y) \cap D \cap (\cup \mathcal{C}) = I(x, y) \cap C \cap J_C$, for some $C \in \mathcal{C}$, with $I(x, y) = [x, y] \cup [y, x]$ and $J_C = (\leftarrow a_C]\backslash\{a_C\}$ or $J_C = [a_C \rightarrow)\backslash\{a_C\}$. So, as in the finite case, it is possible to construct a complete alternating path. Any path of minimal length will be clean, and this will enable us to construct $k + 1$ pairwise disjoint maximal chains, contradicting the assumption $k = \text{disj}(M(\mathcal{P}))$.

To conclude, let us give another cutset-property for chain-complete ordered sets.

Proposition 2.5 [10] *Let $\mathcal{P}$ be a chain-complete ordered set and k be a finite integer, then $M(\mathcal{P})$ has a cutset of size $\leq k$ if and only if, for each finite subset F of $\mathcal{P}$, there is a subset X_F of P of size $|X_F| \leq k$ which is a cutset of $M(\mathcal{P} \mid F \cup X_F)$.*

The proof is by induction, using a limit with respect to an ultrafilter.

2.5 More results on ordered sets $\mathcal{P} = \chi(\mathcal{C})$ where $\mathcal{C} \subseteq M(\mathcal{P})$ is finite

As we already emphasized, the proof of the fact that $M(\mathcal{P})$ is compact if $\mathcal{P} = \chi(\mathcal{C})$ (where $\mathcal{C}$ a finite subfamily of $M(\mathcal{P})$) uses the equivalence of two properties of Bell and Ginsburg, in order to employ an induction. Therefore nothing can be deduced from this proof about the possible number of pairwise disjoint maximal chains or the size of the minimal cutsets of $M(\mathcal{P})$, beyond the fact that they are finite. For example, is it true that for any $x \in P$, there is $F \subseteq \text{inc}(x)$ such that $\{x\} \cup F$ is a cutset of $M(\mathcal{P})$ of size $\leq |\mathcal{C}|$, as it is the case for finite ordered sets? Note that the size of a minimal cutset of $M(\mathcal{P})$ can be any finite integer, in a finite ordered set $\mathcal{P} = \chi(\mathcal{C})$ with $|\mathcal{C}| = 3$. This would mean that $\mathcal{P} = \chi(\mathcal{C})$ has a finite cutset-number $\leq |\mathcal{C}|$ (the ordered set $\mathcal{P}$ is said to have a finite *cutset-number $\leq k$* if for any $x \in P$ there is $F \subseteq \text{inc}(x)$ such that $\{x\} \cup F$ is a cutset of $M(\mathcal{P})$ of size $\leq k$). In fact, the proof of the compactness of $M(\mathcal{P})$ does not even ensure that $\mathcal{P} = \chi(\mathcal{C})$ has a finite cutset-number, bounded or not by some function of $|\mathcal{C}|$. Actually the first question has a negative answer; nevertheless there is a positive result. Recall that I is an interval of the ordered set $\mathcal{P}$ if for any $x, y \in I$ and $z \in P$ such that $x \leq z \leq y$, then $z \in I$. The result, due to the first author is then

Proposition 2.6 [9] *There exist two functions $f, g : \omega\backslash\{0\} \rightarrow \omega\backslash\{0\}$ such that for any ordered set $\mathcal{P} = \chi(\mathcal{C})$ where $\mathcal{C}$ is a finite subfamily of $M(\mathcal{P})$, we have:*

1. $\mathrm{disj}(M(\mathcal{P})) \leq f(|\mathcal{C}|)$ and

2. $\mathcal{P}$ has a finite cutset-number $\leq g(|\mathcal{C}|)$

(and, moreover, each cutset of $M(\mathcal{P})$ which is an interval contains a cutset of size $\leq g(|\mathcal{C}|)$). However, these functions grow faster than any polynomial function and can be bounded by an exponential function (more precisely by a Fibonacci sequence).

The finite case is straightforward since the maximal elements of a cutset of $M(\mathcal{P})$ which is an interval provide a cutset of size $\leq |\mathcal{C}|$. In the general case, we replace the notion of maximal elements by the notion of *up-equivalence* between chains (C and D are up-equivalent if for any $c \in C$, there is $d \in D$ such that $c \leq d$ and conversely). The proof is by induction on $|\mathcal{C}|$ and deals simultaneously with $\mathrm{disj}(M(\mathcal{P}))$ and with this disjointedness when we restrict our attention to an arbitrary interval. Then it uses the compactness of $M(\mathcal{P})$ to translate these results about disjointedness in terms of cutsets. Of course the existence of f follows a posteriori from the existence of g, with $f \leq g$.

3 The Gallai-Milgram theorem for infinite directed graphs

3.1 The classical Gallai-Milgram theorem

If $\mathcal{G} = (V, E)$ is a finite directed graph with no independent set of size $k + 1, k < \omega$, the Gallai-Milgram theorem, [13], says that V can be covered by $\leq k$ pairwise disjoint paths.

For a finite ordered set $\mathcal{P} = (P, \leq)$ we get Dilworth's theorem, [12]: if $\mathcal{P}$ has no antichain of size $k + 1(k < \omega)$, then P can be covered by $\leq k$ (pairwise disjoint) chains.

Dilworth's theorem can be extended in this form to any infinite ordered set, but we need the finiteness assumption on k. The proof is by a standard compactness argument, using the fact that the restriction of a chain is a chain, and so using the transitivity of the order. Now an infinite chain can have quite a complicated order-structure. Therefore it is clear that the classical notion of paths is not adequate for an extension, in a similar way, of the Gallai-Milgram theorem to infinite directed graphs (not even in the denumerable case).

3.2 The infinite case, [11]

3.2.1 Notion of path in infinite graphs

A graph is *acyclic* if its transitive closure is an ordering (that really means that there is no cycle).

Definition 3.1 A graph is a *path* if its transitive closure is a linear ordering. A *path of a graph* is a subgraph (not necessarily induced) which is a path; if the graph is acyclic we can restrict our attention to induced subgraphs. A *vertex-path* of a graph $\mathcal{G}$ is the set of vertices of a path of $\mathcal{G}$ (the distinction between path and vertex-path of a graph $\mathcal{G}$ is meaningless if $\mathcal{G}$ is acyclic). The *order-type* of a path is defined in the obvious way.

Remark. The family of the maximal vertex-paths of a graph usually does not satisfy the same weaker Menger's property as the family of the maximal chains of an ordered set. In the acyclic graph shown in figure 7, the maximal vertex-paths pairwise intersect but the

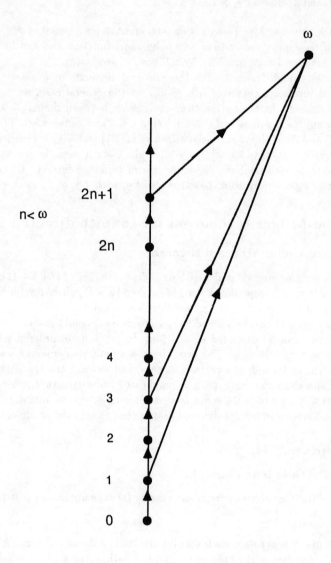

Figure 4: Path of order-type $\omega + 1$

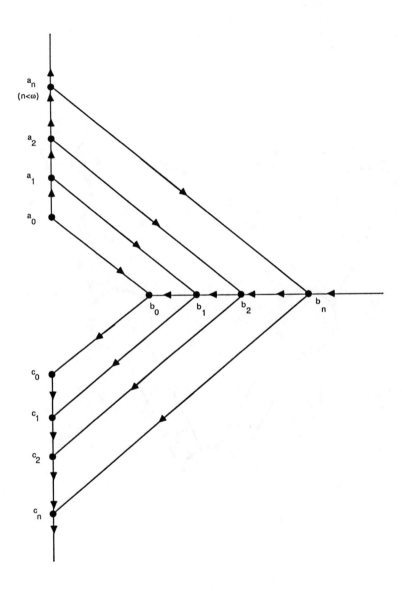

Figure 5: Path of order-type $\omega + \omega^* + \omega$

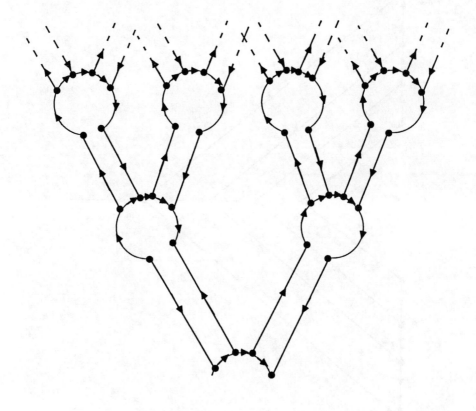

Figure 6: Path of order-type $\omega + (\omega^* + \omega) \cdot \eta + \omega^*$

three maximal vertex-paths $\{x_n : n < \omega\} \cup \{z\}, \{y_n : n < \omega\} \cup \{z\}, \{x_{2n} : n < \omega\} \cup \{y_{2n+1} : n < \omega\}$ have no common element.

We can note that this notion of path generalizes the classical finite notion. It is also very easy to check that, with this definition, the Gallai-Milgram theorem holds in the very particular case where the maximal size of an independent set is 1 (everything is trivial if the graph is transitive or acyclic). More precisely the following obvious fact holds.

Fact. *If $\mathcal{G} = (V, E)$ is a graph such that there is no independent set of size 2, then there is $E' \subseteq E$ such that $\mathcal{G}' = (V, E')$ is a path with the additional property: For any $x, y \in V$ with x below y in the transitive closure of $\mathcal{G}'$, either $(x, y) \in E'$ or there is $z \in V$ such that $(x, z) \in E'$ and $(z, y) \in E'$.*

More generally, the following question arises for graphs with cycles.

Question. How to know when a graph $\mathcal{G} = (V, E)$ admits a spanning path; (i.e. when the vertex-set of a graph is a vertex-path)?

An obvious necessary condition is that V is a weak vertex-path of $\mathcal{G}$ (see definition below) but the condition is not sufficient for graphs with cycles as we shall see in (3.2.2).

Definition 3.2 A *weak vertex-path* of a graph $\mathcal{G} = (V, E)$ is a subset P of V such that any finite subset F of P is contained in a finite vertex-path $F' \subseteq P$ of $\mathcal{G}$ (this means that P is the union of an up-directed family of finite vertex-paths of $\mathcal{G}$).

This distinction is meaningless for acyclic graphs. We remark that P is a weak vertex-path of $\mathcal{G}$ if and only if it is also a weak vertex-path of the induced graph $\mathcal{G} \mid P = (P, E \cap (P \times P))$, and also that a finite weak vertex-path is a vertex-path. From the definition it follows also that each weak vertex-path is contained in a maximal one (which is not true for vertex-paths, as we shall see later).

Besides the notions of path, it is also possible to introduce a notion of chain.

Definition 3.3 A *chain* of a graph $\mathcal{G} = (V, E)$ is a subset C of V such that any finite subset of C is contained in a finite vertex-path of $\mathcal{G}$ (in particular two elements of C are related in the transitive closure of $\mathcal{G}$).

By a standard compactness argument, there is a linear ordering of C respecting the above property (where we replace "vertex-path" by "path"). For acyclic graphs, this notion reduces to the classical notion of chain in the transitive closure of the graph, which is an order.

For transitive graphs, this notion, like the notion of vertex-paths, reduces to the notion of a subset without an independent set of size 2.

A subset of a chain is a chain and each chain is contained in a maximal one. A (weak) vertex-path is a chain but the converse does not hold for infinite graphs (even if acyclic). In the acyclic graph depicted in figure 8, $\{x_n : n < \omega\} \cup \{z\}$ is a maximal chain which is not a vertex-path.

The three notions of maximal chain, maximal weak vertex-path, maximal vertex-path, are identical if and only if each chain is contained in a maximal vertex-path. For example this is the case (by compactness) if the set of maximal vertex-paths of $\mathcal{G} = (V, E)$ is closed in 2^V (i.e. compact) and if each finite vertex-path is contained in a maximal vertex-path (this last condition always holds for acyclic graphs).

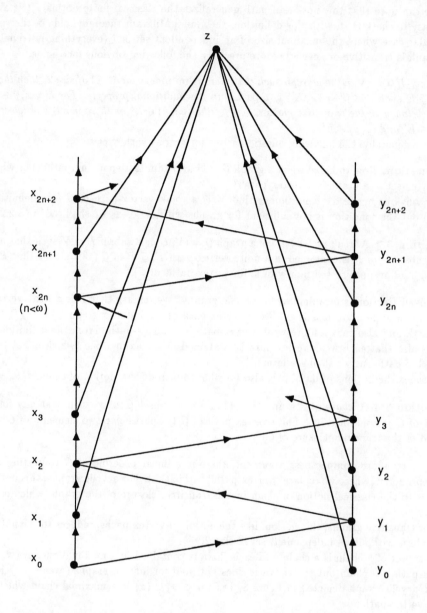

Figure 7

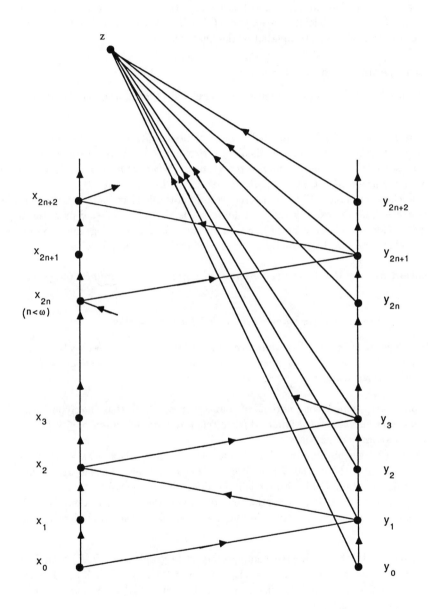

Figure 8

Thus the notion of chain (even for acyclic graph) does not seem to be very helpful to approach the problem (see 3.2.2) of the extension of the Gallai-Milgram theorem to the infinite. For example, we shall see in (3.2.2) an example of an acyclic graph where the transitive closure has no antichain of size 3, but the vertex-set cannot be covered by finitely many (even non disjoint) vertex-paths (cf. example 3). Therefore the width of the transitive closure is not directly related to the question.

3.2.2   Conjectures and a few results

The main problem is whether the extension of the Gallai-Milgram theorem holds for vertex-paths.

Problem. If $\mathcal{G} = (V, E)$ is a directed graph with no independent set of size $k+1 (k < \omega)$, is then V the union of $\leq k$ (pairwise disjoint) vertex-paths (or weak vertex-paths)?

We can solve this problem for undirected graphs: *If there is no independent set of size $k + 1$, V is the union of $\leq k$ pairwise disjoint vertex-paths.*

By the classical Gallai-Milgram theorem, the hypothesis that there is no independent set of size $k + 1$ means exactly that any finite set is the union of $\leq k$ pairwise disjoint vertex-paths (for any directed graph). So, in the case of undirected graphs, the previous property (which can also be proved directly) is deduced from the following result.

Proposition 3.1 *Let $\mathcal{G} = (V, E)$ be an undirected graph with no infinite independent set. Then*

1. *V is an union of finitely many pairwise disjoint vertex-paths.*

2. *If each finite subset of V is contained in the union of $\leq k$ (respectively $\leq k$ pairwise disjoint) finite vertex-paths, then V is the union of $\leq k$ (respectively $\leq k$ pairwise disjoint) vertex-paths.*

The proof uses some strong notions of connectedness. Note that the hypothesis of (2) means that V *is an up-directed union of finite sets which are an union of $\leq k$ (respectively $\leq$ pairwise disjoint) vertex-paths.*

For $k = 1$ (2) says nothing but that V is a vertex-path if it is a weak vertex-path.

If $k \geq 2$, the hypothesis that $\mathcal{G}$ is undirected is essential for (2) as shown by the graph depicted in figure 9 (and (1) is also trivially false for directed graphs).

If $k = 1$ (2) is trivial if $\mathcal{G}$ is acyclic, but otherwise the fact that there is no infinite independent set is essential, as shown by the next example or the graph depicted in figure 9.

Example 1. In the undirected complete bipartite graph $\mathcal{G} = (V, E)$ with $V = X \cup Y, X \cap Y = \emptyset, X = \{x_n : n < \omega\}, Y = \{Y_\alpha : \alpha < (2^\omega)^+\}, E = (X \times Y) \cup (Y \times X), V$ is a weak vertex-path but is not the union of $< (2^\omega)^+$ vertex-paths. We can also note that there is no maximal vertex-path.

Example 2. $\mathcal{G} = (V, E)$ is a countable undirected (bipartite) graph such that V is a weak vertex-path but V is not the union of finitely many pairwise disjoint vertex-paths (and V is the union of the two vertex-paths $\{x_n : n < \omega\}$ and $\{x_{2n+1} : n < \omega\} \cup \{y_n : n < \omega\}$.

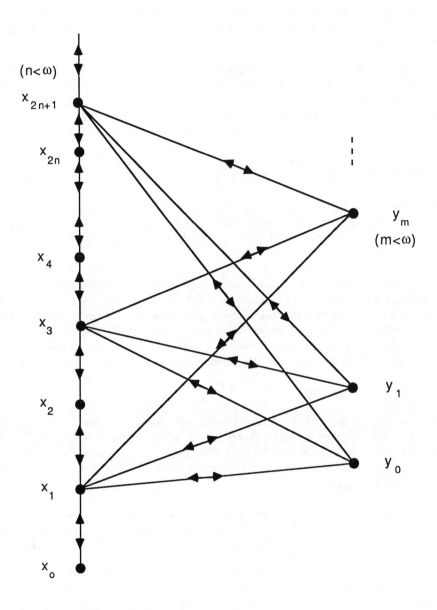

Figure 9

We can also remark that the vertex-path $\{x_{2n+1}\} \cup \{y_n : n < \omega\}$ is not contained in a maximal one. But each finite vertex-path is contained in a maximal vertex-path.

Using this type of idea, we can also build a countable undirected (bipartite) graph such that V is a weak vertex-path but V is not the union of finitely many (even non disjoint) vertex-paths.

Example 3. Let $\mathcal{P} = (P, \leq)$ be the ordered set such that: $P = X \cup Y$ with $X = \{x_n : n \in \omega^* + \omega\}$, $Y = \{y_n : n \in \omega^* + \omega\}$, $x_n < x_m, y_n < y_m$ if $n < m, y_n < y_m$ if $n < m, y_n$ and x_m are incomparable if $m \leq n$.

Let $\mathcal{G}_k = (V_k, E_k)$ be the directed graph ($k \in \omega \backslash \{0\}$), such that $V_k = P \times k$, $((z, \ell), (z', \ell')) \in E$ if and only if $\ell = \ell'$ and $z < z'$ or $\ell' = \ell + 1$ and $z \in X, z' \in Y$ $(z, z' \in P, \ell, \ell' \in k)$.

Let $\mathcal{G} = (V, E)$ be the directed graph such that: $V = \cup\{V_k \times \{k\} : k \in \omega \backslash \{0\}\}$ and $((t, k), (t', k')) \in E$ if and only if $k = k'$ and $(t, t') \in E_k$ or $k < k'(k, k' \in \omega \backslash \{0\}, t \in V_k, t' \in V_{k'})$.

The graphs $\mathcal{G}$ and $\mathcal{G}_k$ are countable and acyclic, with no infinite independent set and such that each finite subset of V (respectively V_k), is contained in the union of two finite pairwise disjoint vertex-paths of V (respectively V_k). The sizes of the independent sets are finitely bounded in the graphs $\mathcal{G}_k(k \in \omega \backslash \{0\})$. Nevertheless V_k is not the union of $k+1$ (even non pairwise disjoint) vertex-paths, ($k \in \omega\{0\}$), and V is not the union of finitely many (even non pairwise disjoint) vertex-paths. We can also note that the transitive closures of these graphs are orderings with no antichain of size 3.

Additionally, let us mention that the Gallai-Milgram theorem holds for transitive graphs (by a standard compactness argument or by an easy reduction to Dilworth's theorem); it also holds when $k = 2$ for the graphs where the undirected dual graph of the transitive closure is connected. In this last case, the decomposition into ≤ 2 pairwise disjoint paths is unique.

We present now a last result.

Proposition 3.2 *Let* $\mathcal{G} = (V, E)$ *be a countable path (we assume that* $(x, x) \notin E$ *for* $x \in V$*)* $(H, \cdot)$ *a finite group, and let a function* $\theta : E \to H$ *be given. For any finite path* $p = (x_0, x_1, \ldots, x_n)$*, we define* $\tilde{\theta}(p) = \theta(x_0, x_1) \cdot \theta(x_1, x_2) \cdot \ldots \cdot \theta(x_{n-1}, x_n) \in H$*. Then there is* $E' \subseteq E$ *such that* $\mathcal{G}' = (V, E')$ *is a path and for any finite path* p *of* $\mathcal{G}'$*,* $\tilde{\theta}(p)$ *depends only upon the extremities of* p*.*

We first prove that we can partition V in maximal intervals of the path where the required property is satisfied and accordingly we replace E by a smaller set of edges (the set of the edges with extremities in two different intervals remains unchanged). Then, if there are two distinct intervals, taking $x < y$ in two different intervals such that the number of different $\tilde{\theta}(p)$, where p is a finite path of the new graph with extremities x and y, is minimum (amongst the values different from 1), we are able to construct by induction (using the denumerability of V) a subset $E' \subseteq E$ with the required property on the interval $[x, y]$; this contradicts the fact that x and y are in two different classes.

We can give another interpretation of this proposition, when $(H, \cdot)$ is a permutation group, to show that this proposition is related to the Gallai-Milgram property. Let $\mathcal{G} = (V, E)$ be a countable path and for each $x \in V$, let B_x be a set $\{x_i : i \in n\}$ of size n, where n is a fixed integer. For each $x, y \in V$ such that $(x, y) \in E$, let $\theta(x, y)$ be a bijection from

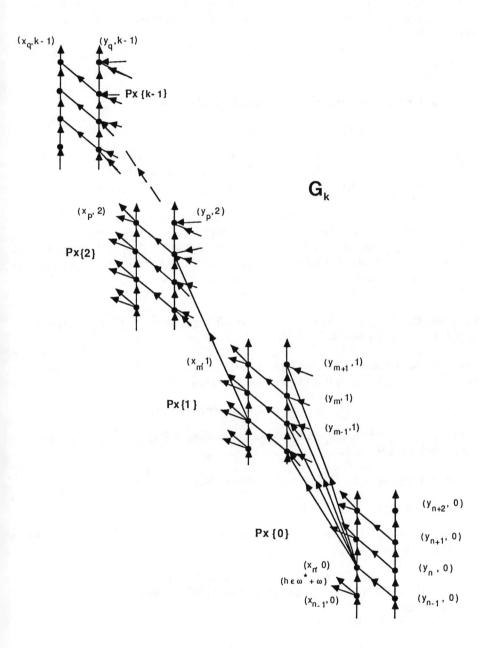

Figure 10

B_x to B_y (or a bijection of n), which can be interpreted as a "linkage" between B_x and B_y. Then the previous property says simply in this case that there exists a "coherent linkage" between all the "boxes" B_x.

Finally, let us mention the problem of the extension of the previous proposition to uncountable paths.

4   Two more examples in combinatorics

4.1   Unfriendly partitions of a graph

For an integer $n \in \omega \setminus \{0\}$, an *n-partition* of a set X is a map $\pi : X \to n$. An *unfriendly n-partition* of an undirected graph $\mathcal{G} = (V, E)$ is an n-partition π of V such that $|\{y \in V : \{x, y\} \in E \text{ and } \pi(x) \neq \pi(y)\}| \geq |\{y \in V \setminus \{x\} : \{x, y\} \in E \text{ and } \pi(x) = \pi(y)\}|$, for any $x \in V$.

It is easy to prove that *any finite graph has an unfriendly 2-partition* (if $\mathcal{G} = (V, E)$, take $\pi : V \to 2$ such that $|\{\{x, y\} \in E : \pi(x) \neq \pi(y)\}|$ is maximum). Thus by a standard compactness argument, *any locally finite graph* (i.e. each element has a finite degree), *has an unfriendly 2-partition*. Cowan and Emerson asked if this was the case for any graph, in particular for a graph having a single vertex with infinite degree.

The following positive results are due to Aharoni, Milner, Prikry.

Theorem 4.1 [7] *If $\mathcal{G} = (V, E)$ has only finitely many vertices of infinite degree, then there is an unfriendly 2-partition of $\mathcal{G}$.*

Theorem 4.2 [7] *Let $k < \omega$ and $m_1 < m_2 < \ldots < m_k$ be infinite cardinals which are regular except maybe m_1. If $\mathcal{G} = (V, E)$ is a graph such that*

$$|\{x \in V : x \text{ is of finite degree}\}| < m_1$$

and such that any vertex with infinite degree has a degree $\in \{m_1, \ldots, m_k\}$, then there is an unfriendly 2-partition of $\mathcal{G}$.

But in a negative sense, Milner and Shelah recently proved the following result:

Theorem 4.3 [17]

1. *It is consistent that there is a graph of size $\aleph_\omega$ with no unfriendly partition and such that each vertex has an infinite degree $\aleph_0, \aleph_1$ or $\aleph_\omega$.*

2. *There is a graph of size $(2^\omega)^{(+\omega)}, (where (2^\omega)^{(+\omega)} = \aleph_{a+\omega}$ if $2^\omega = \aleph_a)$ with no unfriendly partition where all the vertices have infinite degree.*

Nevertheless they also proved:

Theorem 4.4 [17] *Any graph has an unfriendly 3-partition.*

The main question is the following one.

Problem. Is there an unfriendly 2-partition for any denumerable graph?

4.2 The profile function

A *relational structure* of arity $\eta = (n_i)_{i \in I}$ on the set E, where n_i are strictly positive integers, is a family $\mathcal{R} = (\mathcal{R}_i)_{i \in I}$ where each $\mathcal{R}_i$ is an n_i-ary relation on the set E, i.e. an ordered pair (E, R_i), where R_i is a subset of E^{n_i}. For a subset A of E, the *restriction* of $\mathcal{R}$ to A, is the relational structure $\mathcal{R} \,|\, A = (R_i \,|\, A)_{i \in I}$ where $\mathcal{R}_i \,|\, A = (A, R_i, \cap A^{n_i})$. An *isomorphism* between two relational structure of same arity is defined in an obvious way.

For a given relational structure $\mathcal{R} = (\mathcal{R}_i)_{i \in I}$ and a cardinal $\kappa \leq |E|$, we denote by $\varphi_{\mathcal{R}}(\kappa)$ the number of different restrictions, up to isomorphism, of $\mathcal{R}$ to subsets of E of size κ, and we call $\varphi_{\mathcal{R}}$ the *profile function* (of $\mathcal{R}$).

Using elementary linear algebra, the second author proved in [19] that for every integer n, $\varphi_{\mathcal{R}}(n) \leq \varphi_{\mathcal{R}}(n+1)$ provided that $2n + 1 \leq |E|$.

In particular *if all the restrictions of $\mathcal{R}$ to subsets of size $p < \omega$ are isomorphic (i.e. $\varphi_{\mathcal{R}}(p) = 1$), this is also the case for the restrictions of size $p' \leq p$, $(\varphi_{\mathcal{R}}(p') = 1)$, provided that $2p - 1 \leq |E|$.*

The linear algebra argument does not work if we study infinite restrictions, $K < |E|$ and we can ask if $\varphi_{\mathcal{R}}$ is non decreasing. First we have to restrict our attention to cardinals $\kappa < |E|$, since for example, if $\mathcal{R} = (E, \leq)$ where $\leq$ is a linear ordering of E of type ω_1, then $\varphi_{\mathcal{R}}(\aleph_1) = 1$, while $\varphi_{\mathcal{R}}(\aleph_0) = \aleph_1$.

Pouzet and Woodrow obtained that $\varphi_{\mathcal{R}}(\kappa') \leq \varphi_{\mathcal{R}}(\kappa)$ *provided that $\kappa' \leq \kappa < |E|$ and $\varphi_{\mathcal{R}}(\kappa)$ is finite.* They used a result of Kierstead und Niykos for the case where $\varphi_{\mathcal{R}}(\kappa) = 1$ and $\mathcal{R}$ is an m-regular hypergraph (i.e. an ordered pair (E, R), where R is a subset of $[E]^m = \{F \subseteq E : |F| = m\}$), where $m \in \omega$.

Then Macpherson, Mekler and Shelah proved in [18] that $\varphi_{\mathcal{R}}(\kappa') \leq \varphi_{\mathcal{R}}(\kappa)$ if $\kappa' \leq \kappa$ for relational structures of bounded arity, provided that κ or $\varphi_{\mathcal{R}}(\kappa)$ is sufficiently small with respect to $|E|$. The techniques use indiscernible sets and are quite complicated comparatively to the proof of the case where κ is finite.

The question $\varphi_{\mathcal{R}}(\kappa') \leq \varphi_{\mathcal{R}}(\kappa)$ for $\kappa' \leq \kappa < |E|$ still arises.

References

[1] U. Abraham, A note on Dilworth's theorem in the infinite case, *Order* 4(1987), 108-125.

[2] R. Aharoni, König's duality theorem for infinite bipartite graphs, *J. London Math. Soc.* (2) **29**(1984), 1-12.

[3] R. Aharoni, Menger's theorem for graphs containing no infinite paths, *Europ. J. Combinatorics* 4(1983), 201-204.

[4] R. Aharoni, Menger's theorem for countable graphs. To appear in *J. Combin. Th.* Ser. B.

[5] R. Aharoni, C. St. J. A. Nash-Williams, S. Shelah, A general criterion for the existence of transversals, *Proc. London Math. Soc.* (3) 47(1983), 43-68.

[6] R. Aharoni, J-M. Brochet, M. Pouzet, The Menger property for infinite ordered sets. To appear in *Order*.

[7] R. Aharoni, E. C. Milner, K. Prikry, *Unfriendly partitions of a graph,* Preprint 1987.

[8] M. Bell, J. Ginsburg, Compact spaces and spaces of maximal complete subgraphs, *Trans. Amer. Math. Soc.* 283, 1(1982), 329-338.

[9] J. M. Brochet, *On the size of disjoint families of maximal chains and the cutset-number of infinite ordered sets spanned by finitely many chains.* Preprint 1987.

[10] J. M. Brochet, M. Pouzet, Maximal chains and cutsets of an ordered set: A Menger type approach. To appear in *Order.*

[11] J. M. Brochet, M. Pouzet, *Gallai-Milgram properties for infinite graphs.* Preprint 1988.

[12] R. P. Dilworth, A decomposition theorem for partially ordered sets, *Ann. of Math. (2)* **51**(1950), 161-166.

[13] T. Gallai, A. N. Milgram, Verallgemeinerung eines graphentheoretischen Satzes von Rédei, *Acta Sc. Math.* **21**(1960), 181-186.

[14] H. Kierstead, A minimax theorem for chain-complete ordered sets. To appear in *Order.*

[15] K. Menger, Zur allgemeinen Kurventheorie, *Fund. Math.* **16**(1927), 96-115,

[16] E. C. Milner, K. Prikry, The cofinality of a partially ordered set, *Proc. London Math. Soc.* (3) **46**(1983), 454-470.

[17] E. C. Milner, S. Shelah, *Graphs with no unfriendly partition.* Preprint 1988.

[18] D. Macpherson, A. Mekler, S. Shelah, *The number of infinite substructures.* Preprint 1987.

[19] M. Pouzet, Application d'une propriété combinatoire des parties d'un ensemble aux groupes et aux relations, *Math. Z.* **150**(1976), 117-134.

[20] F. P. Ramsey, On a problem of formal logic, *Proc. London Math. Soc.* **30**(1930), 284-286.

[21] S. Todorcevic, *A chain decomposition theorem,* Preprint 1986.

[22] N. Zaguia, *Schedules, cutsets and ordered sets,* Ph.D. thesis, University of Calgary. July 1985.

CHVÁTAL-ERDŐS THEOREM FOR DIGRAPHS*

N. CHAKROUN, D. SOTTEAU
LRI
UA 410 CNRS
Bât. 490
Université Paris-Sud
91405 Orsay Cedex
France

Abstract

This paper deals with two generalizations of a theorem of Chvátal and Erdős which states that if the connectivity of a graph is at least equal to its stability number α, then the graph is hamiltonian. First we prove that under the same conditions, in the case $\alpha \leq 3$, the graph is pancyclic except for two exceptions. Then, in the directed case, if $\kappa(D)$ is the connectivity of a digraph D and $\alpha_2(D)$ the maximum order of a subdigraph of D without 2-cycle, we prove that if $\kappa(G) \geq \alpha_2(D)$, then D is pancyclic for the cases $\alpha_2(D) = 2$ or 3 except for a few exceptions .

1 Introduction

In what follows G will always denote a non directed graph without loop or multiple edge, $\kappa(G)$ its connectivity and $\alpha(G)$ its stability number. In the directed case, D will denote a digraph without loop or multiple arc (but with possibly cycles of length 2 that we will call 2-cycle), $V(D)$ its vertex set and $E(D)$ its arc set. The connectivity number of a digraph D, denoted by $\kappa(D)$, is well defined as the minimum cardinality of a subset S of vertices such that $V(D) - S$ is not strongly connected. On the other hand there are several possible generalizations of the stability number for digraphs (see [8] or [10]). Here we will only work with one of them, $\alpha_2(D)$, which is the maximum order of an induced subdigraph of D without 2-cycle.

When it is clear, we will often use A, instead of $V(A)$, for the set of vertices of a graph or a digraph A. Similarly, we will often use X for the sub(di)graph induced by a subset X of vertices of A.

In a digraph D, we will often work on the partial graph G induced by the 2-cycles of D, and consider G and its subgraphs, either as non directed graphs with edges instead of 2-cycles, or as partial (sub)digraphs of D without specifying, when it is clear by the context.

In the graph G , for any subset X of vertices, $\Gamma(X)$ will denote the set of vertices of $G - X$ adjacent to at least one vertex of X. For any vertex disjoint subgraphs A and B of G, $\Gamma_A(B)$ will denote the set of vertices of A adjacent to at least one vertex of B.

*Research partially supported by PRC Math Info.

G. Hahn et al. (eds.), Cycles and Rays, 75–86.
© 1990 by Kluwer Academic Publishers. Printed in the Netherlands.

To avoid repetitions, we will always use the same notation $u \xrightarrow{A} v$ to denote a hamiltonian path from u to v in the graph (or digraph) A, possibly reduced to one vertex if $|A| = 1$ and, therefore, $u = v$. We will denote a hamiltonian cycle by the sequence of its vertices or subpaths.

C_n (C_n^*) will denote a (directed symmetric) cycle of length n.

For any two complete symmetric digraphs A and B, $EI(A, B) \geq p$ will mean that there exist at least p internally vertex disjoint paths in $A \cup B$ from any vertex of A to any vertex of B. If A and B have cardinality at least p this implies that there are at least p independant arcs (i. e. vertex disjoint arcs) from A to B (see theorem 11.7 of [3]).

For definitions and notations not given here, see [3] or [10].

In [5] Chvátal and Erdős proved , for any graph G, that if $\kappa(G) \geq \alpha(G)$, then G is hamiltonian. This result has given rise to many other sufficient conditions involving the connectivity and the stability number for graphs and digraphs to have some path or cyclic properties. The reader interested in all the problems and results on this subject can refer to the recent well structured survey of Jackson and Ordaz [10]. Here we will just give the proofs of three specific results, one for graphs of stability number at most 3, the other ones for digraphs D with $\alpha_2(D) \leq 3$.

2   Non-directed case

Let us first recall the following result which will be used later.

Theorem 2.1 (Amar, Fournier, Germa [1]) *A graph G such that $\alpha(G) = 2$ and $\kappa(G) \geq \alpha(G)$ is either pancyclic or isomorphic to C_4 or C_5.*

The following remark will also be used several times. Its proof, which is easy, is left to the reader.

Remark 2.2 *If G is a non 2-connected graph with $\alpha(G) = 2$ then $V(G)$ can be covered by two disjoint complete graphs.*

Our first result is only an extension of a theorem of Amar, Fournier and Germa [1] . The proof given here is due to W. McCuaig. It is much nicer and shorter than the one we had before.

Theorem 2.3 *A graph G such that $\alpha(G) = 3$ and $\kappa(G) \geq \alpha(G)$ is either pancyclic or isomorphic to $K_{3,3}$ or $C(8, 1, 4)$ (the circulant graph with vertices the elements of $\mathcal{Z}/8\mathcal{Z}$ and edges $i, i + 1$ and $i, i + 4$ for $0 \leq i \leq 7$) .*

Proof: Let us assume that G is not isomorphic to $K_{3,3}$. Amar, Fournier and Germa proved, in their theorem, that G contains cycles of all lengths between 4 and $|V(G)|$. Let us assume that G contains no triangle. We will show that G is necessarily isomorphic to $C(8, 1, 4)$.

Since G contains no clique of size 3 and no independant set of size 4, we can conclude that $|V(G)| < r(3, 4) = 9$, where $r(3, 4)$ is the classical Ramsay number, and that no vertex of G is adjacent to 4 vertices. So, as G is 3-connected, G is cubic. The only cubic 3-connected triangle free graphs on less than 9 vertices are $K_{3,3}$, $C(8, 1, 4)$ and the 3-dimensional cube. The cube has $\alpha = 4$, and so G is $C(8, 1, 4)$.

Figure 1: D_2 a family of non hamiltonian 2-connected digraphs with $\alpha_2(D) = 2$.

If G has a stability number at least 4 and $\kappa(G) \geq \alpha(G)$, there are only partial results concerning the pancyclicity of G (see [1] or [10]). To end this section, we just recall the following conjecture.

Conjecture 2.4 (Amar, Fournier, Germa) *A graph G, which satisfies $\kappa(G) \geq \alpha(G)$, has cycles of all lengths between 4 and $|V(G)|$ except if G is isomorphic to $K_{\kappa,\kappa}$.*

3 Directed case : $\alpha_2 = 2$

In [2] Bondy suggested to generalize Chvátal and Erdős theorem to digraphs and conjectured that, if $\kappa(D) \geq \alpha_2(D)$, then D is hamiltonian except if it is isomorphic to one of a finite set of graphs. This is not true, as was first noticed by Thomassen [11] who exhibited an infinite family D_2 of counterexamples for $\alpha_2 = 2$ (see Figure 1). However the conjecture maybe true if we allow well characterized infinite families of exceptionnal digraphs. It is supported by Theorems 3.1 and 4.1 for the cases $\alpha_2(D) = 2$ or 3. The following result improves a previous one of Jackson and Ordaz [8] who assumed their digraphs to be 3-connected . Our proof is the same as theirs, pushed a little bit further. As it is not difficult we repeat it here.

Theorem 3.1 *If D is a 2-connected digraph with $\alpha_2(D) \leq 2$ then D is pancyclic except if D is isomorphic to a graph of the family D_2 or C_4^* or C_5^* (see figure 1).*

Proof: If $\alpha_2(D) = 1$, D is a complete symmetric digraph and the result follows. The case $|D| \leq 4$ is obvious. So let us assume that $\alpha_2(D) = 2$ and $|D| \geq 5$. Let us call G the non directed graph with the same set of vertices as D and in which the edge set consists of all pairs of vertices which are joined by a 2-cycle in D. Clearly $\alpha(G) = \alpha_2(D) = 2$. If $\kappa(G) \geq 2$ the result follows from theorem 2.1 since either G and therefore D is pancyclic or G is a C_4 or a C_5 and then, either D is pancyclic or D is a C_4^* or a C_5^*. So we can assume that G is not 2-connected. Then, by Remark 2.2, $V(G)$ can be covered by two complete graphs A and B. The rest of the proof is divided into two cases.
(a) $|A| = 1$ (or similarly $|B| = 1$).
Then D contains cycles of all lengths between 2 and $|B| = |V(D)| - 1$. Moreover, as $\kappa(D) \geq 2$, D contains one arc from A to B and one arc from B to A with different ends in B. So D is hamiltonian.
(b) $|A| \geq 2$ and $|B| \geq 2$.
As $|D| \geq 5$, A or B has cardinality at least 3 and D contains a triangle. As D is 2-connected, there exist two disjoint arcs (a_1, b_1) and (a_2, b_2) from A to B. But also there exist two disjoint arcs from B to A and then necessarily, for some $i = 1$ or 2, there exists one arc ba with $b \neq b_i$ and $a \neq a_i$ except if D is one of the graphs of the family D_2 . If

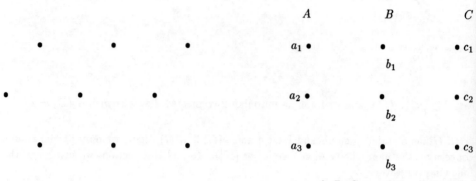

A, B, C complete subdigraphs

F_3 D_3

Figure 2: F_3 non pancyclic and D_3 non hamiltonian, 3-connected digraphs with $\alpha_2(D) = 3$.

$D \neq D_2$ then, using the arc (b, a) and the arc (a_i, b_i) with paths of A and B, we get cycles of all lengths between 4 and $|D|$.

Corollary 3.2 *If D is a 2-connected digraph with $\alpha_2(D) \leq 2$ then D is hamiltonian except if D is isomorphic to a digraph of the family D_2.*

Let us notice that this corollary is a weaker version of the same result with $\alpha_1(D) \leq 2$ where $\alpha_1(D)$ is the maximum order of an induced subdigraph of D without directed cycle. As pointed by Y. Mannoussakis, this stronger result (only for hamiltonicity) is an easy consequence of a theorem of M.C. Heydemann [7] (see [10] for more details).

4   Directed case : $\alpha_2 = 3$

We will now state the most important result of this paper.

Theorem 4.1 *If D is a 3-connected digraph with $\alpha_2(D) \leq 3$, then D is pancyclic except if D is isomorphic to a digraph of the family D_3 or to $C^*(8, 1, 4)$, $K_{3,3}^*$ or F_3 (see Figure 2).*

If $\alpha_2(D) \leq 2$ the result follows from Theorem 3.1; D is pancyclic since D_2, C_4^* and C_5^* are not 3-connected. So we can assume that $\alpha_2(D) = 3$.

Let us first state a proposition which is a particular case of the theorem and is useful to make the proof of the theorem more clear.

Proposition 4.2 *If D is a 3-connected digraph such that $V(D)$ can be covered by three complete symmetric digraphs then either D is pancyclic or D is isomorphic to D_3, $K_{3,3}^*$ or F_3.*

The complete proof of the theorem is very long and involves a lot of cases. It can be found in [4]. Here, by sake of shortness, we will only give a direct proof of its corollary which uses the same techniques but is simpler.

Corollary 4.3 *Every 3-connected digraph* D *with* $\alpha_2(D) \leq 3$ *is hamiltonian except if it is isomorphic to a digraph of the family* D_3.

This last result was independently proved for 5-connected graphs only by Guia, Jackson and Ordaz [6] . To give the proof of Corollary 4.3., we only need the following corollary of Proposition 4.2, that we will prove directly.

Corollary 4.4 *If* D *is a 3-connected digraph such that* $V(D)$ *can be covered by three complete symmetric digraphs then either* D *is hamiltonian or* D *is isomorphic to a digraph of the family* D_3.

Proof: Let us call A, B and C three vertex disjoint complete symmetric digraphs which cover $V(D)$. Let us first introduce two definitions which will be useful.

We will say that D *has a configuration (1)* if D satisfies the hypothesis of the corollary and contains three arcs (a, b), (b', c) and (c', a') with $\{a, a'\} \in A$, $a \neq a'$ if $|A| > 1$, $\{b, b'\} \in B$, $b \neq b'$ if $|B| > 1$ and $\{c, c'\} \in C$, $c \neq c'$ if $|C| > 1$ (with possibly a permutation on A, B and C).

We will say that D *has a configuration (2)* if D satisfies the hypothesis of the corollary and contains four arcs (a, b), (b', c), (c', d) and (d', a') with $\{a, a'\} \in A$, $a \neq a'$ if $|A| > 1$, $\{c, c'\} \in C$, $c \neq c'$ if $|C| > 1$ and $\{b, b', d, d'\} \in B$, $|B| \geq 2$, $\{b, b'\} \cap \{d, d'\} = \emptyset$ and $b \neq b'$ or $d \neq d'$ if $|B| > 2$ (with possibly a permutation on A, B and C).

Let us first notice that if D has a configuration (1) or (2) then it is hamiltonian. We consider two cases.

Case 1: Two of the complete graphs, say A and C, are without arc from A to C.

Then, as $\kappa(D) \geq 3$, we have $|B| \geq 3$, $EI(A, B) \geq 3$ and $EI(B, C) \geq 3$. We may assume that D contains no arc from C to A, otherwise it is not difficult to see that D has a configuration (1) and therefore is hamiltonian. Then $A \cup B$ and $B \cup C$ are 3-connected, $EI(C, B) \geq 3$ and $EI(B, A) \geq 3$.

If $1 \leq |A| \leq 2$ (or similarly if $1 \leq |C| \leq 2$) let $A = \{a_1, a_2\}$ with $a_1 = a_2$ if $|A| = 1$. Then, using the fact that $\kappa(A \cup B) \geq 3$ and therefore that $d_B^+(a_i) \geq 2$ and $d_B^-(a_i) \geq 2$ for $i = 1, 2$, it is easy to see that there exists a directed hamiltonian path in $A \cup B$ between any two vertices of B except at most between two vertices x and y. As $\kappa(B \cup C) \geq 3$ and $|B| \geq 3$, there exist an arc (b, c) from B to C with $b \neq x$ or y and an arc (c', b') from C to B with $c \neq c'$ if $|C| > 1$ and $b \neq b'$. So D has a hamiltonian cycle : $(b' \overset{A \cup B}{\longrightarrow} b, c \overset{C}{\longrightarrow} c', b')$.

So we can now assume that A, B and C have at least three vertices. We will first show that D contains three arcs (a_1, b_1), (b_2, c_2), (c_1, b_3) with a_1 in A, b_1, b_2, b_3 distinct in B and $c_1 \neq c_2$ in C.

Since $EI(A, B) \geq 3$ and $EI(B, C) \geq 3$, there exist two arcs (a, b) and (b', c) with a in A, c' in C and $b \neq b'$ in B. If there doesn't exist an arc $(c, b")$ with $b"$ different from b and b' in B and $c \neq c'$ in C, then, as $EI(C, B) \geq 3$, the only possible configuration of three independant arcs from C to B is (c, b'), $(c', b")$ and $(c", b)$ with $c \neq c'$, $b"$ different from b and b' and $c"$ different from c and c'. But then, as there exists an arc (α, β) from A to B with β different from b and b', we have three arcs as we wanted by taking $a_1 = \alpha$, $b_1 = \beta$, $b_2 = b'$, $c_2 = c'$, $c_1 = c"$ and $b_3 = b$.

Now, as $EI(B, A) \geq 3$, there exist a_i in A and b'_3 in B different from b_1 and b_2 such that $(b'_3, a_i) \in E(D)$. If $a_i \neq a_1$ then D has a configuration (2) : $(a_1, b_1), (b_2, c_2), (c_1, b_3), (b'_3, a_i)$ and is hamiltonian.

So we now assume that $a_i = a_1$ and that there is no arc (a, b_1) with $a \neq a_1$ in A. Also three independant arcs from B to A are necessarily of the form $(b'_3, a_1), (b_1, a_2)$ and (b_2, a_3) with a_2 and a_3 distinct and different from a_1.

By symmetry between A and C, as there exist three arcs $(c_1, b_3), (b_2, a_3)$ and (a_1, b_1) with b_1, b_2 and b_3 distinct in B and $a_1 \neq a_3$ in A, the only possible choice for three independant arcs from B to C, in order to avoid a configuration (2), is $(b_3, c_3), (b_2, c_2)$ and (b'_1, c_1) with b'_1 different from b_2 and b_3 in B and c_3 different from c_1 and c_2 in C. And also D contains no other arc (c, b_3) with $c \neq c_1$ in C.

Now if there is an arc (a, b) with a different from a_2 and b different from b_1 and b_3 then D has a configuration (2) : $(a, b), (b_2, c_2), (c_1, b_3), (b_1, a_2)$ and therefore is hamiltonian.

If there is an arc (a_1, b) with $b \neq b_1$ or b_3 in B, then D has a configuration 2 : $(a_1, b), (b_2, c_2), (c_1, b_3), (b_1, a_2)$.

If there is an arc (a_2, b) with $b \neq b_2, b_3$ or b'_3, then D has a configuration 2 : $(a_2, b), (b_2, c_2), (c_1, b_3), (b'_3, a_1)$.

We may therefore assume that any three independant arcs from A to B are necessarily of the form $(a_1, b_1), (a_i, b_3)$ and (a_2, b_j) with $i \neq 1, 2$ and $j \neq 1, 3$.

Similarly we may assume that the three independant arcs from C to B are $(c_1, b_3), (c_k, b_1)$ and (c_3, b_l) with $k \neq 1, 3$ and $l \neq 1, 3$ otherwise as above D has a configuration (2) and is hamiltonian.

To end the proof of that case it suffices to notice that, with all these arcs, either D has a configuration (2) and is hamiltonian, or necessarily $b'_1 = b_1$, $b'_3 = b_3$, $c_k = c_2$, $a_i = a_3$, $b_l = b_2$, $b_j = b_2$ and there is no other arc between A, B and C. But then D is isomorphic to D_3 which is the exception of the corollary.

Case 2: There is an arc in both directions between any pair of complete graphs

Subcase 2.1 : $EI(A, B) \geq 2, EI(B, C) \geq 2, EI(C, A) \geq 2.$

It is easy to see that either D has a configuration (1) and is hamiltonian or A, B, and C have cardinality at least 2 and the only arcs from A to B, from B to C and from C to A are respectively $(a_i, b_i), (b_i, c_i)$ and (c_i, a_i) for $i = 1, 2$. Let us assume that we are in this last case.

If $|A| = 2$ (or similarly if $|B| = 2$ or $|C| = 2$), then, as $\kappa(D) \geq 3$ and $EI(C, A) = 2$, if $|B| \geq 3$ there exists an arc (b_3, a_i) from B to A with b_3 different from b_1 and b_2 and then D has a hamiltonian cycle: $(b_3, a_i, b_i, c_i \xrightarrow{C} c_j, a_j, b_j \xrightarrow{B-\{b_i\}} b_3)$ with $j = 1$ if $i = 2$ and $j = 2$ if $i = 1$. The case $|A| = |B| = |C| = 2$, which is not difficult, is left to the reader (see also [4]).

Let us now consider the case where A, B and C have cardinality at least 3. As D is 3-connected and there exist only two (independant) arcs from A to B, B to C or C to A, there exists an arc (b_3, a) from B to A, with b_3 distinct from b_1 and b_2, an arc (c_3, b) from C to B with c_3 different from c_1 and c_2 and an arc (a_3, c) from A to C with a_3 different from a_1 and a_2. It is easy to see that if b is different from b_1 and b_2, (and similarly if a is different from a_1 or a_2 or if c is different from c_1 or c_2) then D has a configuration (2):

$(c_3, b), (b_3, a), (a_i, b_i), (b_j, c_j)$ with $i = 1, j = 2$ if $a = a_2$ and $i = 2, j = 1$ otherwise.

So it remains to consider the case where $a \in \{a_1, a_2\}$, $b \in \{b_1, b_2\}$ and $c \in \{c_1, c_2\}$. If $a = a_i$, $b = b_j$ and $c = c_k$ with $\{i, j, k\} \subset \{1, 2\}$ then D has a configuration (1): $(b_3, a_i), (a_3, c_k), (c_3, b_j)$ and thus is hamiltonian.

Subcase 2.2 : $EI(A, B) \geq 1$, $EI(B, C) \geq 1$ and $EI(C, A) = 1$ (or similar cases with permutation of A, B and C).

Since $EI(C, A) = 1$, as D is 3-connected, we have necessarily $EI(C, B) \geq 2$ and $EI(B, A) \geq 2$. We may assume that $EI(A, C) = 1$ otherwise we are in a case similar to subcase 2.1 and D is hamiltonian. So necessarily $EI(A, B) \geq 2$ and $EI(B, C) \geq 2$. But then it is easy to see that either D has a configuration (1) and is hamiltonian or A, B and C have cardinality at least 2, $EI(A, B) = EI(B, A) = EI(B, C) = EI(C, B) = 2$ and D has exactly one arc from A to C and one arc from C to A. Moreover, without loss of generality we may assume that D contains the following arcs: $(a_1, b_1), (b_1, c_1), (a_2, b'_2), (b_2, c_2), (c_2, a_2)$ with $a_1 \neq a_2$ in A, $c_1 \neq c_2$ in C and $b_2 \neq b_1$, $b'_2 \neq b_1$ in B. The case $|A| = 2$ (or similarly $|C| = 2$) is left to the reader and can be found in [4]. We will now deal with the general case where A and C have cardinality at least 3.

As $EI(A, B) = 2$ and D is 3-connected, there exists an arc (a_3, c_i) from A to C with a_3 different from a_1 and a_2. Similarly, as $EI(B, C) = 2$ there exists an arc (a_j, c_3) from A to C with c_3 different from c_1 and c_2. So, as there is only one arc from A to C, $(a_3, c_3) \in E(D)$. As $EI(C, A) = 1$ and D is 3-connected, there exists an arc (c'_3, b_i) from C to B with c'_3 different from c_1 and c_2. We may assume that $b_i = b_1$ otherwise D has a hamiltonian cycle: $(c'_3, b_i \xrightarrow{B} b_1, c_1 \xrightarrow{C - \{c_3, c'_3\}} c_2, a_2 \xrightarrow{A} a_3, c_3, c'_3)$. Similarly, there exists an arc (b_j, a'_3) from B to A with a'_3 different from a_1 and a_2 and we may assume that $b_j = b_1$.

Now, if $a'_3 \neq a_3$, D has a hamiltonian cycle: $(b_1, a'_3 \xrightarrow{A - \{a_2\}} a_3, c_3 \xrightarrow{C} c_2, a_2, b'_2 \xrightarrow{B} b_1)$, and similarly if $c'_3 \neq c_3$. So we may now assume that (c_3, b_1) and (b_1, a_3) are in $E(D)$. But then we can also assume that $b_2 = b'_2$ and $|B| > 2$ otherwise D has a hamiltonian cycle : $(b_1, a_3 \xrightarrow{A} a_2, b'_2 \xrightarrow{B - \{b_1\}} b_2, c_2 \xrightarrow{C} c_3, b_1)$. As $EI(C, B) = 2$ and D is 3-connected, there exists an arc (c, b) from C to B with c different from c_2 and c_3 and $b \neq b_1$ but we can assume that $b = b_2$ otherwise D has a hamiltonian cycle: $(b_1, a_3 \xrightarrow{A} a_2, b_2, c_2 \xrightarrow{C} c, b \xrightarrow{B - \{b_2\}} b_1)$. Similarly, as $EI(B, C) = 2$, D contains an arc (b_3, a) from B to A with b_3 different from b_1 and b_2 and we can assume that $a = a_2$ otherwise D contains a hamiltonian cycle: $(b_3, a \xrightarrow{A} a_2, b_2, c_2 \xrightarrow{C} c_3, b_1 \xrightarrow{B - \{b_2\}} b_3)$.

But then D contains a configuration (1): $(b_3, a_2)(a_3, c_3)(c, b_2)$ and thus is hamiltonian. This ends the proof of Corollary 4.4.

We can now proceed with the proof of Corollary 4.3.

Proof of Corollary 4.3: Let us recall that if $\alpha_2(D) \leq 2$ then Theorem 4.1 and therefore Corollary 4.3 follow from Theorem 3.1. So we can assume that $\alpha_2(D) = 3$. As in the proof of Theorem 3.1, let us consider the non directed graph G with same set as vertices as D and edges between pairs of vertices which are joined by a 2-cycle in D. Clearly, $\alpha(G) = \alpha_2(D) = 3$. First *we can assume that* $\kappa(G) \leq 2$ otherwise the result follows from Theorem 2.3 (in fact since we only want to prove that D is hamiltonian, it is sufficient to apply Chvátal-Erdős' theorem). So, if $S(G)$ is a minimum cutset of G, we have

$|S(G)| \leq 2$. The graph $G - S$ has at least two connected components and one of them, say K, is a complete graph since $\alpha(G) = 3$. Let $H = G - K - S(G)$. We have $\alpha(H) \leq 2$. If $\alpha(H) = 1$ then, as D is 3-connected, it is easy to see that D is hamiltonian (indeed, as $\alpha_2(D) = 3$ necessarily $\kappa(G) \geq 1$. If $\kappa(G) = 1$ then we can apply Corollary 4.4 and if $\kappa(G) = 2$ obviously G itself is hamiltonian).

So let us now assume that $\alpha(H) = 2$. By Remark 2.2, $V(H)$ can be covered by two complete graphs H_1 and H_2. If $\kappa(G) = 0$, or $\kappa(G) = 1$ and $K \subset \Gamma(S(G))$, the result follows from Corollary 4.4. So we can assume that $\kappa(G) = 1$ or 2 and also, when $\kappa(G) = 1$, that there is a vertex k_0 of K not adjacent to $S(G)$. We will now divide the proof into several parts, depending on the connectivity of H.

Case $\kappa(\mathbf{H}) = \mathbf{0}$.

Then H is the union of two disjoint complete graphs H_1 and H_2.

If $\kappa(\mathbf{G}) = \mathbf{1}$, as $\alpha(G) = 3$ either K, H_1 or H_2 is included in $\Gamma(S)$ and the result follows from Corollary 4.4 .

If $\kappa(\mathbf{G}) = \mathbf{2}$ let $S(G) = \{v, w\}$. Let $\mathcal{H} = \{K, H_1, H_2\}$. As $\alpha(G) = 3$ there exist L and L' in $\mathcal{H}$ such that $L \subseteq \Gamma(v)$ and $L' \subseteq \Gamma(w)$. If $L \neq L'$ or if $L = L'$ and vw is an edge of G then the proof ends with Corollary 4.4. If $L = L'$ and vw is not an edge, let us assume, without loss of generality, that $L = L' = K$.

As $\kappa(G) = 2$ there exist x_1 and y_1 in H_1, x_2 and y_2 in H_2 (with, for each i, $x_i \neq y_i$ if $|H_i| > 1$) such that G contains the edges vx_i and wy_i for $i = 1$ and 2.

As $\kappa(D) = 3$ there exists an arc (h, k) from H to K. We can assume, without loss of generality, that h is in H_1 and that $h \neq y_1$ if $|H_1| > 1$. Then D has a hamiltonian cycle: $(h, k \xrightarrow{K} k', v, x_2 \xrightarrow{H_2} y_2, w, y_1 \xrightarrow{H_1} h)$, where k' is any other vertex of K if $|K| > 1$.

Case $\kappa(\mathbf{H}) = \mathbf{1}$.

Without loss of generality we can assume that $V(H)$ can be covered by two complete graphs H_1 and H_2 such that $\Gamma_{H_2}(H_1)$ is reduced to a vertex h_2 which is a cutvertex of H. In that case $|\mathbf{H_2}| > 1$.

If $\kappa(\mathbf{G}) = \mathbf{1}$ let $S(G) = \{v\}$. If K or H_1 or H_2 is included in $\Gamma(v)$ we conclude with Corollary 4.4. Hence, as $\alpha(G) = 3$ necessarily $H_2 - \{h_2\} \subset \Gamma(v)$ and every vertex of H_1 is adjacent to either v or h_2. Moreover we can assume that $|H_1| \geq 2$ and that there is at least one vertex of H_1 not adjacent to h_2 and therefore adjacent to v, otherwise we conclude again with Corollary 4.4. As $\kappa(D) \geq 3$, there are at least three independant arcs from H to $K \cup v$ and also from $K \cup v$ to H in D. It is not difficult (but too teadious to include the details here, see [4]) to see that D is hamiltonian by examining all the different possible cases depending in which part of H those arcs end.

If $\kappa(\mathbf{G}) = \mathbf{2}$, let $S(G) = \{v, w\}$. Then G contains edges vk_1 and wk_2 with k_1 and k_2 in K and $k_1 \neq k_2$ if $|K| > 1$. Also, as $\kappa(H) = 1$, G contains at least one edge x_1v (without loss of generality) where x_1 is a vertex of H_1 different from one of the neighbours of h_2 in H_1, say h_1, if $|H_1| > 1$, and one edge x_2s where $s \in S(G)$ and x_2 is a vertex of H_2 different from h_2.

If $s = w$ then D contains a hamiltonian cycle : $(x_2, w, k_2 \xrightarrow{K} k_1, v, x_1 \xrightarrow{H_1} h_1, h_2 \xrightarrow{H_2} x_2)$. So let us assume $s = v$ and $\Gamma_{H_2}(w) \subseteq \{h_2\}$. If G contains an edge wh with $h \in H_1 - h_1$, then D has a hamiltonian cycle : $(x_2, v, k_1 \xrightarrow{K} k_2, w, h \xrightarrow{H_1} h_1, h_2 \xrightarrow{H_2} x_2)$. So we can assume that w is adjacent to at most h_1 in H_1 and if wh_1 is an edge, then $|H_1| > 1$ and h_1h_2 is the only edge of G between H_1 and H_2.

So let us assume now that $\Gamma_H(w) \subseteq \{h_1, h_2\}$.
For any vertex k of K, $\{w, x_2, x_1, k\}$ cannot be an independant set in G therefore $K \subset \Gamma(w)$. If w is only incident with one h_i, $1 \leq i \leq 2$, then $\{v, h_i\}$ is a cutset of cardinality 2 of G such that $G - \{v, h_i\}$ is the union of three complete graphs, so that we have the case $\kappa(H) = 0$ already settled before.

So *we can assume now that* $\Gamma_H(w) = \{h_1, h_2\}$.
As $\kappa(D) \geq 3$, there exist an arc (k, h) from K to H and an arc from H to K.
If h is in H_1, $h \neq x_1$, (or similarly if h is in H_2, $h \neq x_2$) then D contains a hamiltonian cycle : $(k, h \xrightarrow{H_1} x_1, v, x_2 \xrightarrow{H_2} h_2, w, k' \xrightarrow{K} k)$ where k' is any vertex of K, different from k if $|K| > 1$. If $h = x_1$ or $h = x_2$ and k is different from one of the neighbours of v in K, say k_1, or if $|K| = 1$, then again D contains a hamiltonian cycle, for example : $(k, x_1 \xrightarrow{H_1} h_1, w, h_2 \xrightarrow{H_2} x_2, v, k_1 \xrightarrow{K} k)$.

So we can assume that $|K| \geq 2$, $\Gamma_K(v) = \{k_1\}$ and that the only possible arcs from K to H are (k_1, x_1) and (k_1, x_2). Similarly we can assume that the only possible arcs from H to K are (x_1, k_1) and (x_2, k_1). Since there is no edge in G between K and H we may assume without loss of generality that the only arcs between H and K are (k_1, x_1) and (x_2, k_1). But then, as $\kappa(D) \geq 3$ there must exist an arc (k_1', v) from K to v with $k_1' \neq v$ and D is hamiltonian : $(k_1', v, x_1 \xrightarrow{H_1} h_1, w, h_2 \xrightarrow{H_2} x_2, k_1 \xrightarrow{K} k_1')$.

Case $\kappa(\mathbf{H}) = \mathbf{2}$.

Let $S(H) = \{h, h'\}$ be a cutset of H. As $\alpha(H) = 2$, the graph $H - S(H)$ is the union of two complete graphs, and necessarily $\Gamma(h)$ and $\Gamma(h')$ each contains one of them. Therefore $|H| \geq 4$ and H has one of the following configurations.
(i) Either $V(H)$ is covered by the vertex disjoint union of two complete graphs H_1 and H_2 with two disjoint edges between them say hh_2 and h_1h' where h and h_1 are distinct vertices of H_1 and h' and h_2 are distinct vertices of H_2. The other edges of H are only between h and H_2 or between h' and H_1.
(ii) Or H is covered by the vertex disjoint union of H_1, a complete graph of cardinality at least 3 minus possibly one edge hh' and of a complete graph H_2. The only other edges of H are only between h or h' and H_2 and there are at least two independant ones if $H_2 \geq 2$.

In case (i) *let us first assume that* $\Gamma_H(S(G)) \not\subseteq \{h, h', h_1, h_2\}$.
Let v be a vertex of $S(G)$, with h_0 a neighbour of v in H not in $\{h, h', h_1, h_2\}$. Let k_1 be a neighbour of v in K.
If $\kappa(G) = 2$ and w is the other vertex of $S(G)$ then there exist edges wh_0' with h_0' in H, $h_0' \neq h_0$ and wk_2 with k_2 in K, $k_2 \neq k_1$ if $|K| > 1$. Then obviously G has a hamiltonian cycle $(k_1, v, h_0 \xrightarrow{H} h_0', w, k_2 \xrightarrow{K} k_1)$ where $h_0 \xrightarrow{H} h_0'$ is a hamiltonian path of H which is easy to find since h_0 is not in $\{h, h', h_1, h_2\}$.
If $\kappa(G) = 1$ then, as $\kappa(D) \geq 3$, there exists at least one arc (t, k) from H to K with $t \neq h_0$ and also an arc (k', t') from K to H with $t' \neq h_0$. If $k \neq k_1$ ($k' \neq k_1$ resp.) then D contains a hamiltonian cycle $(k_1, v, h_0 \xrightarrow{H} t, k \xrightarrow{K} k_1)$ $((t' \xrightarrow{H} h_0, v, k_1 \xrightarrow{K} k', t')$ resp.$)$. If (t, k_1) $((k_1, t')$ resp.$)$ is the only arc from $H - \{h_0\}$ to K (from K to $H - \{h_0\}$ resp.$)$, then necessarily, if $|K| > 1$, D contains an arc $(v, k")$ with $k" \neq k_1$ and again D contains

a hamiltonian cycle $(t' \xrightarrow{H} h_0, v, k" \xrightarrow{K} k_1, t')$.

Now assume that $\Gamma_H(S(G)) \subseteq \{h, h', h_1, h_2\}$.

If $\kappa(G) = 2$, as there exist an edge vx and an edge wy with $x \neq y$ in H, it is easy to see that D is hamiltonian except if vh and wh' (or vh_1 and wh_2) are the only two edges and there is no other arc between $S(G)$ and H. In this last case, as $\kappa(D) \geq 3$, there is an arc in D from K to $H - \{h, h'\}$ (or $H - \{h_1, h_2\}$) and again D is hamiltonian .

If $\kappa(G) = 1$, as, for some k_0 in K, vk_0 is not an edge (as we assumed in the beginnig of the proof), necessarily H_1 or H_2 has cardinality 2 otherwise we get a contradiction with $\alpha(G) = 3$. We can assume that $|H_2| = 2$ (the case $|H_1| = 2$ is similar).

If $|H_1| > 3$ there is no edge vx_1 for x_1 in $H_1 - \{h, h_1\}$. As the set $\{v, k_0, x_1, h_2\}$ is not independant in G, there exists an edge vh_2. Then we can assume that there is no edge vh', otherwise we conclude with Corollary 4.4. Now necessarily $H - h \subset \Gamma(h')$ otherwise $\{v, k_0, h', x_1\}$ is an independant set of G for any x_1 in $H_1 - \{h, h_1\}$. Also the set $\{v, k_0, h, h'\}$ is not independant and therefore either v or h' is adjacent to h and in both cases we conclude with Corollary 4.4.

If $|H_1| = |H_2| = 2$, as $h_1 h_2$ is not an edge, there is necessarily an edge vh_2 or vh_1 otherwise $\{v, k_0, h_1, h_2\}$ would be an independant set. Suppose G contains vh_2 (the other case is symmetric). Then we can assume that vh and vh' is not an edge otherwise we conclude with Corollary 4.4. Thus hh' is an edge otherwise $\{v, k_0, h, h'\}$ would be an independant set. Again $V(D)$ can be covered with three complete symmetric digraphs and we conclude with Corollary 4.4.

Case (ii) can be settled in a very similar way. The technique is the same: consider several cases depending on the connectivity of $\kappa(G)$ and use the fact that D is 3-connected. We omit the details since they bring nothing new (see [4]).

Case $\kappa(\mathbf{H}) \geq \mathbf{3}$.

This case is easy to deal with, since H is hamilton connected. If $\kappa(G) = 2$, the graph G itself is hamiltonian. Otherwise, using the fact that D is 3-connected and the existence of at least two independant arcs from K to H and two independant arcs from H to K, it is not difficult to find a hamiltonian cycle in D.

5 Open problems

Again, since there is already a recent survey on the subject, we will be very brief here and just give the problems which are closely related to the previous results.

First it would be very interesting to answer the following question. Does there exist a non hamiltonian 4-connected graph D with $\alpha_2(D) \leq 4$? Our feeling is that such a graph does not exist and we can state the following conjecture.

Conjecture 5.1 *Every digraph satisfying* $\kappa(D) \geq \alpha_2(D)$ *is hamiltonian except if* $\alpha_2(D) = 2$ *or* $\alpha_2(D) = 3$ *and* D *is isomorphic respectively to a digraph of the family* D_2 *or* D_3.

A weaker conjecture would state the same result for κ-connected digraphs such that $V(D)$ can be covered by κ complete symmetric subdigraphs.

Let us mention the following results of Jackson [9] which are the best known until today (in the sense that the lower bound on $\kappa(G)$ is the smallest known to imply the hamiltonicity of D).

Theorem 5.2 *Let D be a digraph with $\alpha_2(D) \leq m$. If $\kappa(D) \geq 2^m(m+2)!$ then D is hamiltonian.*

Proposition 5.3 *Let D be a digraph such that $V(D)$ can be covered with m complete symmetric subdigraphs. If $\kappa(D) \geq m(m-1)$ then D is hamiltonian.*

If one cannot settle Conjecture 5.1 or its weaker version, maybe one can try to improve the lower bounds on $\kappa(D)$ in the above results. A linear bound in $\alpha_2(D)$ would be already very nice.

Acknowledgements: We would like to thank W. McCuaig for his carefull reading of the original paper and his subsequent suggestions. W. McCuaig also noticed that, using methods similar to those in the paper, it is possible to characterize the nonhamiltonian graphs with $\kappa = 2$ and $\alpha = 3$.

References

[1] D. Amar, I. Fournier and A. Germa, Pancyclism in Chvátal-Erdős graphs, *Graphs and Combinatorics*. Submitted.

[2] A. Bondy, Hamilton cycles in graphs and digraphs, Technical report, University of Waterloo, 1975.

[3] A. Bondy and U. Murty, *Graph Theory with Applications*, Macmillan Press, 1976.

[4] N. Chakroun, *Problèmes de circuits, chemins et diamètres dans les graphes*, PhD. thesis, Université Paris-Sud, 1986.

[5] V. Chvátal and P. Erdős, A note on Hamilton circuits, *Discrete Mathematics* **2**(1977), 111-113.

[6] C. Guia, B. Jackson and O. Ordaz, Chvátal- Erdős Condition for hamilton cycles in digraphs with stability at most three, *Acta Cientifica Venezolana*. Submitted.

[7] M.-C. Heydemann, Minimum number of circuits covering the vertices of a strong digraph, In B.R. Alspach and C.D. Godsil, editors, *Cycles in Graphs*, pp. 287-296, Annals of Discrete Mathematics **27**(1985).

[9] B. Jackson, A Chvátal-Erdős condition for Hamilton cycles in digraphs, *J. Combinatorial Theory Ser. B*. To appear.

[8] B. Jackson and O. Ordaz, A Chvátal- Erdős Condition for 2-Cyclability, Hamiltonicity and Pancyclability in Digraphs, *Ars Combinatoria*. To appear.

[10] B. Jackson and O. Ordaz, Chvátal-Erdős conditions for paths and cycles in graphs and digraphs. A survey, *Preprint*, 1987.

[11] C. Thomassen, Long cycles in digraphs, *Proc. London Math. Soc.* **42**(1981), 231-251.

LONG CYCLES AND THE CODIAMETER OF A GRAPH II

G. FAN
Departement of Combinatorics and Optimization
University of Waterloo
Waterloo, Ontario, N2L 3G1
Canada

Abstract

The codiameter of a graph, introduced and studied in [2], is the minimum, taken over all pairs of vertices x and y in the graph, of the maximum length of an (x, y)-path. Further results on the codiameter, and on long cycles of a graph, are given in this paper.

1 Introduction

As in [2], let x and y be two distinct vertices in a graph G and define the *codistance* $d^*(x, y)$ between x and y to be the maximum length of an (x, y)-path in G (an (x, y)-path is a path connecting x and y). If no (x, y)-path exists, we set $d^*(x, y) = 0$. The *codiameter* of a nontrivial graph G, denoted by $d^*(G)$, is the minimum codistance in G; that is,

$$d^*(G) = min\{d^*(x, y) \; : \; x, \; y \in V(G)\}.$$

The codiameter of the trivial graph is defined to be zero. We shall use $d_H^*(x, y)$ to denote the codistance of x and y in the subgraph H, i.e. the maximum length of an (x, y)-path with all its internal vertices in H.

All graphs considered are finite, undirected, and without loops or multiple edges. The vertex set of a graph G is denoted by $V(G)$ and the edge set by $E(G)$. We define $e(G) = |E(G)|$. The set of neighbours of a vertex $x \in V(G)$ is denoted by $N(x)$, and $d(x) = |N(x)|$ is called the *degree* of x. If H is a subgraph of G, then $N_H(x)$ denotes the set, and $d_H(x)$ the number, of the neighbours of x which are in the subgraph H. The vertex x is said to be *joined* to H if $d_H(x) > 0$. If F is another subgraph of G, the set of the vertices in H which are joined to F is defined by

$$N_H(F) = \bigcup_{x \in V(F)} N_H(x);$$

furthermore, $E(F, H)$ denotes the set, and $e(F, H)$ the number, of edges with one endvertex in H and the other in F. The edge with endvertices x and y is denoted by xy. An xy-*cycle* is a cycle containing xy.

Let C be a cycle in a graph G. If we consider a subset $\{a_1, a_2, ..., a_p\}$ of $V(C)$, we will assume that C has an orientation which is consistent with the increasing order of the

G. Hahn et al. (eds.), Cycles and Rays, 87–94.
© *1990 by Kluwer Academic Publishers. Printed in the Netherlands.*

indices of $a_i (1 \leq i \leq p)$. For each a_i, $i = 1, 2, ..., p$, we define the *successor* of a_i to be the vertex on C immediately after a_i, according to the orientation of C. For $i \neq j$, $C[a_i, a_j]$ denotes the segment of C from a_i to a_j, determined by the orientation of C, and $C(a_i, a_j]$ is the segment obtained from $C[a_i, a_j]$ by deleting the endvertex a_i.

2 Preliminaries

The following theorem, Theorem 2.1, can be obtained by combining Theorems 3 and 4 in [2].

Theorem 2.1 *Let G be a k-connected graph, where $k = 2, 3$, and $xy \in E(G)$. Suppose that C is a longest xy-cycle, of length c, in G and H is a component of $G - C$. If $d(v) \geq d$ for all $v \in V(H)$, then*

$$c \geq (k-1)(d+2-k) + 1,$$

with equality only if H is a complete graph in which every vertex has the same k or $d+3-k$ neighbours on C.

As in [2], let C be a cycle in a graph G and H a component of $G - C$. A *strong attachment* of H to C (in G) is a subset $T = \{u_1, u_2, ..., u_t\} \in N_C(H)$, where $u_i (1 \leq i \leq t)$ are in order around C, such that each ordered pair (u_i, u_{i+1}) (define $u_{t+1} = u_1$) are joined to H in G by two independent edges. For fixed C and H, a strong attachment $T = \{u_1, u_2, ..., u_t\}$ of H to C is *maximum* if it has maximum cardinality over all strong attachments of H to C; if there is no edge from H to C at all, define $t = 0$ and $T = \emptyset$.

Suppose that C is a cycle in a graph G and H is a separable component of $G - C$. Let B_1 and B_2 be two distinct endblocks of H and b_1 and b_2 the (unique) cut vertices of H contained in $V(B_1)$ and $V(B_2)$, respectively. Let $T = \{u_1, u_2, ..., u_t\}$ be a maximum strong attachment of H to C in G. We define two disjoint subsets of the t ordered pairs (u_i, u_{i+1}), $1 \leq i \leq t$. A pair (u_i, u_{i+1}) is called a *best pair (associated with B_1 and B_2)* if there are two independent edges joining one of u_i and u_{i+1} to $B_1 - b_1$ and the other to $B_2 - b_2$, and a *good pair* if exactly one of u_i and u_{i+1} is joined to $(B_1 - b_1) \cup (B_2 - b_2)$. If (u_i, u_{i+1}) is a good pair such that u_i or u_{i+1} is joined to $B_j - b_j$, we say that (u_i, u_{i+1}) is *associated with B_j ($j = 1$ or 2)*. Two good pairs are said to be a *matched couple (associated with B_1 and B_2)* if one of them is associated with B_1 and the other with B_2. Two matched couples are *disjoint* if they have no good pair in common.

The following lemma, Lemma 2.1, is part of Lemma 1 in [2].

Lemma 2.1 *Let C be a cycle, of length c, in a graph G. Suppose that H is a component of $G - C$ and $T = \{u_1, u_2, ..., u_t\}$ is a maximum strong attachment of H to C. Set*

$$S = N_C(H) \setminus T \text{ and } s = |S|.$$

Then the following statements are true.

(a) Every vertex in S is joined to exactly one vertex in H.

(b) If C is a longest cycle, then

$$c \geq \begin{cases} 2(s+1) & \text{if } t \leq 1 \\ \\ \sum_{i=1}^{t} d_H^*(u_i, u_{i+1}) + 2s & \text{if } t \geq 2 \end{cases}$$

(c) If G is k-connected and $|V(C)| \geq k$, then

$$t \geq \min\{k, |V(H)| + |D(T)|\},$$

where

$$D(T) = \{u_i \in T : d_H(u_i) \geq 2\}.$$

Finally, we restate Lemma 2 in [2] as Lemma 2.2 below.

Lemma 2.2 *Suppose that C is a cycle in a graph G and H is a separable component of $G - C$. Let $T = \{u_1, u_2, ..., u_t\}$ be a maximum strong attachment of H to C in G. Let B_1 and B_2 be two endblocks of H and b_1 and b_2 the cut vertices of H contained in $V(B_1)$ and $V(B_2)$, respectively. If there are two distinct vertices $v_1, v_2 \in V(C)$ such that v_1 is joined to $B_1 - b_1$ and v_2 to $B_2 - b_2$, then at least one of the following statements is true.*

(i) T has two best pairs associated with B_1 and B_2.

(ii) T has a best pair and a matched couple associated with B_1 and B_2.

(iii) T has two disjoint matched couples associated with B_1 and B_2.

3   Main Results

We first give the following theorem in which a new type of sufficient condition for a graph to have a large codiameter or a long cycle is introduced.

Theorem 3.1 *Let G be a 2-connected graph on n vertices. If the set of vertices of degree less than d induces a complete subgraph in G, or is empty, then $d^*(G) \geq \min\{d, n-1\}$.*

Proof: Let x and y be any two distinct vertices. If $xy \notin E(G)$, add xy to $E(G)$. We need to prove that $d^*(x, y) \geq \min\{d, n-1\}$. Let C be a longest xy-cycle, of length c. If $c = n$, then $d^*(x, y) = c - 1 = n - 1$, as required. Suppose now that $c < n$, and let K be the complete subgraph induced by the set of vertices of degree less than d. If $G - C$ has more than one component, then there is at least one, say H, such that $V(H) \cap V(K) = \emptyset$. So $d(v) \geq d$ for all $v \in V(H)$, and by Theorem 2.1 with $k = 2$, $c \geq d + 1$, which gives that $d^*(x, y) \geq d$. Therefore, we may suppose that $G - C$ has only one component and $V(G - C) \cap V(K) \neq \emptyset$. Since K is a complete subgraph, for any $u \in V(C) \setminus N_C(G - C)$, we must have that $u \notin V(K)$ and so $d(u) \geq d$. Since C is a longest xy-cycle, there is at least one vertex $u \in V(C) \setminus N_C(G - C)$, and hence

$$c \geq d_C(u) + 1 = d(u) + 1 \geq d + 1,$$

which gives that $d^*(x, y) \geq d$, as required. □

Corollary 3.1 *Let G be a 2-connected graph, at most one vertex of which has degree less than d. Then $d^*(G) \geq d$.*

Proof: Let G be a graph satisfying the hypothesis of the corollary. Then there is a vertex u such that $d(v) \geq d$ for all $v \in V(G) \setminus \{u\}$. Since $V(G) \setminus \{u\} \neq \emptyset$, $n \geq d + 1$. The result follows from Theorem 3.1 with K consisting of the single vertex u.   □

Another consequence of Theorem 3.1 is the following corollary, which is also a consequence of a combination of results of Ore [4] and Pósa [5]. To show the relationship between the Ore's condition and the above-presented condition, we derive it from Theorem 3.1.

Corollary 3.2 *Let G be a 2-connected graph on n vertices. Suppose that $d(u) + d(v) \geq m$ for every pair of independent vertices u and v. Then $d^*(G) \geq min\{\frac{m}{2}, n - 1\}$.*

Proof: Let K be the set of vertices of degree less than $\frac{m}{2}$. The given condition implies that no two vertices in K are independent, that is, K induces a complete subgraph in G. The result follows from Theorem 3.1.   □

In the case when G is 3-connected, Enomoto [1] has proved the following result.

Theorem 3.2 *Let G be a 3-connected graph on n vertices. Suppose that $d(u) + d(v) \geq m$ for every pair of independent vertices u and v. Then $d^*(G) \geq min\{m - 2, n - 1\}$.*

In analogue to Theorem 3.1 we have the following stronger result.

Theorem 3.3 *Let G be a 3-connected graph on n vertices. If the set of vertices of degree less than d induces a complete subgraph in G, or is empty, then $d^*(G) \geq min\{2d - 2, n - 1\}$.*

Remark 3.1 It can be seen from the proof of Corollary 3.2 that Theorem 3.2 is indeed a consequence of Theorem 3.3.

Proof: Let x and y be any two distinct vertices of G. If $xy \notin E(G)$, add xy to $E(G)$. We need to prove that $d^*(x, y) \geq min\{2d - 2, n - 1\}$. Let C be a longest xy-cycle, of length c. If $c = n$, then $d^*(x, y) = c - 1 = n - 1$, as required. Suppose now that $c < n$, and let K be the complete subgraph induced by the set of vertices of degree less than d. If $G - C$ has more than one component, then there is at least one, say H, such that $V(H) \cap V(K) \neq \emptyset$. So $d(v) \geq d$ for all $v \in V(H)$. If H consists of a single vertex, say z, then $c \geq 2d(z) - 1 \geq 2d - 1$; if H contains at least two vertices then, by Theorem 2.1 with $k = 3$, $c \geq 2d - 1$. In either case we have that $c \geq 2d - 1$, which gives $d^*(x, y) \geq 2d - 2$, as required. Therefore we may suppose that $G - C$ has only one component and $V(G - C) \cap V(K) \neq \emptyset$. Let $N_C(G - C) = \{a_1, a_2, ..., a_p\}$, where $a_i(1 \leq i \leq p)$ are in order around C. Since G is 3-connected, $p \geq 3$. Suppose, without loss of generality, that $xy \in E(C[a_p, a_1])$ and y is the successor of x. Let b_1 and b_2 be the two successors of a_1 and a_2 on C, respectively. Since K is a complete graph and $V(G - C) \cap V(K) \neq \emptyset$, we have that $b_1, b_2 \notin V(K)$. Moreover, since $G - C$ has only one component,

$$d_C(b_1) = d(b_1) \geq d \quad \text{and} \quad d_C(b_2) = d(b_2) \geq d.$$

Furthermore, for any $u \in V(C(b_1, b_2])$, if $ub_2 \in E(G)$, then $u^+b_1 \notin E(G)$; for any $u \in V(C(b_2, b_1])$ and $u \neq x$, if $ub_1 \in E(G)$, then $u^+b_2 \notin E(G)$, for otherwise, by a standard technique, we would obtain an xy-cycle longer than C, where u^+ is the successor of u. This gives that

$$|V(C([b_1, b_2]))| \geq e(b_1, C([b_1, b_2])) + e(b_2, C([b_1, b_2]))$$

and

$$|V(C([b_2, b_1]))| \geq e(b_1, C([b_2, b_1])) + e(b_2, C([b_2, b_1])) - 1.$$

Hence,

$$c = |V(C([b_1, b_2]))| + |V(C([b_2, b_1]))| \geq d_C(b_1) + d_C(b_2) - 1 \geq 2d - 1,$$

which gives $d^*(x, y) \geq 2d - 2$, as required. $\qquad\square$

Fraisse and Jung [3] have obtained some results on long cycles in k-connected graphs. We restate two of them below, as Theorem 3.4, in terms of the codiameter.

Theorem 3.4 *Let G be a k-connected graph, where $k \geq 2$, C a longest cycle, of length c, in G and H a component of $G - C$. Suppose that $d^*(H) = l$. Then*

(a) $c \geq k(l + 2)$;

(b) $c \geq (k - 2)l + 2d(z)$ for some $z \in V(H)$.

If H consists of a single vertex, then $d^*(H) = l = 0$, and the results are trivial. We consider the case in which H contains at least two vertices and prove the following generalization.

Theorem 3.5 *Let G be a k-connected graph, where $k \geq 2$, C a longest cycle, of length c, in G and H a component of $G - C$. Suppose that $d^*(H) = l$ and $|V(H)| \geq 2$. Then*

(a) $c \geq k(l + 2)$, with equality only if $|N_C(H)| = k$.

(b) $c \geq (k - 2)l + d(x) + d(y)$ for some two distinct vertices $x, y \in V(H)$, with equality only if $|N_C(H)| = k$.

Proof: Let $T = \{u_1, u_2, ..., u_t\}$ be a maximum strong attachment of H to C. Set $S = N_C(H) \setminus T$ and $s = |S|$. Since $|V(H)| \geq 2$ and $k \geq 2$, we have $t \geq 2$. By (b) of lemma 2.1,

$$c \geq \sum_{i=1}^{t} d_H^*(u_i, u_{i+1}) + 2s. \tag{1}$$

Clearly, for every pair (u_i, u_{i+1}), $1 \leq i \leq t$, we have

$$d_H^*(u_i, u_{i+1}) \geq d^*(H) + 2 \geq l + 2. \tag{2}$$

It follows from (1) that

$$c \geq t(l + 2) + 2s. \tag{3}$$

Moreover, from (c) of lemma 2.1,

$$t \geq min\{k, |V(H)| + |D(T)|\}$$

where $D(T) = \{u_i \in T : d_H(u_i) \geq 2\}$. Set $h = |V(H)|$ and $t' = |D(T)|$. Then

$$t \geq h + t' \quad \text{if } t < k. \tag{4}$$

Since G is k-connected, $d(v) \geq k$ for all $v \in V(H)$. Hence

$$e(H, C) = \sum_{v \in V(H)} (d(v) - d_H(v)) \geq h(k - h + 1).$$

On the other hand, by (a) of lemma 2.1 and the definition of $D(T)$,

$$e(H, C) \leq t'h + (t - t') + s.$$

Combining the two inequalities above gives

$$s \geq h(k - h + 1) - t'(h - 1) - t = (k - t)h + (t - h - t')(h - 1).$$

Since $h \geq 2$ and by (4),

$$s \geq (k - t)h \quad \text{if } t < k. \tag{5}$$

We now prove (a) and (b) separately.

Proof of (a): If $t \geq k$, it follows from (3) that $c \geq k(l + 2)$, with equality only if $t = k$ and $s = 0$, that is, $|N_C(H)| = k$, as required. Suppose now that $t < k$. We may assume that $c \leq k(l + 2)$, otherwise there is nothing to prove. This latter inequality, together with (3), gives

$$(k - t)(l + 2) \geq 2s. \tag{6}$$

Combining (5) and (6) yields

$$l + 2 \geq 2h.$$

Since $l = d^*(H) \leq |V(H)| - 1 = h - 1$, this contradicts our assumption that $h \geq 2$.

Proof of (b): We consider two cases.

(i) H is nonseparable. We first claim that there are two vertices x and y in $V(H)$ such that

$$d_H(x) + d_H(y) \leq 2l. \tag{7}$$

If H consists of exactly two vertices, say x and y, then $d_H(x) = d_H(y) = 1 = l$ and so (7) holds. If $h \geq 3$, then H is 2-connected, and by Corollary 3.1, there must be at least two vertices in $V(H)$, say x and y, such that $d_H(x) l + 1$ and $d_H(y) l + 1$, that is, $d_H(x) \leq l$ and $d_H(y) \leq l$. This proves the claim (7). Furthermore, by (a) of lemma 2.1,

$$d_C(x) + d_C(y) \leq 2t + s.$$

Combining this inequality with (7) yields

$$d(x) + d(y) \leq 2l + 2t + s.$$

Rewrite (3) as

$$c \geq (t - 2)l + (2l + 2t + s) + s.$$

Then we have

$$c \geq (t - 2)l + d(x) + d(y) + s. \tag{8}$$

If $t \geq k$, it follows from (8) that $c \geq (k - 2)l + d(x) + d(y)$, with equality only if $t = k$ and $s = 0$, that is, $|N_C(H)| = k$, as required. If $t < k$, then, by (5) and (8),

$$c \geq (t - 2)l + d(x) + d(y) + (k - t)h = (k - 2)l + d(x) + d(y) + (k - t)(h - l).$$

Since $l \leq h - 1$,

$$c \geq (k-2)l + d(x) + d(y) + (k-t) > (k-2)l + d(x) + d(y),$$

which completes (i).

(ii) H is separable. Let B be an endblock of H and b the cut vertex of H contained in $V(B)$. Suppose that $d^*(B) = l'$. We claim that there is a vertex $z \in V(B - b)$ such that

$$d_B(z) \leq l'. \tag{9}$$

If B has only two vertices, let z be the unique vertex of $B - b$, then $d_B(z) = 1 = l'$ and so (9) is true. Otherwise B is 2-connected, and by Corollary 3.1, there must be at least one vertex $z \in V(B - b)$ such that $d_B(z) < l' + 1$, i.e. $d_B(z) \leq l'$, as claimed. Now, let B_1 and B_2 be any two endblocks of H. Suppose that $d^*(B_1) = l_1$ and $d^*(B_2) = l_2$. Since (9) is true for any endblock, we have two vertices $x \in V(B_1 - b_1)$ and $y \in V(B_2 - b_2)$ such that

$$d_{B_1}(x) \leq l_1 \quad \text{and} \quad d_{B_2}(y) \leq l_2.$$

Again, by (a) of lemma 2.1,

$$d_C(x) + d_C(y) \leq 2t + s.$$

Therefore

$$d(x) + d(y) \leq l_1 + l_2 + 2t + s. \tag{10}$$

From the definition, if (u_i, u_{i+1}) is a best pair, associated with the two endblocks B_1 and B_2,

$$d_H^*(u_i, u_{i+1}) \geq l_1 + l_2 + 2, \tag{11}$$

and if (u_p, u_{p+1}), (u_q, u_{q+1}) is a matched couple,

$$d_H^*(u_p, u_{p+1}) + d_H^*(u_q, u_{q+1}) \geq l_1 + l_2 + 4. \tag{12}$$

However, by lemma 2.2, T has either a best pair or a matched couple. In the former case, applying (11) and (2) to (1), we have

$$c \geq (l_1 + l_2 + 2) + (t-1)(l+2) + 2s,$$

and by (10),

$$c \geq (t-1)l + d(x) + d(y) + s.$$

In the latter case, applying (12) and (2) to (1), we have

$$c \geq (l_1 + l_2 + 4) + (t-2)(l+2) + 2s,$$

and, again, by (10),

$$c \geq (t-2)l + d(x) + d(y) + s.$$

Thus, in either case, (8) holds, and the proof of (ii) follows that of (i). This proves (b), and so the theorem. $\square$

Acknowledgement

This paper is part of the author's Ph.D. thesis supervised by Professor J.A. Bondy, whose many suggestions have improved the presentation of this work.

References

[1]   H. Enomoto, Long paths and large cycles in finite graphs, *J. Graph Theory* **8**(1984), 287-301.

[2]   G. Fan, Long cycles and the codiameter of a graph I, *J. Combin. Theory Ser. B.* To appear.

[3]   P. Fraisse and H.A. Jung, Oral presentation, Workshop on Cycles and Rays, Montréal 1987.

[4]   O. Ore, Hamilton connected graphs, *J. Math. Pures Appl.* **42**(1963), 21-27.

[5]   L. Pósa, On the circuits of finite graphs, *Magyar Tud. Akad. Mat. Kutató Int. Kózl.* **8**(1963), 355-361.

COMPATIBLE EULER TOURS IN EULERIAN DIGRAPHS

H. FLEISCHNER
Institut für Informationsverarbeitung
Akademie der Wissenschaften
A1010 Wien
Austria

B. JACKSON
Mathematical Sciences Department
Goldsmiths' College
London, SE14 6NW
United Kingdom

Abstract

We show that if D is an Eulerian digraph of minimum degree $2k$ then D has a set S of $[\frac{1}{2}k]$ Euler tours such that each pair of adjacent arcs of D is consecutive in at most one tour of S. We conjecture that our bound of $[\frac{1}{2}k]$ may be improved to $k-2$, and obtain some partial results in support of this conjecture.

1 Definitions

In the following D will denote a finite Eulerian digraph without loops, but possibly containing multiple arcs. For v a vertex and H a subgraph of D let $A^+(v, H)$ and $A^-(v, H)$ denote the sets of arcs of D from v to $V(H)$ and from $V(H)$ to v, respectively, and put $A(v, H) = A^+(v, H) \cup A^-(v, H)$. Let $V_4(D)$ be the set of vertices of D of degree at least four and $\delta_4(D) = \min\{d(v) : v \in V_4(D)\}$. Choose $v \in V_4(D)$. By a *transition at* v we shall mean a pair of arcs (a, b) with $a \in A^-(v, D)$, and $b \in A^+(v, D)$. A *transition system at* v is a partition, $T(v)$, of $A(v, D)$ into transitions. If $T(v)$ is defined for every $v \in V_4(D)$ then $T = \{T(v) : v \in V_4(D)\}$ is a *transition system for* D. We shall say two transition systems for D, $T_1 = \{T_1(v)\}$ and $T_2 = \{T_2(v)\}$ are *compatible* if $T_1(v) \cap T_2(v) = \phi$ for each $v \in V_4(D)$.

Given an Euler tour E of D, we may define, in an obvious way, the transition system $T(E)$ corresponding to E. We shall say that E is *compatible* to T if T is compatible to $T(E)$. Similarly, if for $i = 1$ and 2, E_i is an Euler tour of D, then we shall say that E_1 is compatible to E_2 if $T(E_1)$ and $T(E_2)$ are compatible.

G. Hahn et al. (eds.), Cycles and Rays, 95–100.

2   Compatible Euler Tours

The problem of determining the maximum number of pairwise compatible Euler tours in an undirected graph was first suggested by Kotzig [4], for the special case of complete graphs. Some partial results for arbitrary graphs have been obtained in [3] and [1]. In the present note we shall consider the analogous problem for Eulerian digraphs. We shall prove:

Theorem 1 *Let T_1, T_2, ..., T_h be transition systems for an Eulerian digraph D with $\delta_4(D) \geq 2k$. If $h \leq \frac{1}{2}k - 1$ then D has an Euler tour which is compatible to T_i for all i, $1 \leq i \leq h$.*

In order to prove Theorem 1 we shall need to extend a result of Häggkvist [1] on arbitrary graphs, for the special case of bipartite graphs.

Theorem 2 *Let G be a simple bipartite graph with bipartition $V(G) = X \cup Y$ where $|X| = |Y| = m$. If $d_G(x) + d_G(y) \geq m + 2$ for all pairs of non-adjacent vertices $x \in X$, $y \in Y$ then every 1-factor of G is contained in a Hamilton cycle of G.*

Proof: We proceed as in [1]. Let F be a 1-factor of G. Let H be the digraph obtained by orienting the edges of G from X to Y and then contracting the edges of F. Using the hypothesis that $d_G(x) + d_G(y) \geq m + 2$ for all non-adjacent $x \in X$, $y \in Y$, it follows that H is simple, strongly connected and that $d_H(u) + d_H(v) \geq 2m = 2|H|$ for all pairs of non-adjacent vertices u, v of H. By Meyniel's Theorem [5], H has a Hamiltonian cycle C. Let F_1 be the edges of G corresponding to the arcs of C. Then $F \cup F_1$ is the required Hamilton cycle of G containing F.

We shall also use the concept of *vertex splitting*. Given a vertex v of degree $2m$ in an Eulerian digraph $D = (V, A)$ and a transition system $T(v)$ at v we shall say that the digraph $H = ((V - \{v\}) \cup \{v_1, v_2, \ldots, v_m\}, A)$ is obtained from D by *splitting v along $T(v)$* if:

(i) $A^-(x, H) = A^-(x, D)$ and $A^+(x, H) = A^+(x, D)$ for all $x \in V - \{v\}$, and

(ii) $A(v_i, H) \in T(v)$, for $1 \leq i \leq m$.

Intuitively H is obtained from D by 'splitting' v into m vertices of indegree and outdegree one.

Proof of Theorem 1: We proceed by induction on $|V_4(D)|$. If $V_4(D) = \emptyset$ the theorem is trivially true. Hence suppose $V_4(D) \neq \emptyset$ and choose $v \in V_4(D)$. Let $A^-(v, D) = \{a_1, a_2, \ldots, a_m\}$, $A^+(v, D) = \{b_1, b_2, \ldots, b_m\}$, and let $H_1, H_2, \ldots, H_s$ be the (weakly connected) components of $G - v$. Since G is Eulerian, $A^-(v, H_i) = A^+(v, H_i)$ for all $i, 1 \leq i \leq s$. Thus we may choose a transition system at v, $T(v)$, such that each transition of $T(v)$ is either disjoint from or contained in $A(v, H_i)$ for all $i, 1 \leq i \leq s$.

Let G be the bipartite graph with vertex bipartition $A^-(v, D) \cup A^+(v, D)$ in which a_i and b_j are adjacent for $1 \leq i \leq j \leq m$ if and only if either $\{a_i, b_j\} \in T(v)$, or $\{a_i, b_j\} \notin T(v)$ and $\{a_i, b_j\} \notin T_g(v)$ for all $g, 1 \leq g \leq h$. Thus each vertex of G has degree at least $m - h$. Our aim is to construct a new digraph D^* from D by 'splitting' v into m vertices of degree two $v_1, v_2, \ldots, v_m$ such that $A(v_i, D^*) \notin T_j(v)$ for all i, j, $1 \leq i \leq m$ and $1 \leq j \leq h$, and such that D^* is connected. Considering the $A(v_i, D^*)$ as edges of G, it can be seen that this is equivalent to finding a 1-factor F of G such that if M is any matching properly contained in F and $V(M)$ is the set of vertices of $V(G)$ $(= A(v, D))$ incident with edges of M then $V(M)$ cannot be expressed as a union of sets $A(v, H_i)$, $i \in S$ for some $S \subset \{1, 2, \ldots, s\}$. Noting that:

(a) each $A(v, H_i)$ is a union of elements of $T(v)$, and

(b) $T(v)$ may be considered as a 1-factor of G,

it follows that it will be sufficient to find a 1-factor F of G such that $F \cup T(v)$ is a Hamilton cycle of G. (To see this, note that if F is a 1-factor of G disjoint from $T(v)$ and M is a proper subset of F such that $V(M) = \underset{i \in S}{\cup} A(v, H_i)$ then $F \cup T(v)$ is a disconnected 2-factor of G.)

Since $m \geq k \geq 2h + 2$, we have that G is a bipartite graph on $2m$ vertices of minimum degree at least $m - h \geq \frac{1}{2}(m + 2)$. By Theorem 2, we deduce that any 1-factor of G is contained in a Hamilton cycle of G. Thus we may construct the required 1-factor F of G by choosing a Hamilton cycle C of G containing $T(v)$ and putting $F = C - T(v)$.

We now use F to construct D^* from D by splitting v along the transition system defined by F. By the definition of G, and since $F \cap T(v) = \emptyset$, it follows that $A(v_i, D^*) \notin T_j(v)$ for all $1 \leq i \leq m$ and $1 \leq j \leq h$. Moreover, as noted above, D^* is connected and hence Eulerian. For each j, $1 \leq j \leq h$, let T_j^* be the system of transitions for D^* obtained from T_j by deleting $T_j(v)$.

Since each vertex of $V_4(D^*)$ has degree at least $2k$ it follows by induction that D^* has an Euler tour E^* which is compatible to T_j^* for all j, $1 \leq j \leq h$. Clearly E^* induces an Euler tour E of D which is compatible to T_j for all j, $1 \leq j \leq h$. □

Corollary 3 *If D is an Eulerian digraph and each vertex of $V_4(D)$ has degree at least $2k$ then D has a set of $[\frac{1}{2}k]$ pairwise compatible Euler tours.*

Proof: Use Theorem 2 and the greedy algorithm (start with an Euler tour E_1 of D and let $T_1 = T(E_1)$). □

Examples with $|V_4(D)| = 1$ illustrate that the bound $\frac{1}{2}k - 1$ in Theorem 2 cannot be improved to $\frac{1}{2}(k = 1)$. Consider for example the digraph D of Figure 1.

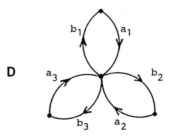

Figure 1

Putting $T_1 = (\{a_1, b_3\}, \{a_2, b_2\}, \{a_3, b_1\})$ we have $k = 3$ and $h = 1$. Moreover D has no Euler tour compatible to T_1. We feel however that Corollary 3 may be strengthened to:

Conjecture 1 *If D is an Eulerian digraph and each vertex of $V_4(D)$ has degree at least $2k$, then D has a set of $k - 2$ pairwise compatible Euler tours.*

Since the Eulerian digraphs with $|V_4(D)| = 1$ seem to be, in some sense, extremal with respect to Theorem 1, we have tested Conjecture 1 by verifying it for this family of digraphs. This follows immediately from the following more general result (in which a *block* of D is a subgraph induced by a block in the underlying undirected graph of D).

Theorem 4 *Let D be an Eulerian digraph such that $\delta_4(D) = 2k \geq 4$, and such that each block of D is a cycle. Then the maximum number of pairwise compatible Euler tours of D is equal to $k - 1 - \epsilon$, where $\epsilon = 1$ if $k = 4$ or 6 and $\epsilon = 0$ otherwise.*

Theorem 4 also has the following Corollary which gives an obvious necessary condition for Conjecture 1 to hold.

Corollary 5 *Let D be an Eulerian digraph, $v \in V_4(D)$, and $d(v) = 2k$. Then there exist h transition systems $T_1(v), T_2(v), \ldots, T_h(v)$, pairwise compatible at v, such that the digraph D_i, obtained from D by splitting v along $T_i(v)$ is connected for all i, $1 \leq i \leq h$, where $h = k - 1 - \epsilon$, and $\epsilon = 1$ if $h = 4$ or 6 and $\epsilon = 0$ otherwise.*

Our proof of Theorem 4 uses techniques developed in [1]. We shall need:

Theorem 6 *Let F_0 be a 1-factor of the complete bipartite graph $K_{m,m}$. Then the maximum number of disjoint 1-factors $F_1, F_2, \ldots, F_h$ of $K_{m,m}$ such that $F_0 \cup F_i$ is a Hamilton cycle for all i, $1 \leq i \leq h$, is $h = m - 1 - \epsilon$, where $\epsilon = 1$ if $m = 4$ or 6 and $\epsilon = 0$ otherwise.*

Proof: Let (X, Y) be the bipartition of $K_{m,m}$ and let K_m^* be the complete digraph obtained by orienting each edge of $K_{m,m}$ from X to Y and then contracting the edges of F_0. Associating each edge of $K_{m,m} - F_0$ with the corresponding arc of K_m^* gives a bijection between the 1-factors F of $K_{m,m}$ such that $F_0 \cup F$ is a Hamilton cycle, and the Hamilton cycles of K_m^*. The theorem now follows from the result of Tillson [6] that the maximum number of arc-disjoint Hamilton cycles in K_m^* is $m - 1 - \epsilon$. $\square$

Proof of Theorem 4: We first show that there exist $k - 1 - \epsilon$ pairwise compatible Euler tours of D. Our proof is by induction on $|V_4(D)|$ but in order to push through the induction we need a rather complicated inductive hypothesis. We shall say that two digraphs $D_1 = (V_1, A_1)$ and $D_2 = (V_2, A_2)$ are *equivalent* if $V_1 = V_2$, $A_1 = A_2$, $V_4(D_1) = V_4(D_2)$, $A^-(v, D_1) = A^-(v, D_2)$ and $A^+(v, D_1) = A^+(v, D_2)$ for all $v \in V_4(D_1)$. Two Euler tours E_1 of D_1 and E_2 of D_2 are *compatible* if $T(E_1)$ and $T(E_2)$ have no common transition at any $v \in V_4(D_1)$. We shall prove: (*) Let $D_1, D_2, \ldots, D_h$ be pairwise equivalent Eulerian digraphs such that

(1) each block of D_i is a cycle and for each $v \in V_4(D_i)$, the transition system $T_i(v)$ induced by the cycles through v is the same for all i, $1 \leq i \leq h$, and

(2) $d(v) \geq 2k$ for $v \in V_4(D_i)$ and $h = k - 1 - \epsilon$.

Then D_i has an Euler tour E_i such that E_i and E_j are compatible for al i, j, $1 \leq i < j \leq h$. Note that Theorem 4 will follow from (*) taking $D_i = D$ for all i, $1 \leq i \leq h$.

We proceed by induction on $|V_4(D_1)|$. If $V_4(D_1) = \emptyset$ then (*) is trivially true. Hence suppose $V_4(D_1) \neq \emptyset$ and choose $v \in V_4(D_1)$. Let $d(v) = 2m \geq 2k$ and let $K_{m,m}$ be the complete bipartite graph with bipartition (X, Y) where $X = A^-(v, D_i)$ and $Y = A^+(v, D_i)$, $1 \leq i \leq h$. Let F_0 be the 1-factor of $K_{m,m}$ corresponding to the transition system $T_i(v)$ defined in (1). By Theorem 6, there exist disjoint 1-factors $F_1, F_2, \ldots, F_h$ of $K_{m,m}$ such that $F_0 \cup F_i$ is a Hamilton cycle for all $1 \leq i \leq h$, and $h = k - 1 - \epsilon$. Now let D_i^* be the digraph obtained from D_i by splitting v along the transition system defined by F_i for all i, $1 \leq i \leq h$. Since $F_0 \cup F_i$ is a Hamilton cycle, D_i^* is connected and hence Eulerian. Moreover $D_1^*, D_2^*, \ldots, D_h^*$ are pairwise equivalent and satisfy (1) and (2). Since $|V_4(D_i^*)| < |V_4(D_i)|$ it follows from the inductive hypothesis that D_i^* has an Euler tour E_i^* such that E_i^* and E_j^* are compatible for all $i, j, 1 \leq i < j \leq h$. Let E_i be the Euler tour of D_i induced by E_i^*. Then E_i and E_j are clearly compatible at every vertex of $V_4(D_1)$ other than v, and are also compatible at v since $F_i \cap F_j = \emptyset$. Thus $E_1, E_2, \ldots, E_h$ are the required Euler tours of $D_1, D_2, \ldots, D_h$.

Finally we show that the maximum number of pairwise compatible Euler tours of D is indeed equal to $k - 1 - \epsilon$. Choose $v \in V_4(D)$ with $d(v) = 2k$ and suppose that $E_1, E_2, \ldots, E_t$ are pairwise compatible Euler tours of D. Let $T_0(v)$ be the system of transitions at v corresponding to the cycles containing v, and $T_i(v)$ be the system of transitions at v corresponding to E_i for $1 \leq i \leq t$. Let $K_{k,k}$ be the complete bipartite graph with bipartition (X, Y) where $X = A^-(v, D)$ and $Y = A^+(v, D)$. Then $T_1(v), T_2(v), \ldots, T_t(v)$, correspond to disjoint 1-factors of $K_{k,k}$ such that $T_0(v) \cup T_i(v)$ is a Hamilton cycle of $K_{k,k}$. By Theorem 6, $t \leq k - 1 - \epsilon$.

Proof of Corollary 5: Choose an Euler tour E_0 of D and let $T(v)$ be the transition system defined at v by E_0. Define a new digraph $H = (\{v, x_1, x_2, \ldots, x_k\}, A(v, D))$ where $j = A^-(v, D), A^+(v, H) = A^+(v, D)$ and $A(x_i, H) \in T(v)$ for all $i, 1 \leq i \leq k$. Then H satisfies the hypothesis of Theorem 4 (with $V_4(H) = \{v\}$), so H has h pairwise compatible Euler tours $E_1, E_2, \ldots, E_h$. Let $T_i(v)$ be the transition system at v defined by $E_i, 1 \leq i \leq h$. Then $T_i(v), T_2(v), \ldots, T_h(v)$ are as required.   □

Finally we give a result which strengthens Conjecture 1 and Theorem 1 for the special cases when $\delta_4(D) = 6$ and, in Theorem 1, when T_1 induces an Euler tour of D.

Theorem 7 *Let D be an Eulerian digraph such that $\delta_4(D) \geq 6$ and let E_1 be an Euler tour of D. Then D has an Euler tour E_2 compatible to E_1.*

Proof: We proceed by induction on $|V_4(D)|$. If $V_4(D) = \emptyset$ then the theorem is trivially true. Hence suppose $V_4(D) \neq \emptyset$. If $d(v) \geq 8$ for all $v \in V_4(D)$ then the theorem follows by Theorem 1. Hence we may assume $d(v) = 6$ for some $v \in V_4(D)$. Let

$$T_1(v) = \{(a_1, b_2), (a_2, b_3), (a_3, b_1)\}$$

be the transition system defined at v by E_1, where $A^-(v, D) = \{a_1, a_2, a_3\}$ and $A^+(v, D) = \{b_1, b_2, b_3\}$. Let

$$T_2(v) = \{(a_1, b_3), (a_3, b_2), (a_2, b_1)\},$$

and let E_1' be the Euler tour of D defined by the transition system $(T(E_1) - T_1(v)) \cup T_2(v)$. Let D^* be the Eulerian digraph obtained by splitting v along $T_2(v)$ and let E_1^* be the Euler tour of D^* induced by E_1'. Since $|V_4(D^*)| < |V_4(D)|$, D^* has an Euler tour E_2^* compatible to E_1^*. Let E_2 be the Euler tour of D induced by E_2^*. Then E_2 is compatible to E_1 at every vertex of $V_4(D) - \{v\}$ by the compatibility of E_2^* and E_1^*, and is compatible at v since $T_2(v) \cap T_1(v) = \emptyset$.   □

3   Closing Remarks

3.1   One possible way to verify Conjecture 1 would be to proceed as in the proof of Theorem 4, but removing condition (1) from the inductive hypothesis (∗). The problem being that when we choose $v \in V_4(D_i)$ we cannot assume that the transition systems $T_i(v)$ induced by the 'separating' transitions at v in D_i are the same for all i, $1 \leq i \leq m - 2$. To overcome this, one would need to replace Theorem 6 by:

Conjecture 2 *Let* $F'_1, F'_2, \ldots, F'_{m-2}$ *be 1-factors of* $K_{m,m}$. *Then there exist disjoint 1-factors* $F_1, F_2, \ldots, F_{m-2}$ *such that* $F'_i \cup F_i$ *is a Hamilton cycle for all* i, $1 \leq i \leq m - 2$.

An analogous conjecture for K_{2m} was formulated in [1].

3.2   In view of Theorem 6, we feel it would be an interesting problem to characterise the 4-regular Eulerian digraphs which have two compatible Euler tours. An obvious necessary condition is that the digraph be 2-connected. The fact that this condition is not also sufficient follows by considering the complete digraph K_3^*.

References

[1]   H. Fleischner, A.J.W. Hilton, B. Jackson, On the maximum number of pairwise compatible euler trails, *J. Graph Theory*. To appear.

[2]   R. Häggkvist, On F-hamiltonian graphs, in: *Graph Theory and Related Topics* (J.A. Bondy, U.S.R. Murty, eds.), Academic Press (1979), 219-231.

[3]   B. Jackson, Compatible Euler tours for transitions systems in eulerian graphs, *Discrete Math.* **66**(1987), 127-131.

[4]   A. Kotzig, Problem session, Proc. 10th S.E. Conf. Combinatorics, Graph Theory, Computing, Congr. Numer. XXIV(1979), 914-915.

[5]   M. Meyniel, Une condition suffisante d'existence d'un circuit hamiltonien dans un graphe orienté, *J. Combin. Theory Ser. B* **14**(1973), 137-147.

[6]   T.W. Tillson, A hamiltonian decomposition of $K_{2m}^*, 2m \geq 8$, *J. Combin. Theory Ser. B* **29**(1980), 68-74.

EDGE-COLOURING GRAPHS AND EMBEDDING PARTIAL TRIPLE SYSTEMS OF EVEN INDEX

A. J. W. HILTON
Department of Mathematics
University of Reading
P.O. Box 220
Whiteknights
Reading RG6 2AX
United Kingdom

C. A. RODGER
Department of Algebra, Combinatorics and Analysis
Mathematical Annex
Auburn University
Auburn, Alabama 36849
U.S.A.

Abstract

We show that a conjecture about edge-colouring certain graphs implies a conjecture about embedding partial triple systems of even index. We give some evidence to support each of these conjectures.

1 Introduction

A *partial triple system* of order r and index λ, a PTS(r, λ) for short, is a collection of triples of elements of an r-set such that each pair of elements is in at most λ of the triples. If each pair of elements is in exactly λ triples, then we have a *triple system* of order r and index λ, or a TS(r, λ) for short. It is well-known that these exist if and only if $r \geq 3$ and the following two conditions are satisfied:

$$\lambda r(r - 1) \equiv 0 \pmod 6$$

and

$$\lambda(r - 1) \equiv 0 \pmod 2.$$

When r and λ obey these, we call r λ-*admissible*.

A well-known conjecture of Lindner [13, 14] is that any PTS$(r, 1)$ can be extended to a TS$(n, 1)$ by the introduction of further elements and triples if n is admissible and $n \geq 2r+1$. The number $2r + 1$ here is best possible. It is natural to extend Lindner's conjecture to all values of λ. Thus we have the following more general conjecture.

G. Hahn et al. (eds.), Cycles and Rays, 101–112.
© *1990 by Kluwer Academic Publishers. Printed in the Netherlands.*

Conjecture 1 *Any PTS(r, λ) can be extended to a TS(n, λ) whenever n is λ-admissible and $n \geq 2r + 1$.*

Often the PTS(r, λ) is said to be *embedded* in the TS(n, λ).

Early results on Lindner's conjecture were by Treash [21] and Lindner [13]. Conjecture 1 was proved for $\lambda = 1$ and $n \geq 4r + 1$ by Andersen, Hilton and Mendelsohn [3], and more recently for any value of λ and $n \geq 4r + 1$ (except possibly if $r \leq 14$ and λ is even) by Rodger and Stubbs [18] and Stubbs [20]. The most recent progress has been by Hilton and Rodger [10] in a paper to which this is a sequel. We showed that Conjecture 1 is true if $4 | \lambda$.

Our recent result uses the methods discussed in this paper, and so is in itself strong evidence for the merits of the approach of this paper. The method is similar to that used for latin squares, timetables and Hamiltonian decompositions by Andersen, Hilton, Nash-Williams and Rodger [5, 6, 7, 8, 11, 12, 16, 17].

2   Two conjectures

In this section we explain briefly the two conjectures which we believe incorporate the real truth about embedding maximal PTS(r, λ)'s when λ is even. There is quite a lot of evidence in the rest of this paper to support these conjectures and we explain in this paper why Conjecture 2* on edge-colourings implies Conjecture 3 on embedding PTS(r, λ)'s when λ is even.

First we need to explain a number of graph-theoretical concepts, some of which are non-standard.

A graph consists of a set $V(G)$ of *vertices*, a set $L^{1/2}(G)$ of *half-loops*, and a set $E(G)$ of *edges*. Each edge is incident with two distinct vertices, and contributes one to the degree of each of the vertices it is incident with. Each half-loop is incident with one vertex, and contributes one to the degree of that vertex. A *loop* at a vertex is the union of two half-loops at that vertex, and so contributes two to the degree of the vertex. The set of all loops of G is denoted by $L(G)$. The set of all half-loops of G which are not part of a loop of G will be denoted by $H(G)$. Thus $L^{1/2}(G) = \left(\bigcup_{\ell \in L(G)} \ell \right) \cup H(G)$. A graph in which all half-loops occur in pairs, each pair forming a loop, is a *normal graph*. A graph is regular of degree d if the sum of the degrees due to the edges and half-loops incident with each vertex is d. Clearly each normal graph G with maximum degree $\Delta = \Delta(G)$ can be turned into a regular graph of degree Δ by the introduction of the appropriate number of half-loops at each vertex.

An *edge-colouring* of a graph G is a function $\gamma : E(G) \cup L(G) \cup H(G) \to \mathcal{C}$, where $\mathcal{C}$ is a set of colours. Note that this implies that each loop receives one colour, and so the two corresponding half-loops receive the same colour.

We now introduce a variant of the idea of an edge-colouring. A *split-loop colouring* is a function $\gamma : E(G) \cup L^{1/2}(G) \to \mathcal{C}$. Thus a split-loop colouring may be thought of as a kind of edge-colouring in which each loop receives two colours (or possibly the same colour twice); in fact, if we 'split' each loop of G by inserting a vertex in each loop, forming a loopless graph G^*, say, then a split-loop colouring of G corresponds to an edge-colouring of G^*.

A λ-*half-loop-factor* of a graph H is a subgraph which is regular of degree λ. A λ-*factor* of a graph H is a subgraph which is normal and is regular of degree λ. Thus a

λ-factor has its usual meaning, and any loop contributes two to the vertex it is on; but a λ-half-loop-factor may have some half-loops.

A λ-*half-loop-factorization* of a regular graph H of degree $x\lambda$ is a decomposition of H into x (edge and half-loop)-disjoint λ-half-loop-factors. Thus a λ-half-loop-factorization of H is a split-loop-colouring of H in which each colour class is a regular half-loop factor of degree λ.

A split-loop-colouring of a graph H with colours $c_1, \ldots, c_k$ is said to be *equalized* (on the edges) if, for each $i, j \in \{1, \ldots, k\}$, $i \neq j$.

$$||C_i| - |C_j|| \leq 1,$$

where, for $1 \leq i \leq k$, C_i is the set of edges of H of colour c_i (so half-loops are *not* included in C_i).

A split-loop-colouring of a graph H is *skew-free* (on the half-loops) at a vertex v if not more than half the half-loops at v have the same colour. It is *skew-free* if, for each $v \in V(H)$, it is skew-free at v.

We are now in a position to state Conjecture 2.

Conjecture 2 *Let λ be even and let $x \geq 2$. Let H be a normal regular connected graph of degree $x\lambda$. Then H has a skew-free equalized λ-half-loop factorization if and only if the following two conditions are satisfied:*

(i) when $x > 2$, H does not have exactly one loop,

(ii) when $x = 2$, the number of loops of H is even.

A simple case where we cannot prove Conjecture 2 are when $\lambda = 2$, $x = 3$ and H is a normal graph with two or three loops.

In Lemmas 5.3 and 5.4 of [10], we showed that if λ and x are both even, and if either H has an even number of loops, or H has an odd number of loops and there is a vertex that is incident with at least two loops, then H has a skew-free equalized λ-half-loop factorization. We also showed in [10] that conditions (i) and (ii) are necessary for the existence of such a factorization.

We shall in fact need a slight extension of Conjecture 2.

Conjecture 2* *Conjecture 2 is true, and moreover, if H satisfies (i) and (ii), if H has degree λx, and if the number ℓ of half-loops of H satisfies $\{\lambda(x-1) - 2\}x \leq \ell \leq \lambda(x-1)x$, then H has a skew-free equalized λ-half-loop-factorization with at least one λ-half-loop factor F such that $H \backslash F$ has the following property. Either the number of half-loops of F is $\lambda(x-1)$, and $H \backslash F$ does not contain some components which, between them, contain exactly one whole loop and all the $\lambda(x-1)$ half-loops which were paired with the half-loops of F; or the number of half-loops of F is $\lambda(x-1) - 2$, and $H \backslash F$ does not contain some components which, between them, contain all the $\lambda(x-1) - 2$ half-loops which were paired with half-loops of F, but no whole loops.*

We now turn to the main conjecture, Conjecture 3, on embedding a PTS(r, λ) when λ is even. Conjecture 3 takes its inspiration from Ryser's theorem [19] on embedding latin rectangles, or more particularly from a development of Ryser's theorem due to Cruse [4]. Before stating Conjecture 3, we need some further preliminaries.

Given a PTS(r, λ) T, let G be the following normal loopless multigraph. The vertex set of G is the set of elements of T; two vertices of G are joined by x edges if they are in

$\lambda - x$ triples of T. We call G the *missing-edge graph* of T. If we are trying to embed T in a TS(n,λ), then the integer n will be known to us; assuming that $\Delta(G) \leq \lambda(n-r)$ and that $\lambda(n-r) - d_G(v)$ is even for each $v \in V(T)$, we define a normal regular graph G^o by adjoining the requisite number of loops at each vertex of G to make the degree $\lambda(n-r)$. For $v \in V(T)$, let $N(v)$ be the number of triples which contain the vertex v.

We can now state Conjecture 3.

Conjecture 3 *Let λ be even. A PTS(r,λ) T can be embedded in a TS(n,λ) without inserting any further triples on the elements of T if and only if the following four conditions are satisfied:*

(i) n is λ-admissible,

(ii) $N(v) \geq \frac{1}{2}\lambda(2r - n - 1)$ *$(\forall v \in V(T))$,*

(iii) $\sum_{v \in V(T)} N(v) \leq \lambda\left\{\binom{n-r}{2} + \binom{r}{2} - \frac{1}{2}r(n-r)\right\}$,

(iv) G^o contains no component with exactly one loop, and if $n - r = 2$, then G^o contains no component with an odd number of loops.

It is shown in Lemma 2.1 of [10] that condition (ii) implies that the definition of G^o is valid. We shall show that Conjecture 3 implies Conjecture 1. It is shown in Theorem 4.3 of [10] that conditions (i)–(iv) are necessary for T to be embeddable in a TS(n,λ) without inserting any further triples on the elements of T. In [10] it is shown that if $4|\lambda$ then conditions (i)–(iii), together with a slightly stronger version of condition (iv), are sufficient to embed T.

3   Amalgamated Triple Systems

Roughly speaking, an amalgamated TS(n,λ) is what you get if you take a TS(n,λ) and amalgamate several of the vertices. Here, and in the rest of this paper, we think of a TS(n,λ) as a λK_n, that is a complete graph on n vertices with each edge replicated λ times, whose edges are coloured with $\frac{1}{6}\lambda n(n-1)$ colours, each colour class forming a K_3.

We now give a formal definition of an amalgamation of a normal graph. Given a normal graph G, let $U \subset V(G)$ and let Q be an element $Q \notin V(G)$. Then H is an amalgamation of G formed by amalgamating U if H has a vertex set $\{Q\} \cup (V(G)\backslash U)$ and if there is a bijection

$$\phi : E(G) \cup L(G) \to E(H) \cup L(H)$$

such that

(Ai) $\phi(e)$ joins two vertices x and y in $V(H)$ if e joins x and y, and $x,y \in V(G)\backslash U$,

(Aii) $\phi(e)$ joins two vertices x and Q in $V(H)$ if e joins x and y, and $x \in V(G)\backslash U$, $y \in U$,

(Aiii) $\phi(e)$ is a loop on x if e is a loop on x and $x \in V(G)\backslash U$,

(Aiv) $\phi(e)$ is a loop on Q if either e joins two vertices $x,y \in U$, or e is a loop on a vertex y and $y \in U$.

The vertex Q will be called the *amalgamated vertex* (or the *source vertex*). It will be understood that if G has an edge-colouring γ, then this will be transferred to H by the bijection ϕ.

With a TS(n, λ) as an edge-coloured λK_n in mind, if $U \subset V(\lambda K_n)$ then the amalgamation W of K_n formed by amalgamating U is called an *amalgamated* TS(n, λ) (or an *amalgamated triple system of order n and index* λ). Thus if $U = V(\lambda K_n)$ then W would be a graph consisting of one vertex (Q) and $\frac{1}{2}\lambda n(n-1)$ loops; the loops would be coloured with $\frac{1}{6}\lambda n(n-1)$ colours and there would be three loops of each colour.

The main object of this paper, to show that Conjecture 2* implies Conjecture 3, will be achieved by considering what is needed to reverse the process of amalgamation of TS(n, λ)'s. To this end we need to study amalgamation itself in some detail. So we start by giving a number of properties satisfied by amalgamated TS(n, λ)'s. Lemmas 3.1, 3.2, 3.3, 3.4 and 3.5 are all proved in [10].

Lemma 3.1 *An amalgamated* TS$(n, \lambda)S$ *with* $n - r$ *vertices amalgamated satisfies the following properties.*

(Bi) Each vertex has degree $\lambda(n-1)$ *except, if* $r < n-1$, *for the source vertex* Q *which has degree* $\lambda(n-r)(n-1)$. *[If* $r = n-1$, *then any vertex can be designated as the source vertex.]*

(Bii) The source vertex has $\frac{1}{2}\lambda(n-r)(n-r-1)$ *loops on it; no other vertex has any loops on it.*

(Biii) The number of edges between two vertices x *and* y, *where* $x \neq y$, *is*

$$d(x)d(y)/\lambda(n-1)^2.$$

Note: $d(x)$ denotes the degree of the vertex x.

Recall that a TS(n, λ) is a λK_n whose edges are coloured with $\frac{1}{6}\lambda n(n-1)$ colours so that each colour class forms a K_3.

Lemma 3.2 *In an amalgamated* TS(n, λ) S *with* $n - r$ *vertices amalgamated, we have the following property:*

(Biv) The three edges of any given colour induce a subgraph of one of the following four types:

Q $\qquad\qquad$ Q $\qquad\qquad$ Q

$\qquad\qquad\qquad\qquad\qquad\qquad\qquad\qquad\qquad\qquad\qquad$ z

$\qquad\quad$ x $\qquad\quad$ $x \qquad\quad y$ $\qquad$ $x \qquad\qquad\quad y$

$\qquad\quad$ $x \neq Q$ $\qquad$ $Q \notin \{x, y\}, x \neq y$ $\qquad$ $Q \in \{x, y, z\}$,

$\qquad\qquad\qquad\qquad\qquad\qquad\qquad\qquad\qquad\qquad$ $|\{x, y, z\}| = 3.$

We call these triangles 3-triangles, 2-triangles, 1-triangles and 0-triangles respectively; we sometimes denote them by $\{Q, Q, Q\}$, $\{Q, Q, x\}$, $\{Q, x, y\}$ and $\{x, y, z\}$ respectively.

Associated with an amalgamated TS(n, λ) S with $n - r$ vertices amalgamated, we have a PTS(r, λ) T consisting of all the 0-triangles of S. Based on T we can construct the missing-edge graph G, and the associated normal regular graph G^o as described before Conjecture 3. The missing-edge graph G is then the set of those edges of 1-triangles of S which are not incident with Q. Each loop of G^o corresponds to a 2-triangle of S.

The amalgamated $TS(n, \lambda)$ S induces in a fairly obvious way a skew-free split-loop colouring of the graph G^o. For suppose that vertices $v_1, \ldots, v_r$ of λK_n are amalgamated to form the vertex Q. For each $i \in \{1, \ldots, n - r\}$, let the edges of $\lambda K_r \backslash \{v_1, \ldots, v_r\}$ which are in triples (coloured triangles) of λK_n with one vertex $v_i \in \{v_1, \ldots, v_{n-r}\}$, but with neither of the other two vertices in $\{v_1, \ldots, v_r\}$, be coloured c_i; let those edges of λK_n which are in triples of λK_n with two vertices v_i and v_j both in the set $\{v_1, \ldots, v_{n-r}\}$, but the third vertex $v \notin \{v_1, \ldots, v_{n-r}\}$, which join v to v_i, be coloured c_i also. Transferring this partial edge-colouring of λK_n to S, we have that the edges and half-loops of G^o are coloured $c_1, \ldots, c_{n-r}$; each edge and each half-loop receives one colour. Also each loop of G^o receives two colours, since each loop of G^o corresponds to a triangle $\{v, v_i, v_j\}$ of λK_n. Thus G^o has a skew-free split-loop colouring with $c_1, \ldots, c_{n-r}$.

Lemma 3.3 [10] *In an amalgamated $TS(n, \lambda)$ S, the induced skew-free split-loop-colouring of G^o is a λ-half-loop-factorization of G^o.*

Lemma 3.4 [10] *Let λ be even. An amalgamated $TS(n, \lambda)$, with $n - r$ vertices amalgamated, has the following property:*
(Bv) If $n - r = 2$, then each connected component of G^o has an even number of loops.

Lemma 3.5 [10] *Let λ be even. An amalgamated $STS(n, \lambda)$, with $n - r$ vertices amalgamated, has the following property:*
(Bvi) If $n - r > 2$, then no connected component of G^o has exactly one loop.

4 Quasi STS's

In this section we define quasi $TS(n, \lambda)$'s for λ even. From the definition it is clear that an amalgamated $TS(n, \lambda)$ is a quasi $TS(n, \lambda)$. We showed in [10] that Conjecture 3 can be reformulated so as to state that, for λ even, a quasi $TS(n, \lambda)$ is an amalgamated $TS(n, \lambda)$; this reformulation seems to be a more illuminating version of Conjecture 3.

For λ even, a *quasi* $TS(n, \lambda)$ is a normal graph H with $\frac{1}{2}\lambda n(n - 1)$ edges (counting parallel edges and loops); it has an edge-colouring with $\frac{1}{6}\lambda n(n - 1)$ colours with three edges of each colour; it contains a special vertex Q; and it satisfies (Bi)–(Bvi).

Since for λ even an amalgamated $TS(n, \lambda)$ satisfies (Bi)–(Bvi) it follows that, for λ even, an amalgamated $TS(n, \lambda)$ is a quasi $TS(n, \lambda)$.

We now state the converse as a conjecture. We showed in [10] that this is equivalent to Conjecture 3.

Conjecture 4 *Let λ be even and n be λ-admissible. Then a quasi $TS(n, \lambda)$ is an amalgamated $TS(n, \lambda)$.*

Thus, if Conjecture 4 is true, then every quasi $TS(n, \lambda)$ can be "undone" to produce a proper $TS(n, \lambda)$.

5 A proof that Conjecture 2* implies Conjecture 3

Since conditions (i)–(iv) in Conjecture 3 are known to be necessary for the embedding of T into a $TS(n, \lambda)$, what needs to be proved is that they are sufficient. Our $TS(n, \lambda)$ T

corresponds to a quasi $TS(n, \lambda)$ S in which the vertex Q has degree $\lambda(n - 1)(n - r)$. Moreover we are supposing that n is λ-admissible and λ is even. Since Conjectures 3 and 4 are equivalent (see [10]), in fact what we have to do is to show that the quasi $TS(n, \lambda)$ S is an amalgamated $TS(n, \lambda)$.

This process itself is done in stages. We take a quasi $TS(n, \lambda)$ S with a vertex Q with degree $\lambda(n - 1)(n - r)$, and from it we produce a quasi $TS(n, \lambda)$ S' with a vertex Q' with degree $\lambda(n-1)(n-r-1)$ such that S is an amalgamation of S'. We repeat this until we have a quasi $TS(r, \lambda)$ S^* with all vertices having degree $\lambda(n - 1)$; but any such quasi $TS(n, \lambda)$ is actually a $TS(n, \lambda)$. Furthermore S^* contains $n - r$ vertices whose amalgamation produces S. Thus our main task will be to produce S' from S.

Let G be the missing edge-graph of S and let G^o be the associated normal regular graph of degree $(n - r)\lambda$ formed by adjoining $\frac{1}{2}\{(n - r)\lambda - d_G(v)\}$ loops to each vertex $v \in V(G)$. Since S is a quasi $TS(n, \lambda)$, it follows from (Bv) and (Bvi) that no connected component of G^o has exactly one loop, and, if $n - r = 2$, that each connected component of G^o has an even number of loops. By Conjecture 2, it follows that G^o has a skew-free equalized λ-half-loop factorization. Let F be one of these λ-half-loop-factors of G^o.

We now go on to describe the procedure we adopt if $n - r \geq 3$ in forming the quasi $TS(n, \lambda)$ S'. First we describe this procedure, and then we justify it. If $n - r = 2$, the procedure is simpler, and is described later.

We 'split off' a vertex u from Q; that is, we introduce a further vertex u, place some triangles on it, remove some triangles from Q and alter other triangles, in a way we now describe. For each edge $v_1 v_2$ of F we form a new 3-triangle $\{v_1, v_2, u\}$. For each edge of $E(G) \backslash F$ we retain the 1-triangle $\{v_1, v_2, Q\}$. For each half-loop of F on a vertex v we form a new 1-triangle $\{v, u, Q\}$. If $h(F, v)$ denotes the number of half-loops of F on the vertex v and $h(G^o, v)$ denotes the number of half-loops of G^o on the vertex v, then we retain $\frac{1}{2}\{h(G^o, v) - 2h(F, v)\}$ 2-triangles incident with v. If $w(u)$ denotes the number of new 1-triangles incident with u, then we form $\frac{1}{2}\{\lambda(n - r - 1) - w(u)\}$ new 2-triangles on u. Finally we remove this number, $\frac{1}{2}\{\lambda(n - r - 1) - w(u)\}$, of 3-triangles from Q.

We now justify the procedure for forming S' explained above. We start by justifying the various numerical assumptions that were made in our description of the procedure. Since the λ-half-loop-factorization was skew-free, for each $v \in V(T)$ it follows that $h(F, v) \leq h(G^o \backslash F, v)$, and so $h(G^o, v) \geq 2h(F, v)$. The number of 1-triangles incident with each $v \in V(T)$ is at most $\lambda(n - r)$; if y is the number of 2-triangles incident with v, then $d_G(v) = \lambda(n - r) - 2y$. Since $2y = h(G^o, v)$, it follows that $h(G^o, v)$ is even. Therefore $\frac{1}{2}\{h(G^o, v) - 2h(F, v)\}$ is a non-negative integer.

Let $h(F)$ denote the number of half-loops of F. Then $h(F) = \lambda r - 2|E(F)|$. Thus $w(u)$, the number of 1-triangles in S' on u, satisfies $w(u) - h(F) = \lambda r - 2|E(F)|$. We therefore need to show that

$$\lambda(n - r - 1) - w(u) = \lambda(n - r - 1) - \lambda r + 2|E(F)| = \lambda(n - 2r - 1) + 2|E(F)|$$

is even and positive. It is clearly even. Since the λ-half-loop-factorization of G^o is equalized,

$$\left\lfloor \frac{1}{n - r}|E(G)| \right\rfloor \leq |E(F)| \leq \left\lceil \frac{1}{n - r}|E(G)| \right\rceil.$$

Therefore we need to show that

$$\frac{\lambda}{2}(n - 2r - 1) + \left\lfloor \frac{1}{n - r}|E(G)| \right\rfloor \geq 0.$$

Since $\left[\sum_{v \in V(T)} N(v)\right] + |E(G)| = \lambda\binom{r}{2}$, the bound in (iii) is equivalent to

$$\frac{\lambda}{2}(n - 2r - 1) + \frac{1}{n-r}|E(G)| \geq 0,$$

which implies the inequality above since $\frac{\lambda}{2}(n - 2r - 1)$ is an integer.

Finally we need to know that the number of 2-triangles we place on u, namely $\frac{\lambda}{2}(n - r - 1) - \frac{1}{2}w(u) \leq \frac{\lambda}{2}(n - 2r - 1) + \left[\frac{1}{n-r}|E(G)|\right]$, is not more than the original number of 3-triangles. But the number of 3-triangles was

$$\begin{aligned}
&\tfrac{\lambda}{6}n(n-1) - \tfrac{\lambda}{2}r(n-r) - \tfrac{1}{3}\sum_{v \in v(T)} N(v)\\
&= \tfrac{\lambda}{6}n(n-1) - \tfrac{\lambda}{2}r(n-r) - \tfrac{\lambda}{3}\binom{r}{2} + \tfrac{1}{3}|E(G)|\\
&= \tfrac{\lambda}{3}\binom{n-2}{2} - \tfrac{\lambda}{6}r(n-r) + \tfrac{1}{3}|E(G)|\\
&= \tfrac{\lambda}{6}(n-r)(n-2r-1) + \tfrac{1}{3}|E(G)|,
\end{aligned}$$

so we need that

$$(1) \qquad \frac{\lambda}{2}(n - 2r - 1) + \left[\frac{1}{n-r}|E(G)|\right] \leq \frac{\lambda}{6}(n-r)(n-2r-1) + \frac{1}{3}|E(G)|.$$

Since $\frac{\lambda}{2}(n - 2r - 1)$ and the right hand side of (1) are both integers, this inequality would follow provided that

$$\frac{\lambda}{6}(2r + 1 - n)(n - r - 3) \leq \frac{1}{3}|E(G)| - \frac{1}{n-r}|E(G)|$$

is true. This can be rewritten as

$$\frac{\lambda}{2}(2r + 1 + n)(n - r - 3) \leq \frac{(n - r - 3)}{(n - r)}|E(G)|.$$

If $n - r > 3$ this is equivalent to

$$\frac{\lambda}{2}(n - r)(2r + 1 - n) \leq |E(G)|.$$

But this inequality is true, as it is equivalent to the bound in (iii). The right-hand side of (1) is an integer (it is the number of 3-triangles). If $n - r = 3$ then $\frac{\lambda}{6}(n - r)(n - 2r - 1)$ is an integer, and so $\frac{1}{3}|E(G)|$. Therefore (1) is true also when $n - r = 3$. This completes the demonstration that the numerical manoeuvres described in the procedure for constructing S' are possible.

We now go on to show that S' is in fact a quasi $TS(n, \lambda)$. It is apparent from the description of the procedure that (Biv) is satisfied, and furthermore that each edge of S' is in a colour class (a triangle).

First consider a vertex $v \in V(G)$. For each edge in F incident with v, a 1-triangle is removed and a new 0-triangle involving v and u is formed. Thus the number of edges from v to Q decreases for this reason by the number, say $e(F, v)$, of edges of F incident with v; the number of (new) edges from v to u equals this amount. For each half-loop in F incident with v, a 2-triangle is removed and replaced by a new 1-triangle involving v and u; each loop on v in G corresponds to a 2-triangle on v, and this process replaces $h(F, v)$

such 2-triangles by $h(F, v)$ (new) 1-triangles. Therefore the number of edges from v to Q decreases for this second reason by $h(F, v)$, and the further number of (new) edges from v to Q equals $h(F, v)$. Therefore altogether the number of edges from v to Q decreases by $e(F, v) + h(F, v) = \lambda$, and the number of (new) edges from v to u is similarly λ.

Now consider u. We have seen that the number of edges from v to u is λ. The number of new 1-triangles involving u is $w(u)$, so this accounts for $w(u)$ edges between u and Q. $\frac{1}{2}\{\lambda(n - r - 1) - w(u)\}$ new 2-triangles are placed on u, which accounts for $\lambda(n - r - 1) - w(u)$ further edges between u and Q. Thus the total number of edges between u and Q is $\lambda(n - r - 1)$.

Finally consider Q. The total number of edges between Q and other vertices is $\lambda(n - r - 1)(r + 1)$. For each half-loop of F a 2-triangle is removed and replaced by a 1-triangle on u; thus a loop is removed from Q, the total number of such loops being $w(u)$. However $\frac{1}{2}\{\lambda(n - r - 1) - w(u)\}$ 2-triangles are placed on u; each such 2-triangle contains a loop on Q, and so this increases the number of loops on Q by $\frac{1}{2}\{\lambda(n - r - 1) - w(u)\}$. Finally $\frac{1}{2}\{\lambda(n - r - 1) - w(u)\}$ 3-triangles are removed from Q. The final number of loops on Q is therefore

$$\lambda \binom{n - r}{2} - \frac{3}{2}\{\lambda(n - r - 1) - w(u)\} + \frac{1}{2}\{\lambda(n - r - 1) - w(u)\} - w(u)$$

$$= \lambda \binom{n - r - 1}{2}.$$

From all this, (Bi)–(Biii) now follow (with r replaced by $(r + 1)$).

We now go on to show that condition (iv) of Conjecture 3 is satisfied by S' when $n - r > 3$. We have to show that no component of $(G')^{\circ}$ contains exactly one loop.

Consider a component C of $(G')^{\circ}$ which does not contain u. Then no half-loop of F was incident with a vertex of C.

If C contained exactly one loop, on v^* say, then in the skew-free λ-half-loop factorization of G° containing F as a λ-half-loop-factor, another λ-half-loop-factor, say F^*, would contain one half-loop on v^*. Let F_C^* be the λ-half-loop-factor F^* restricted to C. Since C has no other loops, F_C^* is a regular graph of degree λ with exactly one half-loop; but removing this one half-loop from F_C^* gives a normal graph with exactly one vertex, namely v^*, of odd degree — which is impossible. Therefore C cannot contain exactly one loop.

Let C_u denote the component of $(G')^{\circ}$ which contains u. To show that we can arrange that C_u does not contain exactly one loop, we have to use Conjecture 2^* (rather than Conjecture 2); this is the only place in the proof where Conjecture 2 itself will not suffice. The number of loops on u is $\frac{1}{2}\{\lambda(n - r - 1) - w(u)\} = \frac{1}{2}\{\lambda(n - r - 1) - h(F)\}$. This number is at least two unless $\frac{1}{2}\{\lambda(n - r - 1) - h(F)\} = 0$ or 1, i.e. unless $h(F) = \lambda(n - r - 1)$ or $\lambda(n - r - 1) - 2$. So suppose that in fact $h(F) = \lambda(n - r - 1)$ or $\lambda(n - r - 1) - 2$. Suppose moreover that no other λ-half-loop factor F' from our λ-half-loop-factorization would yield two loops on u. Since the λ-half-loop-factorization is equalized, the number ℓ of half-loops of G° satisfies

$$\{\lambda(n - r - 1) - 2\}(n - r) \leq \ell \leq (n - r - 1)(n - r).$$

Thus, by Conjecture 2^*, we can choose F so that $G^{\circ} \backslash F$ has the following property. Either $h(F) = \lambda(n - r - 1)$ and $G^{\circ} \backslash F$ does not contain some components which, between them, contain exactly one loop and all the $\lambda(n-r-1)$ half-loops which were paired with half-loops of F; or $h(F) = \lambda(n - r - 1) - 2$, and $G^o\backslash F$ does not contain a component with all the $\lambda(n - r - 1) - 2$ half-loops which were paired with half-loops of F, but no whole loops. In the first case it follows that u has no loops on it in C_u and C_u does not have exactly one loop. In the second case it follows that u has one loop on it, but C_u has at least one other loop. Thus we can select F so that C_u does not contain exactly one loop.

We now show that condition (iv) of Conjecture 3 is satisfied by S' when $n - r = 3$. This time we have to show that no component of $(G')^o$ contains an odd number of loops.

Let the skew-free equalized λ-half-loop factorization of G^o be F, F_1 and F_2, where F is used to form G'. Consider a component C of $(G')^o$ which does not contain u. Then in G^o no half-loop of F was incident with a vertex of C. If C contained an odd number of loops, then $E(C) = (E(C)\cap F_1)\cup(E(C)\cap F_2)$. Also, each loop of C is split into two half-loops, one in F_1 and the other in F_2. Therefore F_1 contains an odd number of half-loops. Removing these half-loops from F_1 gives us a normal graph with an odd number of vertices of odd degree — which is impossible. Therefore C contains an even number of loops.

Let p denote the number of 2-triangles in S and q the number of 3-triangles. Then, since $n - r = 3$,

$$
\begin{aligned}
|E(G)| &= \text{the number of 1-triangles} \\
&= [(\text{the number of 1-triangles}) + p] - p \\
&= \tfrac{1}{2}\lambda r(n - r) - p \\
&= \tfrac{3}{2}\lambda r - [(\text{the number of loops on } Q) - 3q] \\
&= \tfrac{3}{2}\lambda r - 3\lambda + 3q,
\end{aligned}
$$

so $3\,|\,|E(G)|$.

The number of loops in G^o is $\tfrac{1}{2}\{\lambda - r(n - r) - 2|E(G)|\} = \tfrac{1}{2}\lambda r(n - r) - |E(G)|$. The number of loops in $(G')^o$ on u is $\tfrac{1}{2}[\lambda(n - r - 1) - h(F)] = \tfrac{1}{2}[\lambda(n - r - 1) - \lambda r + 2|E(F)|] = |E(F)| - \tfrac{1}{2}\lambda(2r + 1 - n)$. Therefore the number of loops in $(G')^o$ is $\left[\tfrac{1}{2}\lambda r(n - r) - |E(G)|\right] - h(F) + |E(F)| - \tfrac{1}{2}\lambda(2r + 1 - n)$. Since $n - r = 3$, the number of loops in $(G')^o$ is $\left[\tfrac{3}{2}\lambda r - |E(G)|\right] - [\lambda r - 2|E(F)|] + |E(F)| - \tfrac{\lambda}{2}(r - 2) = \lambda - |E(G)| + 3|E(F)|$.

Since the λ-half-loop-factorization we give G^o is equalized, and since $\tfrac{1}{3}\,|\,|E(G)|$, it follows that $3|E(F)| = |E(G)|$, and so the number of loops in $(G')^o$ is λ, and so is even. Since each component of $(G')^o$ other than C_u has an even number of loops, it follows that C_u also has an even number of loops. Therefore condition (iv) of Conjecture 3 is satisfied by S' when $n - r = 3$.

Lastly consider the procedure in forming S' when $n - r = 2$. We give G^o a skew-free equalized λ-half-loop-factorization. Let the two λ-half-loop-factors be F_α and F_β. Then we replace Q and the edges and loops on Q by two vertices v_α and v_β. Corresponding to each edge $w_1 w_2$ of F_α we have a triangle $\{v_\alpha, w_1, w_2\}$, and similarly with each edge of F_β. Corresponding to each loop of G^o on a vertex w, we have a triangle $\{w, v_\alpha, v_\beta\}$. We retain the 0-triangles of S. It is easy to check that S' thus formed is a quasi $TS(n, \lambda)$ [it is in fact an actual $TS(n, \lambda)$].

This concludes our proof that Conjecture 2^* implies Conjecture 3.

6   A proof that Conjecture 2* implies Conjecture 1

In Section 5 we showed that Conjecture 2* implies Conjecture 3. We now show that Conjecture 3 implies Conjecture 1, and thus that Conjecture 2* implies Conjecture 1.

Let n be admissible, $n \geq 2r + 1$, and let T be a $\mathrm{PTS}(n, \lambda)$. If $r \leq 3$ the result follows from the known existence of a $\mathrm{TS}(n, \lambda)$ whenever n is λ-admissible (see for example [15]), so assume that $r \geq 4$. Assume similarly that T contains at least two triples. Then T satisfies the conditions of Conjecture 3 (i) is assumed to be true; (ii) is trivial since $2r - n - 1 \leq 0$; clearly $\sum_{v \in V(T)} N(v) \leq \lambda \binom{r}{2}$, and so condition (iii) is satisfied; finally since G^o is regular of degree $\lambda(n - r)$, and since the number of edges (not counting loops) on each vertex of G^o is at most $\lambda(r - 1)$, there are at least $\frac{1}{2}\{\lambda(n - r) - \lambda(r - 1)\} \geq \frac{\lambda}{2}\{(r + 1)\} \geq \lambda \geq 2$ loops on each vertex, so (iv) is satisfied. Therefore, if Conjecture 3 is true, then T can be embedded in a $\mathrm{TS}(n, \lambda)$, and so Conjecture 1 is true also.

References

[1]   L. D. Andersen and A. J. W. Hilton, Generalized latin rectangles I: construction and decomposition, *Discrete Math.* **31**(1980), 125–152.

[2]   L. D. Andersen and A. J. W. Hilton, Generalized latin rectangles II: embedding, *Discrete Math.* **31**(1980), 225–260.

[3]   L. D. Andersen, A. J. W. Hilton and E. Mendelsohn, Embedding partial Steiner triple systems, *Proc. London Math. Soc.* (3) **41**(1980), 557–576.

[4]   A. B. Cruse, On extending incomplete latin rectangles, *Proc. 5th Southeastern Conf. on Combinatorics, Graph Theory and Computing (1974)*, 333–348.

[5]   A. J. W. Hilton, Hamiltonian decompositions of complete graphs, *J. Combin. Theory Ser. B* **36**(1984), 125–134.

[6]   A. J. W. Hilton, School timetables, Studies on graphs and discrete programming (P. Hansen, ed.), North Holland (1981), 177–188.

[7]   A. J. W. Hilton, Outline latin squares, *Annals of Discrete Math.* **34**(1987), 225–242.

[8]   A. J. W. Hilton, The reconstruction of latin squares with applications to school timetabling and to experimental design, *Math. Programming Study* **13**(1980), 68–77.

[9]   A. J. W. Hilton, Embedding partial triple systems, *J. Combinatorial Math. and Combinatorial Computing* **2**(1987), 77-95.

[10] A. J. W. Hilton and C. A. Rodger, The embedding of partial triple systems when 4 divides λ, *J. Combin. Theory Ser. A.* To appear.

[11] A. J. W. Hilton 226 Transfer complete. htables, *Annals of Discrete Math.* **15**(1982), 239–251.

[12] A. J. W. Hilton and C. A. Rodger, Hamiltonian decompositions of complete *s*-partite graphs, *Discrete Math.* **58**(1986), 63–78.

[13] C. C. Lindner, A partial Steiner triple system of order n can be embedded in a Steiner triple system of order $6n + 3$, *J. Combin. Theory Ser. A* **18**(1975), 349–351.

[14] C. C. Lindner and T. Evans, Finite embedding theorems for partial designs and algebras, SMS 56; Les Presses de l'Université de Montréal (1977).

[15] C. St. J. A. Nash-Williams, Simple constructions for balanced incomplete block designs with block size three, *J. Combin. Theory Ser. A* **13**(1972), 1–6.

[16] C. St. J. A. Nash-Williams, Detachments of graphs and generalized Euler trails, Surveys in Combinatorics (I. Anderson, ed.), Cambridge Univ. Press (1985), 137–151.

[17] C. St. J. A. Nash-Williams, Amalgamations of almost regular edge-colourings of simple graphs, *J. Combin. Theory Ser. B* **43**(1987), 322-342.

[18] C. A. Rodger and S. J. Stubbs, Embedding partial triple systems, *J. Combin. Theory Ser. A* **44**(1987), 241–253.

[19] H. J. Ryser, A combinatorial theorem with an application to latin squares, *Proc. Amer. Math. Soc.* **2**(1951), 550–552.

[20] S. J. Stubbs, Embedding partial triple systems, Ph.D. dissertation, Auburn University (1986).

[21] C. Treash, The completion of finite incomplete Steiner triple systems with applications to loop theory, *J. Combin. Theory Ser. A* **10**(1971), 259–265.

ON THE RANK OF FIXED POINT SETS OF AUTOMORPHISMS OF FREE GROUPS

W. IMRICH[*]
Institut für Mathematik und Angewandte Geometrie
Montanuniversität Leoben
A-8700 Leoben
Austria

S. KRSTIC
Matematički Institut
YU-11000 Beograd
Yugoslavia

E. C. TURNER
Department of Mathematics and Statistics
State University of New York at Albany
Albany, NY 12222
USA

Abstract

Let α be a homomorphism of a free group G into itself. We derive a bound on the rank of the fixed point group Fix(α) of α. For automorphisms of f. g. groups we also obtain a bound on the rank of the fixed point set, i. e. the set consisting of all words of finite or infinite length in a free generating set of G which are fixed under the action of α.

1 Introduction

S. M. Gersten [5] showed that the fixed point subgroup Fix(α) of an automorphism α of a finitely generated free group G is itself finitely generated. D. Cooper [2] extended this result by showing that the set Fp(α) of points and ends of G fixed by α is also finitely generated.

We derive a bound on the rank of Fix(α) which also holds for infinitely generated groups. The proof heavily depends on a method introduced by R. Z. Goldstein and E. C. Turner [6] for proving an extension of Gersten's result. This bound was also found independently by M. M. Cohen and M. Lustig. See [1] for this and related results.

---

[*]Supported by NSERC grant A5367.

G. Hahn et al. (eds.), Cycles and Rays, 113–122.
© 1990 by Kluwer Academic Publishers. Printed in the Netherlands.

For automorphisms of finitely generated groups we give a bound on the rank of $\mathrm{Fp}(\alpha)$. In this case we make use of the so-called bounded cancellation-property (see [2]) of automorphisms of f. g. groups.

For generalizations of these results to homomorphisms and monomorphisms of subgroups of free groups we refer to a forthcoming extension of this paper [7].

2   Rank of the fixed point subgroup

In all known examples the rank of the fixed point subgroup of an automorphism α of a free group G is less than or equal to the rank of G. This is trivial if G has infinite rank. For f. g. free groups this has been shown to hold for periodic automorphisms (see J. L. Dyer and G. P. Scott [3]), for so-called change of maximal tree automorphisms (see S. M. Gersten [4]) and for automorphisms realizable by surface autohomeomorphisms (see W. Jaco and P. B. Shalen [9]).

S. M. Gersten's proof [5] also provides a bound on the rank of $\mathrm{Fix}(\alpha)$, but it depends on α and can be arbitrarily large. Below we deduce a better bound that also holds for homomorphisms of free groups into themselves. However, this bound too can be arbitrarily large.

Theorem 1 (See also [1]) *Let G be a free group freely generated by S and let α be a homomorphism of G into itself. Then*

$$\mathrm{rank}\,\mathrm{Fix}(\alpha) \leq \sum_{s \in S} n_\alpha(s),$$

where $n_\alpha(s)$ denotes the number of occurrences of s in $\alpha(s)$.

Proof: We consider the graph Γ on the vertices of G with the edge-set

$$E(\Gamma) = \{[w, \alpha(s)^{-1}ws]_s \mid w \in G, s \in S\}.$$

The subscript is used to distinguish the edges

$$[w, \alpha(s_1)^{-1}ws_1]_{s_1} \text{ and } [w, \alpha(s_2)^{-1}ws_2]_{s_2}$$

if $\alpha(s_1)^{-1}ws_1 = \alpha(s_2)^{-1}ws_2$ and will usually be omitted. For the sake of convenience we shall also admit $[w, \alpha(s)^{-1}ws]_s$ for $s \in S^{-1}$. We consider $[w, \alpha(s)^{-1}ws]_s$ and $[\alpha(s)^{-1}ws, \alpha(s^{-1})^{-1} \cdot \alpha(s)^{-1}ws \cdot s^{-1}]_{s^{-1}} = [\alpha(s)^{-1}ws, w]_{s^{-1}}$ to be the same edge, but equipped with different orientations.

The graph Γ was introduced by Goldstein and Turner [6], who proved that the rank of the component Γ_1 of Γ containing the vertex 1 is the rank of the fixed subgroup $\mathrm{Fix}(\alpha)$.

In fact, this is not hard to see. For, let $H = \mathrm{Fix}(\alpha)$ and $Y = \Gamma(G, S)/H$, i. e. the coset graph of the subgroup H of G with respect to the generating set S. The vertices of Y are the cosets Hx, $x \in G$, and the edges are the oriented pairs $(Hx, Hxs)_s, s \in S$, subject to the same conventions as above. Then the mapping $Hx \mapsto \alpha(x)^{-1}x$ and $(Hx, Hxs) \mapsto [\alpha(x)^{-1}x, \alpha(xs)^{-1}xs]$ is an isomorphism of Y onto Γ_1 and it is well known that the rank (or cyclomatic number) of Y is the rank of H.

We now reorient every edge

$$[w, \alpha(s)^{-1}ws]_s$$

from w to $\alpha(s)^{-1}ws$ if (the reduced form of) w ends in s^{-1}. This way not every edge will be oriented. The exception are those edges $[w, \alpha(s)^{-1}ws]$ for which w does not end in s^{-1} and (the reduced form of) $\alpha(s)^{-1}ws$ does not end in s. Furthermore, some edges will be oriented both ways. This happens, when w ends in s^{-1} and $\alpha(s)^{-1}ws$ ends in s.

In this way every vertex of Γ becomes the origin of at most one oriented edge.

Let Γ^* denote the reoriented graph Γ and let F be the subgraph of Γ^* induced by the oriented edges. It is easy to see that every connected component of F contains at most one cycle.

The edge $[w, \alpha(s)^{-1}ws]$ remains unoriented if s does not cancel in ws and if ws is an initial segment of $\alpha(s)$. Because the number of such initial segments of $\alpha(s)$ is $n_\alpha(s)$, exactly that many edges of the form $[w, \alpha(s)^{-1}ws]_s$ remain unoriented. Thus the set U of unoriented edges has

$$\sum_{s \in S} n_\alpha(s)$$

elements.

Let U_1 be the set of those edges of U contained in Γ_1^*. Thus Γ_1^* consists of finitely many components of F, together with the edges of U_1. Since every component of F contains at most one cycle the minimal connected subgraph $\mathrm{core}(\Gamma_1^*)$ of Γ_1^* containing 1, U_1 and these cycles must be finite. Let n be the number of vertices of $\mathrm{core}(\Gamma_1^*)$. Since no oriented edge originates at 1 and at most one at every other vertex, $\mathrm{core}(\Gamma_1^*)$ contains at most $n-1$ oriented edges. The rank of $\mathrm{core}(\Gamma_1^*)$ is therefore bounded by

$$n - 1 + |U_1| - (n-1).$$

Since $|U_1| \leq |U| = \sum_{s \in S} n_\alpha(s)$ and since

$$\mathrm{rank}\, \mathrm{Fix}(\alpha) = \mathrm{rank}(\Gamma_1) = \mathrm{rank}(\Gamma_1^*) = \mathrm{rank}\, \mathrm{core}(\Gamma_1^*),$$

the assertion of the theorem follows.□

3 Rank of the fixed point set

In this section we wish to investigate the action of automorphisms of finitely generated free groups on words of infinite length. To fix ideas, let G be a f.g. free group freely generated by S. Furthermore, let $|w|$ denote the length of a (reduced) word w in $S \cup S^{-1}$ and for reduced words w_1 and w_2 let $w_1 \wedge w_2$ denote the common initial segment of w_1 and w_2.

If w_1 is an initial segment of w_2 we shall write $w_1 \nearrow w_2$. Thus, $w_1 \wedge w_2 \nearrow w_1, w_2$.

We say an infinite reduced word w is the limit of a sequence $w_1, w_2, \ldots$ if $\lim_{n \to \infty} |w_n \wedge w| = \infty$.

It is easy to see (compare [2]) that for any reduced infinite word $s_1 s_2 s_3 \ldots$ in $S \cup S^{-1}$

$$\lim_{n \to \infty} \alpha(s_1 \ldots s_n)$$

exists and that, freely reduced, it is an infinite word. Thus α acts on the set Ω of infinite reduced words. We set $W = G \cup \Omega$.

We shall use the notation $\overline{\text{Fix}(\alpha)}$ for the set $\text{Fix}(\alpha)$ together with the set of all limit points and shall consider $\overline{\text{Fix}(\alpha)}$ to be generated by $\text{Fix}(\alpha)$. It is in this sense that we speak of a generating set of $\text{Fix}(\alpha)$ as a generating set of $\overline{\text{Fix}(\alpha)}$.

In the sequel we shall make use of the fact that there exists a constant N_α which only depends on α such that $w_1 \wedge w_2 = 1$ implies

$$|\alpha(w_1) \wedge \alpha(w_2)| \leq N_\alpha.$$

This has been shown in [2]. The constant N_α is called the cancellation bound. It is also mentioned in [2] that $N_\alpha \leq 2S_\alpha^2$, where

$$S_\alpha = \max_{s \in S}\{|\alpha(s)|, |\alpha^{-1}(s)|\}.$$

S_α is called the size of α. For a short combinatorial proof of the stronger inequality

$$N_\alpha \leq \frac{1}{2} \max_{s \in S}\{|\alpha(s)|\} \max_{s \in S}\{|\alpha^{-1}(s)|\}$$

see [8].

Let A be a subset of W. Then the set B generated by A is defined as the smallest set containing A and all limit points of $\langle A \rangle$ such that

$$x, y \in B \text{ and } x \in G \Rightarrow x^{-1}, xy \in B.$$

We say B is finitely generated if A is finite. The smallest cardinality of a generating set of B is called the rank of B.

The fixed point set of α is defined by

$$\text{Fp}(\alpha) = \{w \in W, \alpha(w) = w\}.$$

It has been shown by D. Cooper [2] that it is finitely generated.

We give a different proof, which also provides a bound on the rank of $\text{Fp}(\alpha)$. In order to do this we need a closer look at the components of the graph F introduced in the proof of Theorem 1. We recall that every edge of F is oriented and that every vertex in F is the origin of at most one edge.

To any vertex $v_0 \in F$ there thus exists exactly one path

$$\text{p}(v_0) = [v_0, v_1, v_2, \ldots]$$

of maximal length such that v_i is the origin of an edge ending in v_{i+1}. We call such a path uniquely oriented. This path may be finite or infinite. If it is infinite it may become periodic after an aperiodic initial segment.

Let C be a component of F. Then C clearly satisfies (for a formal proof see [7]) one and only one of the following properties:

(i) C has a sink u, i.e. a vertex which is not the origin of any edge. Every $\text{p}(v)$ of C ends in u.

(ii) C has exactly one doubly oriented edge f. Every $\text{p}(v)$ ends in f.

(iii) C has ecactly one circuit c, i.e. a closed, uniquely oriented path. Every $p(v)$ ends in c.

(iv) C has neither sink nor doubly oriented edge or circuit and any two paths $p(x)$ and $p(y)$ differ in at most finitely many edges. These paths are called rays and the collection of these rays is called a sink at infinity.

We recall that the edges of Γ are of the form $[w, \alpha(s)^{-1}ws]$, where $s \in S \cup S^{-1}$. We now define a mapping σ from $E(\Gamma)$ into $S \cup S^{-1}$ by setting

$$\sigma[w, \alpha(s)^{-1}ws]_s = s.$$

We call s the label of $[w, \alpha(s)^{-1}ws]_s$. This mapping extends to a homomorphism of the fundamental groupoid of Γ into G.

Passing from Γ to Γ^* we do not alter σ, thus the restriction of σ to the fundamental group $\pi_1(\Gamma_1^*)$ of Γ_1^* is a homomorphism into G.

We also note that we can extend σ to infinite reduced paths in Γ_1^*, i.e. to paths in which the labels of successive edges are not inverse to each other. The images of such paths are in Ω.

Theorem 2 *Let α be an automorphism of a f. g. free group G freely generated by S. Then*

$$\text{rank } \text{Fp}(\alpha) \leq \sum_{s \in S}(n_\alpha(s) + n_{\alpha^{-1}}(s)) - \text{rank } \text{Fix}(\alpha),$$

where $n_\alpha(s)$ denotes the number of occurrences of s in $\alpha(s)$.

In the proof of this theorem we shall apply the following lemma.

Lemma 1 *Under the assumptions of Theorem 2, let Γ_1^* be defined as in the proof of Theorem 1 and let $n_\alpha(\infty)$ denote the number of sinks at infinity of Γ_1^*. Then*

$$\text{rank } \text{Fix}(\alpha) + n_\alpha(\infty) \leq \sum_{s \in S} n_\alpha(s).$$

Proof of Lemma 1: Consider $\text{core}(\Gamma_1^*)$ and let e be a sink at infinity of Γ_1^*. Then there exists a uniquely oriented path $p(v)$ for some v in $\text{core}(\Gamma_1^*)$ that ends in e. Let x be the vertex of $p(v)$ with largest distance in $p(v)$ from v that is still in $\text{core}(\Gamma_1^*)$. Then x is not the origin of any edge ending in Γ_1^*. Since uniquely oriented paths ending in different sinks at infinity are distinct the number of oriented edges in the core of Γ_1^* is reduced by 1 for every sink of Γ_1^* at infinity. Thus

$$\text{rank } \text{Fix}(\alpha) \leq \sum_{s \in S} n_\alpha(s) - n_\alpha(\infty),$$

which proves the lemma. $\square$

Proof of Theorem 2:

(A) We show first that every sink at infinity of Γ_1^* corresponds to an infinite word fixed by α. To show this, let $p(v)$ be in such a sink e and let q be a path from 1 to v. Without loss of generality we can assume that $qp(v)$ is reduced. Clearly

$$\sigma qp(v) \in \Omega.$$

Let $v = \alpha(a)^{-1}a$ and $\sigma p(v) = s_1 s_2 s_3 \ldots$. We wish to show that α fixes $as_1 s_2 \ldots$. Setting $p(v) = [v, v_1, v_2, \ldots]$ we note that

$$v_i = \alpha(as_1 \ldots s_i)^{-1} as_1 s_2 \ldots s_i$$

by the definition of $p(v)$ and that every v_i ends in s_{i+1}^{-1}. Thus,

$$s_{i+1} \nearrow s_i^{-1} \ldots s_1^{-1} a^{-1} \alpha(as_1 \ldots s_i)$$

and

$$as_1 \ldots s_{i+1} \nearrow \alpha(as_1 \ldots s_i).$$

But this is equivalent to

$$\lim_{i \to \infty} (as_1 \ldots s_i) = \lim_{i \to \infty} \alpha(as_1 \ldots s_i).$$

(B) Suppose we have two distinct reduced paths of Γ_1^* that originate in 1 and end in the infinite sink e. Since Γ_1^* has finite rank these paths are identical from some vertex v on. Let $q_1 p(v)$ be the first of these paths and $q_2 p(v)$ the other. Since $q_2 q_1^{-1}$ is closed we infer $\sigma(q_2 q_1^{-1}) \in \mathrm{Fix}(\alpha)$. As

$$\sigma(q_2 p(v)) = \sigma(q_2 q_1^{-1}) \sigma(q_1 p(v))$$

this means that $\sigma(q_1 p(v))$ and $\sigma(q_2 p(v))$ are equivalent modulo $\mathrm{Fix}(\alpha)$.

(C) Considering the automorphism α^{-1} we can form a graph $\Gamma(\alpha^{-1})$ as we formed Γ for α. We shall thus refer to Γ henceforth also as $\Gamma(\alpha)$. Just as the mapping ϕ from $\Gamma(G, S)/\mathrm{Fix}(\alpha)$ onto $\Gamma_1(\alpha)$, defined by

$$\mathrm{Fix}(\alpha)x \xmapsto{\phi} \alpha(x)^{-1}x,$$

$$(\mathrm{Fix}(\alpha)x, \mathrm{Fix}(\alpha)xs) \xmapsto{\phi} [\alpha(x)^{-1}x, \alpha(xs)^{-1}xs]$$

is an isomorphism, there exists an analogously defined isomorphism from $\Gamma(G, S)/\mathrm{Fix}(\alpha^{-1})$ onto $\Gamma_1(\alpha^{-1})$. As $\mathrm{Fix}(\alpha) = \mathrm{Fix}(\alpha^{-1})$ we thus have an isomorphism ψ from $\Gamma_1(\alpha)$ onto $\Gamma_1(\alpha^{-1})$ defined by

$$\alpha(x)^{-1}x \xmapsto{\psi} (\alpha^{-1}(x))^{-1}x,$$

$$[\alpha(x)^{-1}x, \alpha(xs)^{-1}xs] \xmapsto{\psi} [(\alpha^{-1}(x))^{-1}x, (\alpha^{-1}(xs))^{-1}xs].$$

Of course, this can also be seen directly.

If we define σ on $E(\Gamma(\alpha^{-1}))$ as we defined it on $E(\Gamma(\alpha))$ it is easy to see that σ commutes with ψ. We also note that σ is usually defined on $\Gamma(G, S)/\mathrm{Fix}(\alpha)$ by

$$\sigma(\mathrm{Fix}(\alpha)x, \mathrm{Fix}(\alpha)xs)_s = s.$$

Hence, we can consider $\Gamma(G, S)/\text{Fix}(\alpha)$, $\Gamma_1(\alpha)$ and $\Gamma_1(\alpha^{-1})$ as the same graphs with the same edge-labelling σ and the same orientation of the edges, but with different vertex-labels. $\Gamma_1^*(\alpha)$ and $\Gamma_1^*(\alpha^{-1})$ differ from these graphs by having different edge-orientations, leaving some edges unoriented.

The arguments in (A) show that σq is fixed by α (and α^{-1}) for any reduced path q in Γ_1 leading from 1 to a sink e at infinity in $\Gamma_1^*(\alpha)$ or $\Gamma_1^*(\alpha^{-1})$. Choosing such a path for every one of the finitely many sinks at infinity of $\Gamma_1^*(\alpha)$ and $\Gamma_1^*(\alpha^{-1})$ we obtain finitely many elements of Ω fixed by α. We wish to show that these elements, together with those in $\text{Fix}(\alpha)$ generate $\text{Fp}(\alpha)$.

By (B) it is clear that it suffices to show that any element w in Ω fixed by α is a σq for some reduced path q in Γ_1 leading from 1 to a sink at infinity of $\Gamma_1^*(\alpha)$ or $\Gamma_1^*(\alpha^{-1})$.

(D) Let $w = s_1 s_2 \ldots \in \Omega$ and suppose $\alpha w = w$. Setting $w_n = s_1 s_2 \ldots s_n$ we note that $\alpha w = w$ is equivalent to

$$\lim_{n \to \infty} |w_n \wedge \alpha(w_n)| = \infty.$$

If there is a length L infinitely often attained by $\alpha(w_n)^{-1} w_n$, then there exists a sequence $\{n_i | i \geq 1\}$ such that all $\alpha(w_{n_i})^{-1} w_{n_i}$ are equal. Set $u = w_{n_1}$. Then

$$\alpha(w_{n_i})^{-1} w_{n_i} = \alpha(u)^{-1} u$$

and $w_{n_i} u^{-1}$ is an element of $\text{Fix}(\alpha)$. Since

$$\lim_{i \to \infty} w_{n_i} u^{-1} = w$$

this means that w is the limit of elements in $\text{Fix}(\alpha)$, i. e. $w \in \overline{\text{Fix}(\alpha)}$.

If no finite length is attained infinitely often by $\alpha(w_n)^{-1} w_n$ we have

$$\lim_{n \to \infty} |\alpha(w_n)^{-1} w_n| = \infty. \tag{1}$$

This implies the existence of constant K, such that

$$|\alpha(w_n)^{-1} w_n| > N_\alpha + S_\alpha + 1$$

for $n > K$.

Let $c_n = w_n \wedge \alpha(w_n)$. Since

$$|\alpha(w_n)^{-1} w_n| = |c_n^{-1} w_n| + |c_n^{-1} \alpha(w_n)|$$

at least one of the sets $\{|c_n^{-1} w_n| \mid n \geq 1\}$ and $\{|c_n^{-1} \alpha(w_n)| \mid n \geq 1\}$ has to be unbounded.

(E) Suppose there is an $n > K$ with

$$|c_n^{-1} w_n| \geq 1 \tag{2}$$

and

$$|c_n^{-1} \alpha(w_n)| > N_\alpha. \tag{3}$$

Since $c_n \nearrow w_n$ we have $c_n \nearrow w_{n+i}$ for all $i \geq 0$. As $|c_i| \to \infty$ there must be a k with $w_n \nearrow c_{n+k}$. By (2) $c_n = \alpha(w_n) \wedge \alpha(w_{n+k})$, and hence

$$\alpha(w_n)^{-1} c_n = \alpha(w_n)^{-1} \wedge \alpha(w_n^{-1} w_{n+k}).$$

Since $|c_n^{-1} \alpha(w_n)| > N_\alpha$ by (3) and as $w_n \wedge w_n^{-1} w_{n+k} = 1$ this contradicts the bounded cancellation property.

Hence, for $n > K$, (2) and (3) cannot both hold.

(F) Suppose (3) holds for some $n > K$. Then $w_n = c_n \nearrow \alpha(w_n)$ since (2) cannot hold. Furthermore

$$
\begin{aligned}
|c_{n+1}^{-1} \alpha(w_{n+1})| &\geq |c_n^{-1} \alpha(w_n)| - 1 - |\alpha(s_{n+1})| \\
&= |w_n^{-1} \alpha(w_n)| - 1 - |\alpha(s_{n+1})| \\
&> N_\alpha + S_\alpha - |\alpha(s_{n+1})| \\
&\geq N_\alpha.
\end{aligned}
$$

But this means (3) holds for $n + 1$ and, a fortiori, for all $m \geq n$.

Hence, $w_m \nearrow \alpha(w_m)$ for all $m \geq n$. This, on the other hand, means that

$$\alpha(w_m)^{-1} w_m$$

ends in s_{m+1}^{-1} and that

$$[\alpha(w_m)^{-1} w_m, \ \alpha(w_{m+1})^{-1} w_{m+1}]$$

is oriented from $\alpha(w_m)^{-1} w_m$ to $\alpha(w_{m+1})^{-1} w_{m+1}$ in $\Gamma_1^*(\alpha)$. Therefore

$$p(\alpha(w_n^{-1}) w_n) = [\alpha(w_n)^{-1} w_n, \ \alpha(w_{n+1})^{-1} w_{n+1}, \ldots].$$

(G) We are left with the case that (3) never holds for $n > K$, i.e. that we always have

$$|c_n^{-1} \alpha(w_n)| \leq N_\alpha$$

for $n > K$. But then $|c_n^{-1} w_n| \to \infty$.

Let us consider $\alpha^{-1}(w_n)$ and $\alpha(\alpha^{-1}(w_n)) = w_n$ now and set

$$d_n = w_n \wedge \alpha^{-1}(w_n).$$

By using the same arguments for α^{-1} as previously used for α we infer the existence of an index L such that

$$|d_n^{-1} w_n| \geq 1$$

and

$$|d_n^{-1} \alpha^{-1}(w_n)| > N_{\alpha^{-1}}$$

cannot both hold for $n > L$.

As $|\alpha^{-1}(w_n)| \geq |w_n|/S_\alpha$ and since

$$
\begin{aligned}
|d_n^{-1} \alpha^{-1}(w_n)| &\geq |\alpha^{-1}(w_n)| - N_{\alpha^{-1}} \\
&\geq |w_n|/S_\alpha - N_{\alpha^{-1}}
\end{aligned}
$$

we can make $|d_n^{-1}\alpha^{-1}(w_n)|$ arbitrarily large. Thus

$$[\alpha^{-1}(w_n)^{-1}w_n,\ \alpha^{-1}(w_{n+1})^{-1}w_{n+1}, \ldots]$$

is a uniquely oriented, infinite path in $\Gamma_1^*(\alpha^{-1})$.

(H) By the above

$$\operatorname{rank} \operatorname{Fp}(\alpha) \leq \operatorname{rank} \operatorname{Fix}(\alpha) + n_\alpha(\infty) + n_{\alpha^{-1}}(\infty).$$

Now an application of Lemma 1 yields the assertion of the theorem.□

Corollary 1 *Let α be an automorphism of a f. g. free group G freely generated by S and suppose $\alpha \neq \iota$ fixes the infinite reduced word $s_1 s_2 \ldots \notin \overline{\operatorname{Fix}(\alpha)}$. Set $w_n = s_1 s_2 \ldots s_n$. Then*

$$\lim_{n\to\infty} |\alpha^{-1}(w_n)w_n| = \infty \qquad (4)$$

and there exists an index N such that either

$$s_1 s_2 \ldots s_n \nearrow \alpha(s_1 s_2 \ldots s_n) \text{ for all } n > N \qquad (5)$$

or

$$s_1 s_2 \ldots s_n \nearrow \alpha^{-1}(s_1 s_2 \ldots s_n) \text{ for all } n > N. \qquad (6)$$

Proof: Equation (4) is the same as equation (1) of (D). For (5) and (6) compare (F) and (G). □

Corollary 2 *Let α be a periodic automorphism of a f. g. free group. Then*

$$\operatorname{Fp}(\alpha) = \overline{\operatorname{Fix}(\alpha)}.$$

Proof: Suppose $\alpha^k = \iota$ and that $s_1 s_2 \ldots \notin \overline{\operatorname{Fix}(\alpha)}$ is fixed by α. By Corollary 2 there is an N such that for all $n > N$ either (5) or (6) holds. Without loss of generality we can assume (5) to hold, if neccessary by replacing the rôle of α and α^{-1}. Then

$$s_1 \ldots s_n \nearrow \alpha(s_1 \ldots s_n) \nearrow \ldots \nearrow \alpha^{k-1}(s_1 \ldots s_n) \nearrow s_1 \ldots s_n$$

for all $n > N$. But this contradicts (4). □

References

[1] M. M. Cohen and M. Lustig, On the dynamics and the fixed subgroup of a free group automorphism, *Inventiones Math.* To appear.

[2] D. Cooper, Automorphisms of free groups have f.g. fixed point sets, *J. Algebra* **111**(1987), 453-456.

[3] J. L. Dyer and G. P. Scott, Periodic automorphisms of free groups, *Comm. Alg.* **3**(1975), 195-201.

[4] S. M. Gersten, On fixed points of certain automorphisms of free groups, *Proc. London Math. Soc.* **48** (1984), 72-90; Addendum, **49**(1984), 340-342.

[5] S. M. Gersten, Fixed points of automorphisms of free groups, *Adv.in Math.* **64**(1987), 51-85.

[6] R. Z. Goldstein and E. C. Turner, Fixed subgroups of homomorphisms of free groups, *Bull. London Math. Soc.* **18**(1986), 468-470.

[7] W. Imrich and E. C. Turner, Fixed subsets of homomorphisms of free groups, *manuscript* (1987).

[8] W. Imrich, Computing generators of bounded length for the fixed subgroup of a free group homomorphism, *manuscript* (1988).

[9] W. Jaco and P. B. Shalen, Surface homeomorphisms and periodicity, *Topology* **16**(1977), 347-367.

ON TRANSITION POLYNOMIALS OF 4-REGULAR GRAPHS

F. JAEGER*

Laboratoire de Structures Discrètes

IMAG

BP 53X

38041 Grenoble Cedex

France

Abstract

Let G be a 4-regular graph. For every vertex v of G, there are three distinct possible ways of splitting v into two vertices of degree two, which we call *transitions at v*. A *transition system of G* is a family $p = (p(v), v \in V(G))$, where $p(v)$ is a transition at v. If we perform all vertex-splittings associated to the transitions of p, we obtain a family of disjoint cycles; we denote the number of these cycles by $c(p)$. We also denote by $P(G)$ the set of transition systems of G.

Let A be a mapping which associates to every transition of G a weight chosen in some set of variables, and let τ be another variable. Then the associated *transition polynomial* $Q(G, A, \tau)$ is defined by:

$$Q(G, A, \tau) = \sum_{p \in P(G)} (\pi_{v \in V(G)} A(p(v))) \tau^{c(p)-1}.$$

Special transition polynomials already appear in some work by Penrose (1971), Martin (1977), Las Vergnas (1978). The Tutte polynomial of a plane graph can be described as a transition polynomial of its medial graph. The Jones polynomial of an oriented link (a new invariant of knot theory) has also been identified by Kauffman with a transition polynomial of the diagram of the link.

We survey these results, providing a unifying frame, and in some cases new proofs and generalizations.

1   Introduction

This paper presents the notion of *transition polynomial* as a natural tool for the unification and generalization of a number of results scattered in the literature. It is intended both as a survey and as a research paper.

All graphs will be finite, and may have loops or multiple edges. We refer the reader to [3] and [4] for usual notions of Graph Theory.

We begin in Section 2 with some formulae of Penrose [30] which were apparently the first results of this type to be published. Then we describe briefly in Section 3 some

*This work has been partially supported by the PRC "Mathématiques et Informatiques".

G. Hahn et al. (eds.), Cycles and Rays, 123–150.

© 1990 by Kluwer Academic Publishers. Printed in the Netherlands.

results obtained by Martin [26, 27] a few years later and subsequent improvements by Las Vergnas [22, 23, 24]. This leads us naturally to the notion of transition polynomial, whose definition and general properties are given in Section 4. We present in Section 5 some recent results on the Tutte polynomial of plane graphs [11] which were motivated by the results of Penrose, Martin and Las Vergnas. Section 6 describes an interpretation of transition polynomials which sheds a new light on one of Penrose's formulae [30] and on a theorem of Fleischner [9]. In Section 7 we generalize some known results relating transition polynomials and weighted enumeration of Eulerian orientations. Section 8 describes some connections between transition polynomials and knot theory and in particular prominent work by Kauffman [17]. Finally in Section 9 we conclude by presenting some prospects of extension.

2   Penrose's Formulae

Penrose presented in [30] several formulae for the number of edge-3-colorings of a plane connected cubic graph. As our first introductory example we now describe briefly two of these formulae. Our presentation is essentially the same as the one given by Penrose.

Let G be a plane connected cubic graph. Let us associate to G a "tensor diagram" constructed as follows. For each face f of G, we draw inside f a simple closed Jordan curve $l(f)$ which is parallel and close to the boundary of f. Then for each edge e of G, which bounds the face f_1 on one side and the face f_2 on the other side (f_1 and f_2 are identical iff e is a bridge of G), we draw a "bar" through an interior point of e which crosses $l(f_1)$ and $l(f_2)$. An example of this construction is depicted in Figure 1.

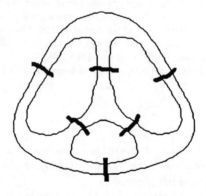

Figure 1

Now let us replace each "bar symbol" in the tensor diagram according to the following scheme:

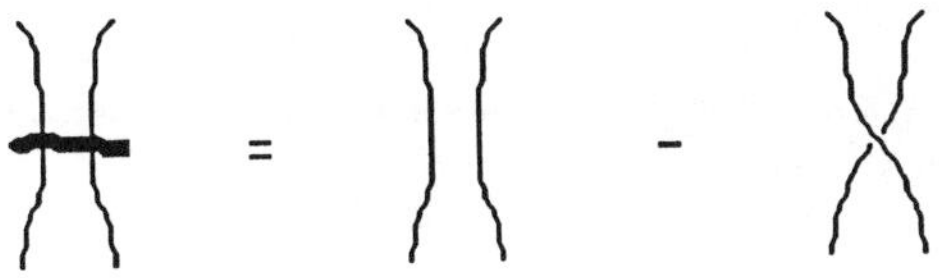

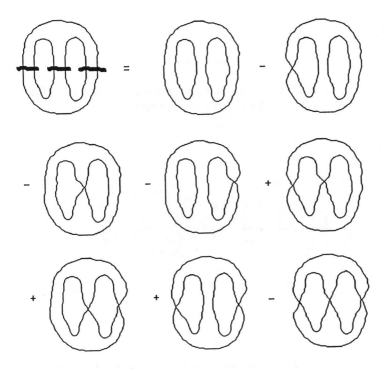

Figure 2

Then the tensor diagram decomposes into a formal signed sum of sets of disjoint curves (here each double point is to be interpreted as a crossing). We illustrate this by an example in Figure 3.

Figure 3

Finally if we replace in this sum each set of n disjoint curves by a monomial τ^n, we obtain a polynomial in one variable τ which we denote by $P(G, \tau)$ and which we call the *Penrose polynomial* of G. For instance, if G has 2 vertices and 3 parallel edges joining these two vertices, Figure 3 shows that $P(G, \tau) = \tau^3 - 3\tau^2 + 2\tau$.

We can now present Penrose's two formulae together in the following result.

Proposition 1 *Let $G = (V, E)$ be a plane connected cubic graph. The number of edge-3-colorings of G is equal to:*

(i)$P(G, 3)$

(ii)$(-1/4)^{|V|/2} P(G, -2)$

If we replace each bar symbol in the tensor diagram associated to G by a vertex according to the following scheme:

Figure 4

we obtain a plane connected 4-regular graph known as the *medial graph* of G (see [29, p.47]) which we shall denote by $M(G)$. See Figure 5 for an example.

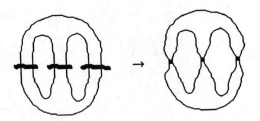

Figure 5

Then clearly the sets of disjoint curves occurring in the expansion defining $P(G, \tau)$ can be interpreted as certain partitions of the edge set of $M(G)$ into cycles. This idea will be made more precise later.

3   Martin Polynomials

These polynomials were introduced by Martin in his thesis [26] in relation with the problem of counting Eulerian cycles or circuits (considered as edge sequences) in undirected or directed Eulerian graphs. We restrict our attention here to 4-regular graphs. In fact it will be convenient to introduce a slightly more general concept of 4-regular graph than the usual one. Let $\mathcal{C}$ by the class of graphs, all vertices of which have degree 2 or 4. Let us consider two graphs of $\mathcal{C}$ as equivalent if one can be obtained from the other by a sequence of operations which can be either the subdivision of an edge or the reverse operation. Then a *4-regular graph* is just an equivalence class of $\mathcal{C}$ under this relation. Thus the equivalence class consisting of the elementary cycles (connected graphs, all vertices of which have degree 2) is considered as a 4-regular graph. We shall call it the *free loop*. It will be denoted by L and represented by a simple closed curve. More generally, the class of graphs consisting of n disjoint elementary cycles will be considered as a set of n free loops and will be denoted by L^n. Every equivalence class will be identified with its unique member which has no vertices of degree two. Thus a 4-regular graph in our sense is a graph, each connected component of which is either a 4-regular graph in the usual sense or a free loop.

Let G be a connected 4-regular graph and v be a vertex of G. If we "split" v according to the following scheme:

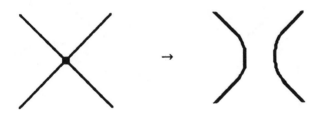

Figure 6

we obtain another 4-regular graph. Such a splitting can be performed in three distinct ways. If one of the three possible splittings yields a disconnected graph (this is the case for instance if there is a loop at v), v is called a *cut-vertex*. Then the resulting graph has exactly two connected components which we call the *pieces with respect to v*.

Martin proved inductively that one can associate to each connected 4-regular graph G a one-variable polynomial, which we denote by $m(G, x)$, in such a way that the following properties hold:

(i) If v is not a cut-vertex of G, $m(G, x) = m(G_1, x) + m(G_2, x) + m(G_3, x)$, where G_1, G_2, G_3 are the results of the three possible splittings at v.

(ii) If v is a cut-vertex of G and G_1, G_2 are the pieces of G with respect to v, $m(G, x) = xm(G_1, x)m(G_2, x)$.

(iii) $m(L, x) = 1$.

These properties can be considered as a recursive definition of the polynomial $m(G, x)$. It is convenient to visualize them as follows:

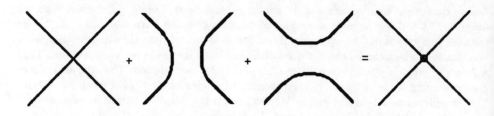

Figure 7(i)

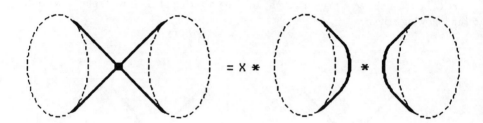

Figure 7(ii)

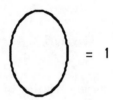

Figure 7(iii)

To interpret a "pictorial equation" such as Figure 7(i) we need only two conventions: first, all diagrams in the equation are to be completed in the same way by a single arbitrary diagram to form the 4-regular graphs which they represent. Second, each diagram stands for the value of the polynomial on the graph which it represents. An additional convention appears in Figure 7(ii): the diagram completions must not intersect the infinite region delimited by the dotted closed curves. For the sake of simplicity we shall use pictorial equations whenever possible.

Martin introduced a similar polynomial, which we denote by m', for 4-regular graphs with a specified Eulerian orientation (that is, an orientation in which every vertex has in-degree and out-degree 2). It can be defined recursively by the following pictorial equations:

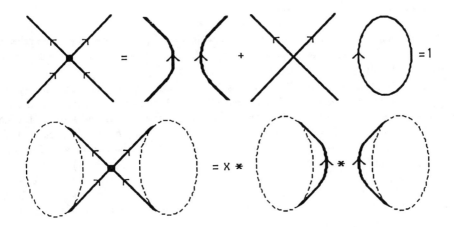

Figure 8

It is not difficult to show by induction that for an undirected connected 4-regular graph G the number of Eulerian cycles of G (taken by convention as equal to the number of ways of splitting the vertices of G which yield a single free loop) is equal to $m(G, 2)$. Similarly if G has received an Eulerian orientation, it is easy to see that $m'(G, 1)$ gives the number of Eulerian circuits of G (with an analogous convention). More generally, Martin showed how his polynomials could be used in the undirected (respectively: directed) case to compute the number $f_k(G)$ (respectively: $f_k'(G)$) of partitions of the edge-set of G into k cycles (respectively: circuits). Again, the definition of $f_k(G)$ and $f_k'(G)$ involves special conventions which will be made clear in the next section.

A few years later, Las Vergnas [23, 24] gave the following closed formulae for the Martin polynomials which completely clarify their relationship with the numbers $f_k(G)$ and $f_k'(G)$ and also provide a simple and convenient definition for them.

Proposition 2 *Let G be a connected 4-regular graph. Then:*
(i)$m(G, x) = \sum_{k \geq 1} f_k(G)(x - 2)^{k-1}$
Moreover if G has received an Eulerian orientation:
(ii)$m'(G, x) = \sum_{k \geq 1} f_k'(G)(x - 1)^{k-1}$

Another result by Martin [26] which will be of interest here concerns the "anticircuits" of a 4-regular graph with an Eulerian orientation. Intuitively, an *anticircuit* is a cycle along which the orientations of the edges alternate. Then the anticircuits form a partition of the edge-set.

Proposition 3 *Let $G = (V, E)$ be a connected 4-regular graph with an Eulerian orientation, and let q be the number of anticircuits of G. Then*

$$m'(G, -1) = (-1)^{|V|}(-2)^{q-1}.$$

Finally in the case of planar graphs Martin related his polynomial $m'(G, x)$ to the classical "Tutte polynomial". We shall present this relation in Section 5.

4   Transition Polynomials

We first need some definitions which will allow us to deal more rigorously with cycles, in particular when loops and free loops are involved. Besides, these definitions appear rather naturally in the study of graphs embedded on surfaces.

Let $G = (V, E)$ be a graph. We shall consider that each edge e of G is subdivided into two *half-edges* (the two *halves* of e), one incident to each end of e. In particular if e is a loop at v, the two halves of e are incident to v. Let us denote by H the set of half-edges of G. Then a *cycle* of G is a sequence $C = (v_1, h_1, e_1.h'_1, \ldots v_k, h_k, e_k, h'_k)$ such that, for $i = 1, \ldots k : v_i \in V$, $e_i \in E$, $h_i \in H$, $h'_i \in H$, h_i and h'_i are the two halves of e_i, h_i is incident to v_i and h'_i is incident to v_{i+1} (with the convention that v_{k+1} is v_1). Cycles will be considered up to initial vertex and direction. The cycle C will be said *simple* if the edges e_i, $i = 1, \ldots k$ are distinct. Then $\{e_i/i = 1, \ldots k\}$ will be called the *edge-set* of C. If this edge-set is E, C is called *Eulerian*. Finally if the vertices v_i, $i = 1, \ldots k$ are distinct, the cycle C will be said *elementary*.

Let $G = (V, E)$ be a 4-regular graph and v be a vertex of G. Let h_1, h_2, h_3, h_4 be the four half-edges incident to v. A *transition at* v is a partition of $\{h_1, h_2, h_3, h_4\}$ into pairs. Thus there are exactly three distinct transitions at v. For $i = 1, \ldots 4$ let e_i be the edge which contains h_i as a half. Consider for instance the transition $t = \{\{h_1, h_2\}, \{h_3, h_4\}\}$. Let us delete from G the vertex v and introduce two new vertices v_1, v_2 to obtain $V' = (V - \{v\}) \cup \{v_1, v_2\}$. Let G' be the graph with vertex-set V' and edge-set E whose vertex-edge incidence relation restricted to $(V - \{v\}) \times E$ is the same as that of G, and such that e_1, e_2 are incident to v_1 and e_3, e_4 are incident to v_2. G' is a graph of C and thus is equivalent to a 4-regular graph which will be said *obtained from G by the splitting of t* and which will be denoted by $G * t$. The transition t will be called *separating* if $G * t$ has more connected components than G. For instance if h_1, h_2 together form a loop at v, the splitting of t creates a free loop and hence t is separating in this case. Note that at most one transition at a given vertex can be separating.

A *transition system* of G is a family $p = (p(v), v \in V(G))$, where $p(v)$ is a transition at v. If we perform all vertex-splittings associated to the transitions of p, we obtain a graph denoted by $G * p$ consisting only of free loops. We denote by $c(p)$ the number of these free loops.

Let $C = (v_1, h_1, e_1, h'_1, \ldots v_k, h_k, e_k, h'_k)$ be a cycle of G. The pair of half-edges $\{h'_i, h_{i+1}\}$ is said to be *consecutive* in C at v_{i+1} for $i = 1, \ldots k$, indices being read modulo k. Let $\mathcal{D}$ be a set of simple cycles of G whose edge-sets form a partition of E (we call such a set $\mathcal{D}$ a *cycle partition of G*). For any vertex v, the set of pairs of half-edges incident to v which are consecutive in some cycle of $\mathcal{D}$ clearly forms a transition at v. The family of all such transitions defines a transition system which is said to be *associated to $\mathcal{D}$*. Then it is easy to see that conversely every transition system is associated to a unique cycle partition. The properties of 4-regular graphs studied in this paper can be formulated in two equivalent ways, either in terms of cycle partitions or in terms of transition systems. We choose the latter for reasons of convenience only. Thus for instance we may define $f_k(G)$ as the number of transition systems p of G such that $c(p) = k$. As another illustration, let us now define an Eulerian orientation as the choice for each edge of one of its halves as *initial* and the other as *terminal* in such a way that each vertex v is incident to two initial half-edges and two terminal half-edges (note that with this definition a loop can be oriented in two distinct ways). The transition at v consisting of the pair of initial half-edges together with the pair of terminal half-edges will be called *anticoherent*, and the two other transitions at v will be said *coherent*. Then, if G has received an Eulerian orientation, we may define $f_k'(G)$ as the number of transition systems p of G with no anticoherent transitions such that $c(p) = k$. On the other hand the transition system consisting of the anticoherent transitions is associated to the cycle partition consisting of the anticircuits informally introduced in the previous section.

Let us denote by $P(G)$ the set of transition systems of the 4-regular graph G. We consider that $P(L)$ contains one (trivial) transition system.

Let A be a mapping which associates to every transition of G a "weight" chosen in some set of variables or constants (A will be called a *weight function for* G). Let τ be another variable (which we shall call the *cycle variable*). The pair (G, A) will be called a *weighted 4-regular graph*. Then the associated *transition polynomial* $Q(G, A, \tau)$ is defined by:

$$Q(G, A, \tau) = \sum_{p \in P(G)} (\pi_{v \in V(G)} A(p(v))) \tau^{c(p)-1}.$$

Example 1 *A is injective (one specific variable for each transition). Then $Q(G, A, 0)$ gives the list of transition systems associated to Eulerian cycles.*

Example 2 *A assigns to each transition a weight equal to 1. Then by Proposition 2(i), $Q(G, A, \tau) = m(G, \tau + 2)$.*

Example 3 *G has received an Eulerian orientation. A assigns a weight equal to 1 to coherent transitions, and a weight equal to 0 to anticoherent transitions. Then by Proposition 2(ii), $Q(G, A, \tau) = m'(G, \tau + 1)$.*

The following immediate result can be viewed as a recursive definition of the polynomial $Q(G, A, \tau)$. Here for a vertex v, A/v denotes the restriction of A to the transitions at vertices distinct from v.

Proposition 4 *Let (G, A) be a weighted 4-regular graph and v be a vertex of G. Let t_1, t_2, t_3 be the three transitions at v. Then:*

$$Q(G, A, \tau) = \sum_{i=1,2,3} A(t_i) Q(G * t_i, A/v, \tau),$$

where A/v is identified in the obvious way with a weight function on $G * t_i$ ($i = 1, 2, 3$). Moreover, for the graph L^n consisting of n free loops, with trivial weight function A: $Q(L^n, A, \tau) = \tau^{n-1}$.

Let us now present some decomposition results for the transition polynomial. For 4-regular graphs G, G_1, G_2 we write $G = G_1 \cup G_2$ if the graph G is the disjoint union of the graphs G_1 and G_2. Also, we write $G = G_1 \& G_2$ if G can be obtained from disjoint copies of G_1 and G_2 by cutting an edge of G_1 in its two halves h_1 and h_1', cutting an edge of G_2 in its two halves h_2 and h_2' and creating two new edges e with halves h_1, h_2 and e' with halves h_1', h_2'. In both cases any transition of G can be identified with a transition of G_1 or with a transition of G_2. If A_1, A_2 are weight functions for G_1, G_2 we denote by $A_{1,2}$ the weight function for G whose restriction to the transitions of G_i is A_i, $i = 1, 2$.

Proposition 5 *Let $(G_1, A_1), (G_2, A_2)$ be two weighted 4-regular graphs. Then:*
$(i)Q(G_1 \cup G_2, A_{1,2}, \tau) = \tau Q(G_1, A_1, \tau)Q(G_2, A_2, \tau)$
$(ii)Q(G_1 \& G_2, A_{1,2}, \tau) = Q(G_1, A_1, \tau)Q(G_2, A_2, \tau)$

Proof: Denote by G the graph $G_1 \cup G_2$ in case (i) and the graph $G_1 \& G_2$ in case (ii). Every transition system p of G can be identified with a pair (p_1, p_2), where p_i is a transition system of G_i $(i = 1, 2)$, with $\pi_{v \in V(G)} A_{1,2}(p(v)) = (\pi_{v \in V(G_1)} A_1(p_1(v)))(\pi_{v \in V(G_2)} A_2(p_2(v)))$.
Moreover clearly $c(p) = c(p_1) + c(p_2)$ in case (i) and $c(p) = c(p_1) + c(p_2) - 1$ in case (ii). Then the result follows immediately from the definition of the transition polynomial. $\square$
We may describe the content of Proposition 5 by the pictorial equations of Figure 9.

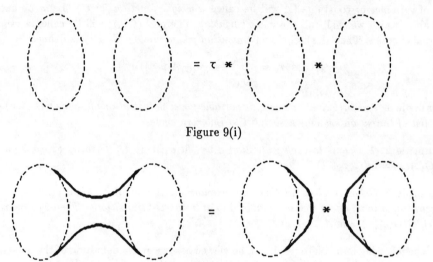

Figure 9(i)

Figure 9(ii)

Proposition 6 *Let (G, A) be a connected weighted 4-regular graph, v be a cut-vertex of G, t_1 be the separating transition at v, t_2, t_3 be the other transitions at v, and G_1, G_2 be the two connected components of $G * t_1$. Then:*

$$Q(G, A, \tau) = (A(t_1)\tau + A(t_2) + A(t_3))Q(G_1, A_1, \tau)Q(G_2, A_2, \tau)$$

where A_i is the restriction of A to the transitions at vertices of G_i, $i = 1, 2$.

Proof: By Proposition 4: $Q(G, A, \tau) = \sum_{i=1,2,3} A(t_i)Q(G * t_i, A/v, \tau)$. Moreover $G * t_1 = G_1 \cup G_2$, $G * t_2 = G_1 \& G_2$ and $G * t_3 = G_1 \& G_2$. Then Proposition 5 yields:

$$Q(G * t_1, A_{1,2}, \tau) = \tau Q(G_1, A_1, \tau)Q(G_2, A_2, \tau) \text{ and}$$

$$Q(G * t_2, A_{1,2}, \tau) = Q(G * t_3, A_{1,2}, \tau) = Q(G_1, A_1, \tau)Q(G_2, A_2, \tau).$$

The result follows since clearly $A/v = A_{1,2}$. $\square$
Proposition 6, together with Proposition 4 in the case where v is not a cut-vertex and the number n of free loops is 1, yields another recursive definition of the transition polynomials in the restricted context of connected 4-regular graphs. For instance, in Examples 2 and 3 it is easy to see that we obtain Martin's definition of the polynomials $m(G, x)$ and $m'(G, x)$.

5 Tutte Polynomials of Plane Graphs

Let $G = (V, E)$ be a graph. For $E' \subseteq E$, we denote by $k(E')$ the number of connected components of the subgraph (V, E').

The *Tutte polynomial* of G, introduced in [36, 37], is the polynomial in two variables

$$t(G, x, y) = \sum_{E' \subseteq E} (x - 1)^{k(E') - k(E)} (y - 1)^{|E'| - |V| + k(E')}.$$

This polynomial satisfies the following properties:

(1) If e is an edge of G which is neither a bridge nor a loop:

$$t(G, x, y) = t(G - e, x, y) + t(G.e, x, y),$$

where $G - e$ (respectively: $G.e$) denotes the graph obtained from G by deleting (respectively: contracting) e.

(2) If e is a bridge (respectively: loop) of G, $t(G, x, y) = xt(G.e, x, y)$ (respectively: $t(G, x, y) = yt(G - e, x, y)$), with the convention that $t(G, x, y) = 1$ if G has no edges.

Note that properties (1), (2) can be viewed as a recursive definition of the Tutte polynomial.

The Tutte polynomial contains as a special case (up to a normalization factor) the classical chromatic polynomial. We shall be interested here with a dual form of this result (see for instance [37]). For a connected plane graph G and a positive integer n we denote by $F(G, n)$ the number of face-colorings of G with n colors $1, \ldots n$ such that adjacent faces receive different colors and the unbounded face receives the color 1.

Proposition 7 *Let $G = (V, E)$ be a connected plane graph. Then for every positive integer n, $F(G, n) = (-1)^{|E| - |V| + 1} t(G, 0, 1 - n)$. Moreover, if G is cubic, $F(G, 4)$ is equal to the number of edge-3-colorings of G.*

Martin [26] established an interesting relationship between his polynomial for 4-regular graphs with an Eulerian orientation and the Tutte polynomial of plane graphs.

Let G be a connected plane graph and let $M(G)$ be its medial graph (see Section 2). We may color the faces of $M(G)$ in two colors, black and white, in such a way that adjacent faces have different colors and that all vertices of G lie inside black faces of $M(G)$. We shall always assume that $M(G)$ has received such a face-coloring. Then we may identify G with the *graph of black faces* of $M(G)$, which has one vertex $v(f)$ in the interior of each black face f of $M(G)$, and one edge with ends $v(f_1), v(f_2)$ for each vertex of $M(G)$ incident to the black faces f_1, f_2. Note that we may define similarly the *graph of white faces of $M(G)$*, which can be identified with the dual graph G^* of G. Consider a transition of $M(G)$ at a given vertex v. If each pair of the transition is formed of half-edges opposite at v, the transition is said to be *crossing*. If each pair of the transition forms an angle of a black (respectively: white) face of $M(G)$, the transition is said to be *black* (respectively: *white*). See Figure 10 where $\{\{h_1, h_2\}, \{h_3, h_4\}\}$ is black, $\{\{h_1, h_4\}, \{h_2, h_3\}\}$ is white and

$\{\{h_1, h_3\}, \{h_2, h_4\}\}$ is crossing.

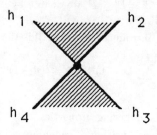

Figure 10

Let us orient every edge of $M(G)$ in such a way that the incident black face is on its left. This clearly defines an Eulerian orientation of $M(G)$. The anticoherent transitions of this orientation are exactly the crossing transitions. Thus its circuits will be called *non-crossing cycles* and its anticircuits will be called *crossing cycles*. Martin [26] proved that when $M(G)$ receives this Eulerian orientation, his polynomial $m'(M(G), x)$ is equal to the "diagonal" Tutte polynomial $t(G, x, x)$. Note that since $M(G) = M(G^*)$ and $t(G, x, y) = t(G^*, y, x)$ this result is autodual. For $x = 1$, the result has an interesting interpretation: it is well known that $t(G, 1, 1)$ is equal to the number of spanning trees of G [37], and $m'(M(G), 1)$ is equal to the number of Eulerian non-crossing cycles of $M(G)$ (see Section 3); in fact the equality of these two numbers has a simple bijective proof [21]. Proposition 3 yields another interesting case [27]: $t(G, -1, -1)$ is equal to $(-1)^{|E(G)|}(-2)^{q-1}$, where q is the number of crossing cycles of $M(G)$. This result can be generalized to non-planar graphs [33] by interpreting the number q as an algebraic invariant of G [32, 34].

Martin's result, as reformulated by Las Vergnas [22, 23, 24] (see Proposition 2(ii)) expresses $t(G, x, x)$ as a transition polynomial for $M(G)$. We can describe Penrose's formulae in a similar way. Indeed, for a plane connected cubic graph G, let A be the weight function on $M(G)$ which assigns the value 0 to the black transitions, the value 1 to the white transitions and the value -1 to the crossing transitions. Then clearly (see Section 2) the Penrose polynomial $P(G, \tau)$ is equal to $\tau Q(M(G), A, \tau)$. Thus, in view of Proposition 7, Penrose's formulae of Proposition 1 express $t(G, 0, -3)$ in terms of a transition polynomial for $M(G)$.

This was the main motivation for the following result (presented in [11] with a slightly different terminology):

Proposition 8 *Let $G = (V, E)$ be a connected plane graph with set of faces R, and let A be a weight function for $M(G)$ which assigns the value α (respectively: β, γ) to all black (respectively: white, crossing) transitions. Assume that $\lambda \neq 0$, $\mu \neq 0$, α, β, γ and τ satisfy:*

$$\tau\alpha + \beta + \gamma = (1/\lambda) + (\tau/\mu), \alpha + \tau\beta + \gamma = (1/\mu) + (\tau/\lambda), \alpha + \beta + \tau\gamma = (1/\lambda) + (1/\mu).$$

Then $t(G, 1 + (\lambda\tau/\mu), 1 + (\mu\tau/\lambda)) = \lambda^{|V|-1}\mu^{|R|-1}Q(M(G), A, \tau)$.

Note first that for $\lambda = \mu = \alpha = \beta = 1$ and $\gamma = 0$ the conditions of Proposition 8 are satisfied for all values of τ. It follows that when A assigns the value 1 to non-crossing transitions and the value 0 to crossing transitions, $t(G, 1 + \tau, 1 + \tau) = Q(M(G), A, \tau)$. This is essentially the above result by Martin and Las Vergnas on the "diagonal" Tutte polynomial.

It is also easy to check that the conditions of Proposition 8 are satisfied for $\lambda = 1/2$, $\mu = 1$, $\alpha = 0$, $\beta = 1$, $\gamma = -1$, $\tau = -2$. Then, using Proposition 7, we obtain the following result:

Proposition 9 *Let $G = (V, E)$ be a connected plane graph, and let A be the weight function for $M(G)$ which assigns the value 0 (respectively: 1, -1) to all black (respectively: white, crossing) transitions. Then $F(G, 4) = (-1)^{|E|}(-1/2)^{|V|-1}Q(M(G), A, -2)$.*

It is easy to see that when G is cubic Proposition 9 reduces to Penrose's formula of Proposition 1(ii).

Two proofs of Proposition 8 are presented in [11]. The first one, based on induction, yields the result in its full generality. The second one deals with two main cases to which the result can be essentially reduced and uses previous results by Las Vergnas [22]. We shall also restrict our attention to these two cases here, with a slightly different approach. The proofs will be only informally sketched.

(i) *The non-crossing case.* We take $\alpha \neq 0$, $\beta \neq 0$, $\gamma = 0$, $\lambda = 1/\beta$, $\mu = 1/\alpha$. Then the conditions of Proposition 8 are satisfied for all values of τ. We must show that $t(G, 1 + (\alpha\tau/\beta), 1 + (\beta\tau/\alpha)) = (1/\beta)^{|V|-1}(1/\alpha)^{|R|-1}Q(M(G), A, \tau)$. A result equivalent to this formula was previously given by Baxter [2] in his study of the Potts model in statistical mechanics. The direct proof by Baxter is similar to the one of [11], Section 6. We now briefly describe a simple inductive proof. Using Propositions 4, 5, 6 we see that $Q(M(G), A, \tau)$ can be recursively defined by the pictorial equations given in Figure 11.

It is then easy to check, interpreting on $M(G)$ the recursive definition of the Tutte polynomial of G, that these pictorial equations also correspond to $\beta^{|V|-1}\alpha^{|R|-1}t(G, 1 + (\alpha\tau/\beta), 1 + (\beta\tau/\alpha))$.

Example: Let us take $\alpha = -1$, $\beta = \tau = 2$. We obtain: $t(G, 0, -3) = (1/2)^{|V|-1}(-1)^{|R|-1}Q(M(G), A, 2)$, or equivalently, by Proposition 7 and Euler's formula: $F(G, 4) = (1/2)^{|V|-1}Q(M(G), A, 2)$.

(ii) *The binor case.* We take $\tau = -2$. Then the determinant of the linear system in α, β, γ which appears in the hypotheses of Proposition 8 vanishes, and the solution vectors (α, β, γ) are defined up to a scalar multiple of $(1, 1, 1)$. This comes from the following result which appears as the "binor equation" in [30] and is also used in [22, 23, 26, 27].

Proposition 10 *Let (G, A) be a weighted 4-regular graph. Let A' be obtained from A by adding the same number to the weights of the three transitions at a given vertex. Then $Q(G, A, -2) = Q(G, A', -2)$.*

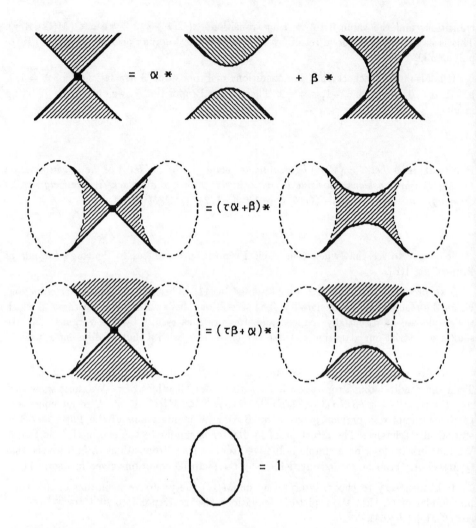

Figure 11

Proof: By induction on $|V(G)|$. We may assume that G is connected. If G has only one vertex v, let t_1, t_2, t_3 be the three transitions at v, and assume that $A'(t_i) = A(t_i) + a$ $(i = 1, 2, 3)$. Then $Q(G, A', \tau) - Q(G, A, \tau) = a(\tau + 2)$ and the result is true. Otherwise we may choose the vertex v such that $A'(t_i) = A(t_i)$ $(i = 1, 2, 3)$. By Proposition 4:

$$Q(G, A, -2) = \sum_{i=1,2,3} A(t_i) Q(G * t_i, A/v, -2) \text{ and}$$

$$Q(G, A', -2) = \sum_{i=1,2,3} A(t_i) Q(G * t_i, A'/v, -2).$$

By induction $Q(G * t_i, A/v, -2) = Q(G * t_i, A'/v, -2)$ $(i = 1, 2, 3)$ and the proof is complete.   □

We may now obtain new results by using Proposition 10 repeatedly. For instance, consider the following generalization of Proposition 3 [22].

Proposition 11 *Let $G = (V, E)$ be a 4-regular graph and q be a transition system of G. Let A be the weight function for G which assigns to all transitions of q the value 0 and to all other transitions the value 1.*
Then $Q(G, A, -2) = (-1)^{|V|}(-2)^{c(q)-1}$.

Proof: The right-hand side of this equality is equal to $Q(G, A', -2)$, where A' is obtained from A by subtracting 1 to the weights of the three transitions at each vertex. □

Assuming again that $G = (V, E)$ is a connected plane graph, the formula of the non-crossing case yields for $\tau = -2$:

$$t(G, 1 - (2\alpha/\beta), 1 - (2\beta/\alpha)) = (1/\beta)^{|V|-1}(1/\alpha)^{|R|-1}Q(M(G), A, -2).$$

By Proposition 10, this will be valid for any weight function A which assigns the value $\alpha + \gamma$ to black transitions, $\beta + \gamma$ to white transitions and γ to crossing transitions, for any number γ. For instance for $\alpha = 1$, $\beta = 2$, $\gamma = -1$ it is easy to see that we obtain the generalized Penrose formula of Proposition 9.

6 Positive Integer Values of the Cycle Variable

Let G be a graph in the class $\mathcal{C}$, that is, a graph all vertices of which have degree 2 or 4. For a positive integer k, an even k-*edge-valuation* of G is a mapping from the set of edges of G to $\{1, \ldots k\}$ such that for $i = 1, \ldots k$ every vertex is incident to an even number of half-edges of value i (with the obvious convention that every edge transmits its value to its halves). There is a natural bijection between the sets of even k-edge-valuations of two equivalent graphs of $\mathcal{C}$. This leads us to the following definition: an *even k-edge-valuation* of a 4-regular graph $G = (V, E)$ is a mapping from the set of edges and free loops of G to $\{1, \ldots k\}$ such that for $i = 1, \ldots k$ every vertex is incident to an even number of half-edges of value i. We shall denote by $S(G, k)$ the set of even k-edge-valuations of G. Let f be an element of $S(G, k)$ and v be a vertex of G. We shall say that *the interaction of f with v is of type v* and we shall write $\theta(f, v) = v$ if f takes the same value on the four half-edges incident to v. Otherwise, the partition of these half-edges according to their f-value is a transition t at v; then we shall say that *the interaction of f with v is of type t* and we shall write $\theta(f, v) = t$. Thus θ is a mapping from $S(G, k) \times V$ to $V \cup P(G)$ which we call the *type mapping*. An *even interaction function for G* is a mapping W which associates to every vertex or transition of G (that is, to every possible interaction type) a "weight" chosen in some set of variables or constants. Then for every f in $S(G, k)$ we define $W(f)$, the weight of f, as equal to $\pi_{v \in V(G)} W(\theta(f, v))$.

Proposition 12 *Let (G, A) be a 4-regular weighted graph and k be a positive integer. Let W be the even interaction function for G defined by:*
(i) for every vertex v, if t_1, t_2, t_3 denote the three transitions at v:

$$W(v) = \sum_{i=1,2,3} A(t_i)$$

(ii) for every transition t, $W(t) = A(t)$.
Then $Q(G, A, k) = (1/k) \sum_{f \in S(G,k)} W(f)$.

Proof: We must show that:

$$\sum_{p \in P(G)} (\pi_{v \in V(G)} A(p(v))) k^{c(p)} = \sum_{f \in S(G,k)} W(f).$$

For $v \in V(G)$ and $f \in S(G, k)$, let us write $t << (f, v)$ if t is a transition at v and $\theta(f, v) \in \{v, t\}$. For $p \in P(G)$ we write $p << f$ if $p(v) << (f, v)$ for every vertex v of G. We observe that for any $v \in V(G)$ and $f \in S(G, k)$:

$$W(\theta(f, v)) = \sum_{t << (f,v)} A(t).$$

Hence $W(f) = \pi_{v \in V(G)} \sum_{t << (f,v)} A(t) = \sum_{p \in P(G), p << f} \pi_{v \in V(G)} A(p(v))$.
Now we may write:

$$
\begin{aligned}
\sum_{f \in S(G,k)} W(f) &= \sum_{f \in S(G,k)} \sum_{p \in P(G), p << f} \pi_{v \in V(G)} A(p(v)) \\
&= \sum_{p \in P(G)} \sum_{f \in S(G,k), p << f} \pi_{v \in V(G)} A(p(v)).
\end{aligned}
$$

Given a transition system p, the elements f of $S(G, k)$ such that $p << f$ are exactly the $k^{c(p)}$ mappings f from the set of edges and free loops of G to $\{1, \dots k\}$ such that any two half-edges forming a pair of some transition of p receive the same f-value. The result now follows immediately.   □

Remark: Using Proposition 4, it is easy to prove Proposition 12 by induction and we leave this as an exercise to the reader.

We now apply Proposition 12 to the study of a generalization of the Penrose polynomial introduced in Section 2. Let G be a plane 4-regular graph. Let q be a transition system of G consisting of non-crossing transitions. An even k-edge-valuation f of G is said to be *compatible with q* if for every vertex v, the interaction of f with v is not of type v or $q(v)$.

Proposition 13 *Let G be a plane 4-regular graph and let q be a transition system of G consisting of non-crossing transitions. Let A be the weight function which assigns the weight zero to the transitions of q, the weight -1 to the crossing transitions and the value 1 to the other transitions. Then for every positive integer k, the number of even k-edge-valuations of G compatible with q is equal to $kQ(G, A, k)$.*

Proof: Let W be the even interaction function for G such that $W(v) = 0$ for every vertex v and $W(t) = A(t)$ for every transition t. By Proposition 12, $kQ(G, A, k) = \sum_{f \in S(G,k)} W(f)$. Moreover for $f \in S(G, k)$, $W(f) = 0$ if f is not compatible with q. On the other hand, if f is compatible with q, $W(f) = (-1)^{x(f)}$, where $x(f)$ denotes the number of vertices v such that $\theta(f, v)$ is equal to the crossing transition at v. Now we observe that for each i in $\{1, \ldots k\}$, the edges of G with f-value equal to i form a set C_i of disjoint simple curves in the plane. For distinct i, j in $\{1, \ldots k\}$, let us denote by $x(C_i, C_j)$ the number of vertices where a curve of C_i crosses a curve of C_j. This number is clearly even and hence $x(f) = \sum_{1 \leq i < j \leq k} x(C_i, C_j)$ is also even. Thus $W(f) = 1$ if f is compatible with q, and the proof is complete. □

For a plane 4-regular graph G and a transition system q consisting of non-crossing transitions, we denote by $R(G, q, \tau)$ the polynomial $\tau Q(G, A, \tau)$, where the weight function A is defined as in Proposition 13. Thus when τ is a positive integer, $R(G, q, \tau)$ gives the number of even τ-edge-valuations of G compatible with q. Assume that $q(v)$ is a separating transition for some vertex v.

Let t be the other non-crossing transition at v, and t' be the crossing transition at v. Then Proposition 4 yields:

$$Q(G, A, \tau) = Q(G * t, A/v, \tau) - Q(G * t', A/v, \tau).$$

It now follows easily from Proposition 5(ii) that $Q(G, A, \tau) = 0$. Thus $R(G, q, \tau)$ vanishes whenever q contains a separating transition. The converse turns out to be true, by a difficult result of Fleischner [9].

Proposition 14 *Let G be a plane 4-regular graph and q be a transition system consisting of non-crossing transitions. The polynomial $R(G, q, \tau)$ vanishes if and only if q contains a separating transition.*

Proof: By Proposition 13, it is enough to show that if q contains no separating transition, there exists an even k-edge-valuation of G compatible with q for some positive integer k. By Fleischner's Theorem, there exists a cycle partition $\mathcal{D}$ of G into elementary cycles $C_1, \ldots C_k$ such that the associated transition system contains no transition of q. If we assign to every edge the index of the cycle of $\mathcal{D}$ which contains it, we obtain the required even k-edge-valuation. □

Remark: The special case of Fleischner's Theorem that we use here corresponds to the difficult core of its proof. Thus it is easy to show the equivalence between Proposition 14 and Fleischner's Theorem.

Fleischner pointed out the following connection between his result and the Four Color Problem. Let G be a plane cubic graph and F be a perfect matching of G. Let us contract every edge of F. We obtain a plane 4-regular graph H. The cycles of the 2-factor $E(G) - F$ of G define a cycle partition of H. Let q be the associated transition system, which clearly consists of non-crossing transitions. This defines a bijective correspondence between the pairs (G, F) (G plane and cubic, F a perfect matching of G) and the pairs (H, q) (H plane and 4-regular, q a system of non-crossing transitions of H). Clearly G is bridgeless if and only if q has no separating transition. Moreover the edge-3-colorings of G are easily seen to be in bijective correspondence with the even 3-edge-valuations of H compatible with

q. Thus the Four Color Theorem [1] is equivalent to the following statement: for any plane 4-regular graph H and any system q of non-crossing, non-separating transitions of H, $R(H, q, 3)$ is non-zero. On the other hand, as we have just seen, Fleischner's Theorem essentially asserts that under the same hypothesis $R(H, q, \tau)$ is non-zero for some positive integer value of τ.

Penrose's first formula (Proposition 1(i)) appears as a simple consequence of the above discussion. Let G' be a plane connected cubic graph. Replace each vertex of G' by a triangle according to the following scheme:

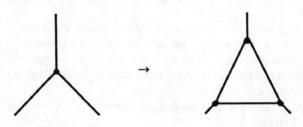

Figure 12

In this way we obtain another plane connected cubic graph G, in which the former edges of G' define a perfect matching F. Associate as above the pair (H, q) to the pair (G, F): H is isomorphic to $M(G')$, q consists of its black transitions (recall that all vertices of G' lie inside black faces of $M(G')$). Then G' and G have the same number of edge-3-colorings, which is equal to $R(H, q, 3)$ by the above discussion. This is easily seen to be equivalent to Proposition 1(i).

7   Ice-Type Models

Let G be a 4-regular graph. We denote by $O(G)$ the set of Eulerian orientations of G. For any Eulerian orientation f and vertex v of G, we shall say that *the interaction of f with v is of type t* and we shall write $\theta(f, v) = t$ if t is the transition at v anticoherent with respect to f. An *Eulerian interaction weight function for G* is a mapping W which associates to every transition t a "weight" $W(t)$ chosen in some set of variables or constants. Now for every f in $O(G)$ we define $W(f)$, the weight of f, as equal to $\pi_{v \in V(G)} W(\theta(f, v))$. Then we obtain the following result, quite similar to Proposition 12.

Proposition 15 *Let (G, A) be a 4-regular weighted graph. For every vertex v and transition t at v, let $W(t)$ be the sum of the values of A on the two transitions at v distinct from t.*

   Then $Q(G, A, 2) = (1/2) \sum\limits_{f \in O(G)} W(f)$.

Proof: We must show that:

$$\sum_{p \in P(G)} (\pi_{v \in V(G)} A(p(v))) 2^{c(p)} = \sum_{f \in O(G)} W(f).$$

For $p \in P(G)$ and $f \in O(G)$ we write $p << f$ if $p(v) \neq \theta(f, v)$ for every vertex v of G. It is then easy to see that:

$$W(f) = \pi_{v \in V(G)} W(\theta(f, v)) = \sum_{p \in P(G), p << f} \pi_{v \in V(G)} A(p(v)).$$

Hence:

$$\sum_{f \in O(G)} W(f) = \sum_{f \in O(G)} \sum_{p \in P(G), p << f} \pi_{v \in V(G)} A(p(v))$$
$$= \sum_{p \in P(G)} \sum_{f \in O(G), p << f} \pi_{v \in V(G)} A(p(v)).$$

Given a transition system p, the elements f of $O(G)$ such that $p << f$ are the $2^{c(p)}$ Eulerian orientations such that all transitions of p are coherent, and the result follows. $\square$

Remarks: + Like Proposition 12, Proposition 15 can easily be proved by induction.

+ A function of the form $\sum_{f \in O(G)} W(f)$ can be viewed as the partition function of a special kind of *ice-type model*, an object of extensive study in statistical mechanics (see [2], Chapter 8). In fact Section 12.3 of [2] presents a result strongly related to Proposition 15.

Proposition 15 can easily be reformulated as follows.

Proposition 16 *Let G be a 4-regular graph and W be an Eulerian interaction weight function for G. For every vertex v, if t_1, t_2, t_3 denote the three transitions at v, let $A(t_i) = -W(t_i) + (1/2) \sum_{j=1,2,3} W(t_j)$.*

Then $\sum_{f \in O(G)} W(f) = 2Q(G, A, 2)$.

We now give three examples of applications of Propositions 15 and 16. The first one appears (in dual form) in [2], Section 12.3.

Proposition 17 *Let $G = (V, E)$ be a connected plane graph. Let W be the Eulerian interaction weight function for $M(G)$ which assigns the value 2 to the black transitions, the value -1 to the white transitions, and the value 1 to the crossing transitions. Then $F(G, 4) = (1/2)^{|V|} \sum_{f \in O(M(G))} W(f)$.*

Proof: Let A be the weight function for $M(G)$ which assigns the value -1 to the black transitions, the value 2 to the white transitions, and the value 0 to the crossing transitions. It follows from Proposition 8 (see the example given for the non-crossing case) that $F(G, 4) = (1/2)^{|V|-1} Q(M(G), A, 2)$. Moreover $Q(M(G), A, 2) = (1/2) \sum_{f \in O(M(G))} W(f)$ by Proposition 15. $\square$

The following result is a special case of Theorem 5.2 of [24].

Proposition 18 *Let $G = (V, E)$ be a 4-regular graph. The number of Eulerian orientations of G is equal to $(1/2)^{|V|-1} m(G, 4)$.*

Proof: Let W be the Eulerian interaction weight function which assigns to every transition the value 1, so that $|O(G)| = \sum_{f \in O(G)} W(f)$. We apply Proposition 16 and consider the weight function A which assigns to every transition the value $1/2$. Then $|O(G)| = 2Q(G, A, 2)$. Hence if A' is the weight function which assigns to every transition the value $1, |O(G)| = (1/2)^{|V|-1} Q(G, A', 2)$. The result now follows from the definition of the Martin polynomial $m(G, x)$ (see Example 2 of Section 4).   □

The following result is a reformulation of Theorem 5.1 of [25].

Proposition 19 *Let G be a 4-regular graph and q be a transition system of G. Let A be the weight function for G which assigns to all transitions of q the value 0 and to all other transitions the value 1. For an Eulerian orientation f of G, let $q.f$ denote the number of anticoherent transitions of f which belong to q. Then:*

(i) $Q(G, A, 2) = \sum_{f \in O(G)} 2^{q.f-1}.$

(ii) $Q(G, A, 2) = k2^{c(q)-1}$ *for some positive odd integer k.*

Proof of (i). Let W be the Eulerian interaction weight function for G which assigns the value 2 to the transitions of q and the value 1 to the other transitions. By Proposition 15, $Q(G, A, 2) = (1/2) \sum_{f \in O(G)} W(f) = \sum_{f \in O(G)} 2^{q.f-1}.$ □

For the proof of part (ii) of Proposition 19 (which is based on part (i)), the reader is referred to [25].

Remark: It follows from Propositions 11 and 19(ii) that $Q(G, A, 2) = kQ(G, A, -2)$ for some odd integer k.

8   Applications to Knot Theory

Recent spectacular developments in knot theory, including the proof of very old conjectures, rely on purely combinatorial methods, some of which involve the concept of transition polynomial. The purpose of this section is to present briefly this fascinating research topic. The interested reader will find more details, and further references, in [10], [18].

Informally speaking, a *link* consists of a finite family of disjoint simple curves (the *components* of the link) in the 3-dimensional Euclidean space $\mathcal{R}^3$ which satisfies a certain "tameness" or "smoothness" property. A *knot* is a link with only one component. An *oriented link* is obtained by choosing a travel direction along each component of the link. Two links are *ambient isotopic* if there exists an isotopic deformation of $\mathcal{R}^3$ which deforms one into the other (this notion is a good mathematical model for the study of the feasible deformations in space of a "physical" link made of elastic string). In the sequel links are considered up to equivalence under ambient isotopy.

A link is usually represented by a *link diagram* (an example is given in Figure 13).

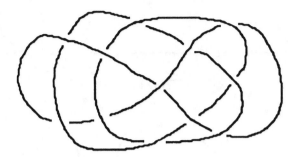

Figure 13

Such a diagram is just a 4-regular plane graph (corresponding to a plane projection of the link with suitable regularity properties), together with a binary information locally visualized at each vertex. This binary information specifies which "branch" overcrosses the other (a branch can be viewed as a pair of opposite half-edges). Each component of the link is represented by a crossing cycle of the diagram. For an oriented link, each crossing cycle is directed into a circuit to represent the travel direction chosen on the corresponding component of the link. Thus the diagram receives an Eulerian orientation for which all crossing transitions are coherent (see Figure 14 where we represent an orientation of the link of Figure 13).

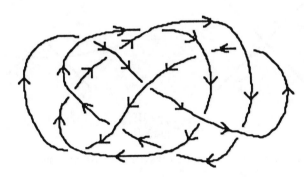

Figure 14

A basic problem in knot theory is the following: given two link diagrams, decide whether or not they represent the same link (in which case we shall say that they are *equivalent*). Reidemeister's Theorem allows us to reformulate this problem in purely combinatorial terms. This theorem asserts that two link diagrams are equivalent if and only if one can be obtained from the other by a finite sequence of elementary transformations, or "moves". Each move transforms only a small part of the diagram, regardless of the remaining part, and belongs to one of the three types depicted in Figure 15.

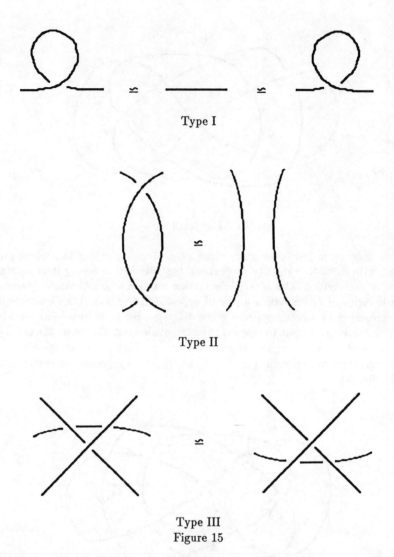

Type I

Type II

Type III
Figure 15

Of course for oriented links one must consider all possible orientations of the portions of diagrams appearing in Figure 15.

Thus Reidemeister's Theorem can be considered as a basis for a purely combinatorial, or "formal", knot theory (as illustrated in [16]).

A basic tool in the study of the equivalence of link diagrams is that of *(oriented) link invariant*: one wants to associate to each (oriented) link diagram a simpler and efficiently computable mathematical object which will be invariant under ambient isotopy of the associated links (or equivalently under Reidemeister moves). As a classical example, we shall mention the fundamental group of the complement of the link in $\mathcal{R}^3$ and the related Alexander polynomial. Conway ([7], see also [16]) proved that a suitable normalization of the Alexander polynomial, which is a polynomial $\nabla(z)$ in one variable z:

(1) is an oriented link invariant
(2) takes the value 1 on the trivial knot (represented by the free loop)
(3) satisfies the property described by the pictorial equation of Figure 16.

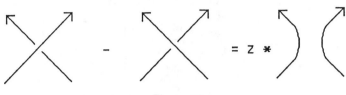

Figure 16

Moreover it can be shown that one can consider (1), (2), (3) as axioms which define $\nabla(z)$ uniquely. There is some similarity with the recursive definition of transition polynomials presented in Proposition 4. However the situation here is much more complicated. Kaufmann [16] designed a combinatorial model for $\nabla(z)$ which is remotely related to the transition polynomial concept, and far more subtle.

More recently, in the course of some studies on von Neumann algebras, Jones discovered a new invariant [15], a one-variable Laurent polynomial satisfying (1), (2), and a property quite similar to (3). This quickly lead several independent research groups [8, 31] to a generalization of both Conway and Jones polynomials through a new invariant: the "homfly" polynomial (whose name comes from the initials of the authors of [8]). This invariant is a Laurent polynomial homogeneous of degree zero in three variables x, y, z which satisfies (1), (2) and the property described in Figure 17.

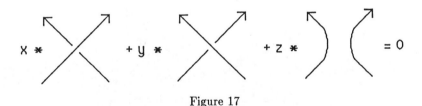

Figure 17

The homfly polynomial can be defined algebraically using representations of oriented links as closed braids and then either Hecke algebra representations of the braid groups (this is the path taken by Jones for his original discovery) or a "term rewriting system" approach in the spirit of theoretical computer science. One can also define the homfly polynomial inductively in a purely combinatorial way on diagrams. But in any case the proof of its existence and unicity is quite involved and no combinatorial model is known in general. We already mentioned that Kauffman found such a model for the Conway polynomial. The only other result in this direction, also due to Kauffman, is the reformulation of the Jones polynomial as a transition polynomial (in his terminology, a *bracket polynomial*). We now present briefly this beautiful and most important work. A detailed account can be found in [10],[17],[18].

Consider a link diagram, that is a 4-regular plane graph G together with a crossing information at each vertex. For a vertex v, let us denote by h_1, h_2, h_3, h_4 the half-edges incident to v in clockwise order around v in such a way that the branch formed by h_1, h_3 overcrosses the branch formed by h_2, h_4.

Then we shall say that the transition $\{\{h_1, h_2\}, \{h_3, h_4\}\}$ is *descending* and the transition $\{\{h_2, h_3\}, \{h_4, h_1\}\}$ is *ascending* (see Figure 18).

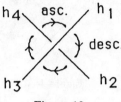

Figure 18

Proposition 20 *Let us associate to every link diagram with underlying graph G the weight function A for G which assigns the value a to all descending transitions, a^{-1} to all ascending transitions, and zero to all crossing transitions. Then $Q(G, A, -a^2 - a^{-2})$ is invariant under Reidemeister moves of types (II) and (III).*

Outline of proof. For a move of type (II) apply Proposition 4 successively to the two vertices of the larger diagram which are to be erased. For a move of type (III) apply Proposition 4 to a well-chosen vertex and use invariance under type (II) moves. □

Let us now consider an oriented link diagram. For each vertex v we can define its *sign* $s(v)$ according to the rule depicted in Figure 19.

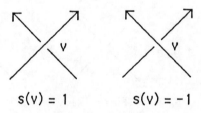

Figure 19

Proposition 21 *Let us associate to every oriented link diagram with underlying graph G the weight function A for G as defined in Proposition 20. Let A' be the weight function which associates to every transition t at the vertex v the value $(-a)^{-3s(v)} A(t)$. Then $Q(G, A', -a^2 - a^{-2})$ is an invariant of oriented links which can be identified (up to a change of variable) with the Jones polynomial.*

Outline of proof. Let $w = \sum_{v \in V(G)} s(v)$. Clearly $Q(G, A', -a^2, -a^{-2}) = (-a)^{-3w} Q(G, A - a^2, -a^{-2})$. The invariance of w under Reidemeister moves of types (II) and (III) is easy to check, and thus the similar invariance for $Q(G, A' - a^2, -a^{-2})$ follows from Proposition 20. The invariance under Reidemeister moves of type (I) is obtained by applying Proposition 4 to the vertex concerned by the move. Finally, Proposition 4 applied to an arbitrary vertex yields the equation described in Figure 20 (compare with Figure 17).

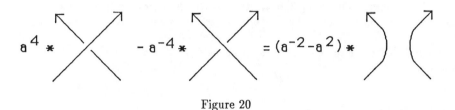

Figure 20

This equation insures the identification with the Jones polynomial.   □

Proposition 21 is the basis of a purely combinatorial proof ([17], [28], [35]) of a conjecture going back to the work by Tait and Little at the end of the last century. Let us call a link diagram *alternating* if when travelling along any component one alternately meets over- and under-crossings.

Proposition 22 *Two equivalent alternating 2-connected link diagrams have the same number of crossings. This number of crossings is minimal among all equivalent connected link diagrams.*

The connection between the Jones polynomial and the Tutte polynomial via the transition polynomial concept (which can be made explicit by putting together Proposition 21 and the non-crossing case of Proposition 8) is already mentioned in [15] (via the Potts model) and further explored in [19]. It is successfully exploited in [35] and this motivated new results on transition polynomials by Kauffman [18].

The Tutte polynomial is also related to the homfly polynomial in the following sense. It is shown in [12] how the Tutte polynomial of a plane graph can be evaluated as the homfly polynomial of an associated oriented link. It follows that the study of the Potts model and of the Four Colour Problem can be approached in terms of the homfly polynomial and that the computation of this polynomial on a diagram is NP-hard.

Other recently discovered link invariants (see [20]) have recursive definitions similar to those of the Jones or homfly polynomials, but no connection with the concept of transition polynomial is known (except of course for the Jones polynomial which is a special case of the Kauffman 2-variable polynomial defined in [20]).

9   Conclusion

As we have seen, transition polynomials play a significant role in various areas, such as knot theory and the study of the Tutte polynomial of plane graphs (related to the Potts model of statistical mechanics and to the Four Colour Problem). The interaction thus allowed between these areas has already been fruitful, and provides a rich research topic. Another promising topic is the generalization of the results presented here. For instance in [23], [24], Las Vergnas extends Martin polynomials to arbitrary Eulerian graphs and generalizes Proposition 18. He also generalizes in [22] the Martin – Las Vergnas formula $t(G, x, x) = m'(M(G), x)$ (see Section 5) from plane graphs to graphs embedded in the projective plane or the torus. The same formula, as well as Propositions 11, 18, and 19(ii) are generalized by Bouchet [6] in terms of isotropic systems (a new concept which unifies 4-regular graphs and binary matroids — see [5]). The extended Penrose formula of Proposition 9 is generalized in [13] to arbitrary matroids in terms of their Tutte polynomials, and the restriction of Proposition 19 to plane graphs is generalized to binary matroids in [14].

We feel that much more could be done in the directions of research that we just briefly indicated.

References

[1] K. Appel, W. Haken, Every planar graph is four colorable, Part I; W. Haken, K. Appel, J. Koch, Every planar graph is four colorable, Part II, *Illinois J. Math.* **21**(1977), 429–567.

[2] R. J. Baxter, *Exactly Solved Models in Statistical Mechanics*, Academic Press, 1982.

[3] C. Berge, *Graphes et Hypergraphes*, Dunod, Paris, 1974.

[4] J. A. Bondy, U. S. R. Murty, *Graph Theory with Applications*, MacMillan, London, 1976.

[5] A. Bouchet, Isotropic systems, *European Journal of Combinatorics*. To appear.

[6] A. Bouchet, Tutte–Martin polynomials and orienting vectors of isotropic systems, preprint.

[7] J. H. Conway, An enumeration of knots and links and some of their algebraic properties, *Computational Problems in Abstract Algebra*, Pergamon Press, New York (1970), 329–358.

[8] P. Freyd, D. Yetter, J. Hoste, W. B. R. Lickorish, K. Millett, A. Ocneanu, A new polynomial invariant of knots and links, *Bull. Amer. Math. Soc.* **12**(1985), 239–246.

[9] H. Fleischner, Eulersche Linien und Kreisüberdeckungen, die vorgegebene Durchgänge in den Kanten vermeiden, *J. Combin. Theory Ser. B* **29**(1980), 145–167.

[10] P. de la Harpe, M. Kervaire, C. Weber, On the Jones polynomial, *L'Enseignement Mathématique* **32**(1986), 271–335.

[11] F. Jaeger, On Tutte polynomials and cycles of plane graphs, *J. Combin. Theory Ser. B* **44**(1988), 127-146.

[12] F. Jaeger, On Tutte polynomials and link polynomials, *Proc. Amer. Math. Soc.* **103**(1988), 647-654.

[13] F. Jaeger, Generalization to matroids of a formula of Roger Penrose, Communication to the 11th British Combinatorial Conference, 1987.

[14] F. Jaeger, On Tutte polynomials of matroids representable over $GF(q)$, *European Journal of Combinatorics*, to appear.

[15] V. F. R. Jones, A polynomial invariant for knots via Von Neumann algebras, *Bull. Amer. Math. Soc.* **12**(1985), 103–111.

[16] L. H. Kauffman, Formal knot theory, Princeton University Press, *Mathematical Notes* **30**(1983).

[17] L. H. Kauffman, State models and the Jones polynomial, *Topology* **26**(1987), 297–309.

[18] L. H. Kauffman, New invariants in the theory of knots, *Amer. Math. Monthly* **95**(1988), 3, 195-242.

[19] L. H. Kauffman, Statistical mechanics and the Jones polynomial, *Proceedings of the 1986 Santa Cruz Conference on Artin's Braid Group, AMS Contemp. Math.* Series (1989).

[20] L. H. Kauffman, An invariant of regular isotopy, To appear in *Trans. Amer. Math. Soc.*

[21] A. Kotzig, Eulerian lines in finite 4-valent graphs and their transformations, Proc. Colloq. Tihany 1966, North Holland, Amsterdam, 1968, 219–230.

[22] M. Las Vergnas, Eulerian circuits of 4-valent graphs imbedded in surfaces, Coll. Math. Soc. J. Bolyai 25 (*Algebraic Methods in Graph Theory*, Szeged, 1978), North-Holland, Amsterdam, 1981, 451–477.

[23] M. Las Vergnas, On Eulerian partitions of graphs, in *Graph Theory and Combinatorics*, R. J. Wilson editor, Research Notes in Mathematics 34, Pitman Advanced Publishing Program, San Francisco, London, Melbourne, 1979, 62–75.

[24] M. Las Vergnas, Le polynôme de Martin d'un graphe Eulérien, Annals of Discrete Mathematics 17, North-Holland, Amsterdam, 1983, 397–411.

[25] M. Las Vergnas, On the evaluation at (3,3) of the Tutte polynomial of a graph, *J. Combin. Theory Ser. B* **45**(1988), 367-372.

[26] P. Martin, Enumérations eulériennes dans les multigraphes et invariants de Tutte – Grothendieck, Thesis, Grenoble, 1977.

[27] P. Martin, Remarkable valuation of the dichromatic polynomial of planar multigraphs, *J. Combin. Theory Ser. B* **24**(1978), 318–324.

[28] K. Murasugi, Jones polynomials and classical conjectures in knot theory, *Topology* **26**(1987), 187–194.

[29] O. Ore, *The Four-Color Problem*, Academic Press, New York, 1967.

[30] R. Penrose, Applications of negative dimensional tensors, in *Combinatorial Mathematics and its Applications*, Proceedings of the Conference held in Oxford in 1969, Academic Press, London, 1971, 221–244.

[31] J. H. Przytycki, P. Traczyk, Invariants of links of Conway type, *Kobe J. Math.* **4**(1987), 115-139.

[32] P. Rosenstiehl, Bicycles et diagonales des graphes planaires, *Cahiers du C.E.R.O.* Vol. 17, no. **2–3–4**(1975), 365–383.

[33] P. Rosenstiehl, R. C. Read, On the principal edge tripartition of a graph, *Annals of Discrete Mathematics* **3**(1978), 195–226.

[34] H. Shank, The theory of left–right paths, in *Combinatorial Mathematics* III, Lecture Notes in Mathematics 452, Springer Verlag, Berlin, 1975, 42–54.

[35] M. Thistlethwaite, A spanning tree expansion of the Jones polynomial, *Topology* **26**(197), 297–309.

[36] W. T. Tutte, A ring in graph theory, *Proc. Cambridge Phil. Soc.* **43**(1947), 26–40.

[37] W. T. Tutte, A contribution to the theory of chromatic polynomials, *Can. J. Math.* **6**(1954), 80–91.

ON INFINITE n-CONNECTED GRAPHS

W. MADER
Universität Hannover
Institut für Mathematik
Welfengarten 1
D-3000 Hannover 1
FRG

Abstract

Every infinite, $(n+1)$-connected graph G contains an infinite subset $S \subseteq V(G)$ such that for all $S' \subseteq S$, $G - S'$ is n-connected. Every infinite, critically $(n + 1)$-connected graph G has a subset $S \subseteq V(G)$ of cardinality $|V(G)|$ such that for all $S' \subseteq S$, $G - S'$ is n-connected.

It was proved in [5] that every infinite, $(n+1)$-connected graph $G = (V(G), E(G))$ must contain vertices $x \neq y$ so that deleting x and y from G, the remaining graph $G - \{x, y\}$ is n-connected. We shall show here that *there is even an infinite set $S \subseteq V(G)$ such that $G - S'$ is n-connected for all $S' \subseteq S$* (also cf. [7, Theorem 4]). For a critically $(n + 1)$-connected, infinite graph we can find such a set S even with $|S| = |V(G)|$. We also touch upon the corresponding problems for graphs of infinite connectivity. At the end of the paper, we show by an example that similar results are not true for directed graphs.

First some definitions and notation. All graphs considered do not have multiple edges nor loops. For $A \subseteq V(G)$ and $A \subseteq G$ (i.e. A a subgraph of G), $G(A)$ denotes the subgraph of G induced by A and $V(A)$, respectively. An $A \subseteq V(G)$ is *independent* in G, if $E(G(A)) = \emptyset$. We write $v \in G$ instead of $v \in V(G)$, and for $A \subseteq V(G)$ and $B \subseteq G$, $A \cap B := A \cap V(B)$. The edge between the vertices x and y is denoted by $[x, y]$, and (x, y) denotes the edge from x to y in a directed graph. For $A \subseteq V(G)$, $N(A; G) := \{x \in G - A :$ there is an $a \in A$ such that $[x, a] \in E(G)\}$, and for $A \subseteq G$, $N(A; G) := N(V(A); G)$. For a set A of vertices in a directed graph G, $N^+(A; G) := \{x \in G - A:$ there is an $a \in A$ such that $(a, x) \in E(G)\}$. For $X, Y \subseteq V(G)$ or $Y \subseteq G$, $d(X, Y; G) := |\{[x, y] \in E(G) : x \in X$ and $y \in Y\}|$ and $d(X; G) := d(X, V(G) - X; G)$, if G is undirected, and $d^+(X; G) := |\{(x, y) \in E(G) : x \in X$ and $y \in G - X\}| =: d^-(V(G) - X; G)$, if G is directed. $\mathcal{C}(G)$ denotes the set of components of the graph G. For disjoint $X, Y \subseteq V(G)$, an X, Y-*path* P has one endvertex in X, the other endvertex in Y, and $|X \cap P| = |Y \cap P| = 1$; for $Y \subseteq G$, an X, Y-path means an $X, V(Y)$-path. For a vertex x, we write x instead of $\{x\}$ in the above notation. The x, y-paths P_i $(i \in I)$ are *openly disjoint*, if $P_i \neq P_j$ and $V(P_i) \cap V(P_j) = \{x, y\}$ for all $i \neq j$. For $x \in G$ and $X \subseteq V(G - x)$ or $X \subseteq G - x$, an x, X-*fan of degree* α is a subgraph which consists of x, X-paths P_i $(i \in I)$ such that $|I| = \alpha$ and the $P_i - x$ $(i \in I)$ are disjoint. By a well known generalization of Menger's theorem (see, for instance, [4, Ch. IV]), an α-connected graph G contains an x, X-fan of degree α for every $x \in G$ and $X \subseteq G - x$ with $|X| \geq \alpha$. A tree T, directed in such a way that $d^-(t; T) = 0$, but $d^-(t'; T) = 1$ for all $t' \in T - t$, is called a t-*branching*.

151

G. Hahn et al. (eds.), Cycles and Rays, 151–160.

N denotes the set of non-negative integers and n is always an element of N. For a set $M, \binom{M}{n} := \{S \subseteq M : |S| = n\}$. A set of mutually incomparable elements in an ordered set is called an *antichain*. The infinite cardinals are denoted by $\aleph_\alpha$, where α is an ordinal. For an ordinal $\lambda, |\lambda|$ denotes the corresponding cardinal number, and $\omega_\alpha := \min\{\lambda : |\lambda| = \aleph_\alpha\}$. The term *"sequence"* always means ω_0-sequence. The axiom of choice is adopted throughout.

We say, that a subset $T \subseteq V(G)$ *separates* $A \subseteq V(G)$ or $A \subseteq G$ in G, if there are at least two $C \in \mathcal{C}(G - T)$ with $V(C) \cap A \neq \emptyset$. For non-adjacent vertices $x \neq y$ in G, we set $\mu(x, y; G) := \min\{|T| : T \text{ separates } \{x, y\} \text{ in } G\}$, and for adjacent vertices $x \neq y$,

$$\mu(x, y; G) := \mu(x, y; G - [x, y]) + 1.$$

For $A \subseteq V(G)$, set $\mu(A; G) := \min_{\{x,y\}\in\binom{A}{2}} \mu(x, y; G)$, if $|A| \geq 2$, and $\mu(A; G) := \aleph_0$, if $|A| \leq 1$. Furthermore, $\mu(G) := \mu(V(G); G)$ for $|G| \geq 2$ and $\mu(G) := |G| - 1$ for $|G| \leq 1$ denotes the *connectivity number of G*, and G is called *α-connected*, if $\mu(G) \geq \alpha$ holds. By a theorem of Menger–Whitney (see, for instance, [4, Ch. IV]), a graph G with $|G| \geq 2$ is α-connected, iff for every $\{x, y\} \in \binom{V(G)}{2}$, there is a system of α openly disjoint x, y-paths in G. If $\mu(G)$ is finite, we define $T(G) := \{T \subseteq V(G) : T \text{ separates } G \text{ and } |T| = \mu(G)\}$; for $\mu(G) \geq \aleph_0$, $T(G)$ be empty. A subset $U \subseteq V(G)$ is called *separable* in G, if there is a $T \in T(G)$ which separates U in G. A subgraph $F \subseteq G$ is called a *fragment of G*, if there is a $T \in T(G)$ such that F is the union of at least one, but not all components of $G - T$. If F is a fragment of G, then

$$\overline{F}^G := G - (V(F) \cup N(F; G))$$

is a fragment of G, as well (we suppress the superscript, if no confusion seems possible). A vertex $x \in G$ is called *critical* in G, if $\mu(G - x) < \mu(G)$, and $Cr(G)$ denotes the set of critical vertices of G. If $Cr(G) \neq \emptyset$, then $\mu(G)$ is finite, and if G is not complete, then $Cr(G) = \bigcup_{T \in T(G)} T$. A graph is called *critically n-connected*, if $Cr(G) = V(G)$ and $\mu(G) = n$.

We search in this paper for vertex sets S in an n-connected graph G such that $B := G - S$ is n-connected, with the vertices of S not just anywhere in G, but "close" to B in G. So we define for any cardinal number α : A subset $S \subseteq V(G)$ is *α-connectivity-preserving* (abbreviated *"α-preserving"*) in G, iff for all $S' \subseteq S$, $\mu(G - S') \geq \alpha$ holds. Of course, if S is α-preserving in G, then every $S' \subseteq S$ is α-preserving in G. It is easy to see that an $S \subseteq V(G)$ is α-preserving in G, iff $\mu(G - S) \geq \alpha$ and $d(s, V(G) - S; G) \geq \alpha$ for all $s \in S$.

In particular, an independent $S \subseteq V(G)$ is α-preserving in an α-connected graph G, iff $\mu(G - S) \geq \alpha$ holds.

We are now able to state precisely the results we shall prove.

Theorem 1 *If G is an $(n + 1)$-connected graph with $|Cr(G)| \geq \aleph_0$, then there exists an independent, n-preserving set S in G with $|S| = |Cr(G)|$. In particular, an infinite critically $(n + 1)$-connected graph G contains an independent, n-preserving set S with $|S| = |G|$.*

Theorem 2 *Every $(n + 1)$-connected, infinite graph G contains an n-preserving, infinite set S. If there is an infinite, independent set in G, then S can be chosen independent, too.*

By Ramsey's theorem (see, for instance, §20 in [1], §1,5 in [2] or Ch. 2.1 in [8]), every infinite graph contains an infinite, independent vertex set or an infinite, complete subgraph. Therefore, if G does not contain an infinite, independent set, then S in Theorem 2 may be chosen so that $G(S)$ is complete. It seems probable that an $(n+1)$-connected, infinite graph G always contains an n-preserving set S with $|S| = |G|$, but I can prove this only in some special cases.

For the proof of Theorems 1 and 2, we need a series of lemmas.

Lemma 1 *If $G - T$ is disconnected, then there is a $\overline{T} \subseteq T$ such that there are $C_1 \neq C_2$ in $\mathcal{C}(G - \overline{T})$ with $N(C_1; G) = N(C_2; G) = \overline{T}$.*

Proof: Choose any $C \in \mathcal{C}(G - T)$ and consider any $C_1 \in \mathcal{C}(G - N(C; G))$ different from C. Then $\overline{T} := N(C_1; G) \subseteq N(C; G) \subseteq T$ and C_1 and the $C_2 \in \mathcal{C}(G - N(C_1; G))$ with $C_2 \supseteq C$ have the property desired.   □

Lemma 2 *If $\mu(G) \geq n$ and there is an infinite n-preserving set S in $G - X$ for an $X \subseteq V(G)$ with $|S| > |X|$, then there is also an n-preserving $S_0 \subseteq S$ in G with $|S_0| = |S|$.*

Proof: For every $x \in X$, choose an $x, (G-X)$-fan F_x of degree n in G. Then $|H| < |S|$ for $H := \bigcup_{x \in X} V(F_x)$ and hence $|S_0| = |S|$ for $S_0 := S - H$. Since $G - (X \cup S_0)$ is n-connected by assumption, $G - S_0$ is n-connected, as well, by construction. Since for all $s \in S$, $d(s, G - (X \cup S); G) \geq n$, S_0 is n-preserving in G.   □

Let G be an infinite, $(n+1)$-connected graph. We shall see later that we have finished, if there is a finite $X \subseteq V(G)$ such that $\mu(G-X) \geq n+1$ and $Cr(G-X)$ is infinite. So let us suppose for the moment that for every finite $X \subseteq V(G)$ with $\mu(G-X) \geq n+1$, $Cr(G-X)$ is finite. Let any infinite $A_0 \subseteq V(G)$ be given. Since $Cr(G)$ is finite, there is an $a_0 \in A_0$ with $\mu(G - a_0) \geq n+1$. Consider $N_0' \subseteq N(a_0; G)$ with $|N_0'| = n+1$. For notational convenience, choose $N_0 \subseteq V(G - a_0)$ with $N_0 \supseteq N_0'$ and $|N_0| = n+2$. For all $x \neq y$ from N_0, choose a system $\mathcal{P}_{x,y}$ of $n+1$ openly disjoint x, y-paths in $G - a_0$ and set $H_0 := \bigcup_{\{x,y\} \in \binom{N_0}{2}} \bigcup_{P \in \mathcal{P}_{x,y}} V(P)$. Then $H_0 \supseteq N_0$ is finite. Since $Cr(G - a_0)$ is finite, there is an $a_1 \in A_0 - (H_0 \cup \{a_0\})$ such that $\mu(G - \{a_0, a_1\}) \geq n+1$. Since $\mu(G - a_0) \geq n+1$, there is an $N_1 \subseteq N(a_1; G - a_0)$ with $|N_1| = n+1$. For all $x \neq y$ from $N_1 \cup H_0$, choose a system $\mathcal{P}_{x,y}$ of $n+1$ openly disjoint x, y-paths in $G - \{a_0, a_1\}$ and set $H_1 := \bigcup_{\{x,y\} \in \binom{N_1 \cup H_0}{2}} \bigcup_{P \in \mathcal{P}_{x,y}} V(P)$. Continuing in this way, we construct sequences $a_0, a_1, a_2, \ldots$ of distinct elements of A_0 and $H_0, H_1, H_2, \ldots$ such that $A := \{a_i : i \in N\}$ and $B := G(\bigcup_{i \in N} H_i)$ are disjoint, $d(a_i, B; G) \geq n+1$ for all $i \in N$, and $\mu(H_i; G(H_{i+1})) \geq n+1$ for all $i \in N$. But then A is $(n+1)$-preserving in $G(A \cup B)$.

So we fix the following *notation for Lemmas 3 and 6*: G is an infinite, $(n+1)$-connected graph, B is an induced subgraph of G with $\mu(B) \geq n+1$ or $|B| = 0$, and $A \subseteq V(G)$ is disjoint from B such that for all $a \in A$, $d(a, B; G) \geq n+1$. This means $A = \emptyset$, if $|B| = 0$, and $A(n+1)$-preserving in $G(A \cup B)$, if $|B| > 0$.

Lemma 3 *If $\mu(G - A) \in N$ and if there are $\aleph_\alpha$ disjoint fragments F of $G - A$ with $F \cap B = \emptyset$, then there is an n-preserving, independent set $S \subseteq V(G) - (A \cup V(B))$ in G with $|S| = \aleph_\alpha$.*

Proof: Let F_λ $(\lambda < \omega_\alpha)$ be a system of disjoint fragments of $G - A$ such that $F_\lambda \cap V(B) = \emptyset$ for all $\lambda < \omega_\alpha$. Choose any $s_\lambda \in F_\lambda$ for every $\lambda < \omega_\alpha$. There is a finite $H_\lambda \subseteq V(G - s_\lambda)$ such that $H_\lambda \supseteq N_\lambda := N(F_\lambda; G - A)$ and $\mu(N_\lambda \cup A \cup B; G(H_\lambda \cup A \cup B)) \geq n$. We get such an H_λ (for $|N_\lambda \cup B| \geq 2$), if in the case $|B| = 0$, we take n openly disjoint

x, y-paths in $G - s_\lambda$ for all $x \neq y$ in N_λ and in the case $|B| \neq 0$, we take an x, B-fan of degree n in $G - s_\lambda$ for every $x \in N_\lambda$.   □

We define a strictly increasing ω_α-sequence κ_λ ($\lambda < \omega_\alpha$) of ordinals $\kappa < \omega_\alpha$ and a monotone ω_α-sequence $\overline{H}_0 \subseteq \overline{H}_1 \subseteq \ldots$ of subsets $\overline{H}_\lambda \subseteq V(G)$ with $|\overline{H}_\lambda| \leq |\lambda|$ or $\overline{H}_\lambda$ finite in the following way: Define $\kappa_0 := 0$ and $\overline{H}_0 := H_0$. Suppose κ_λ and $\overline{H}_\lambda$ are defined for all $\lambda < \nu < \omega_\alpha$. Then $H'_\nu := \bigcup_{\lambda < \nu} \overline{H}_\lambda$ is finite or of cardinality at most $|\nu| < \aleph_\alpha$. Hence there is a least κ_ν such that $F_{\kappa_\nu} \cap (H'_\nu \cup \{s_{\kappa_\lambda} : \lambda < \nu\}) = \emptyset$. Define $\overline{H}_\nu := H'_\nu \cup H_{\kappa_\nu}$.

Since $\overline{H}_\lambda \subseteq \overline{H}_\nu$ for $\lambda < \nu < \omega_\alpha$ and $s_\lambda \in F_\lambda$, the ω_α-sequence κ_λ is strictly increasing. Hence $S := \{s_{\kappa_\lambda} : \lambda < \omega_\alpha\}$ has cardinality $\aleph_\alpha$ and $S \cap (A \cup V(B)) = \emptyset$. Furthermore, S is independent, because $\{s_{\kappa_\lambda} : \nu < \lambda < \omega_\alpha\} \cap (F_{\kappa_\nu} \cup N(F_{\kappa_\nu}; G)) = \emptyset$ for every $\nu < \omega_\alpha$. So it remains only to show that S is n-preserving in G. Let us assume that there is an $S' \subseteq S$ such that $\mu(G - S') < n$. Since $|G - S'| \geq d(s_0; G) > n$, there is a $T' \in T(G - S')$. Then $T := T' \cup S'$ separates G and by Lemma 1, there is a $\overline{T} \subseteq T$ such that there are $C_1 \neq C_2$ in $C(G - \overline{T})$ with $N(C_1; G) = N(C_2; G) = \overline{T}$. Of course, $|\overline{T}| \geq n + 1$ and hence $|\overline{T} \cap S'| \geq 2$, since $|T'| < n$. Set

$$\lambda' := \min\{\lambda < \omega_\alpha : s_{\kappa_\lambda} \in \overline{T} \cap S'\}.$$

Since $N(F_{\kappa_{\lambda'}}; G)$ separates $s_{\kappa_{\lambda'}}$ from all other elements of $\overline{T} \cap S'$, we obtain $C_i \cap N(F_{\kappa_{\lambda'}}; G) \neq \emptyset$ for $i = 1, 2$. Therefore, since $\mu(N(F_{\kappa_{\lambda'}}; G); G(H_{\kappa_{\lambda'}} \cup A \cup B)) \geq n$,

$$|\overline{T} \cap (H_{\kappa_{\lambda'}} \cup A \cup V(B))| \geq n.$$

But this contradicts $|T'| < n$, because $\overline{T} \cap S' \cap (H_{\kappa_{\lambda'}} \cup A \cup B) = \emptyset$, by the definition of $H_{\kappa_{\lambda'}}$ and the choice of s_{κ_λ} for $\lambda > \lambda_0$.

$T_1, T_2 \in T(G)$ are called *parallel* (in G), if there are $C_1 \in C(G - T_1)$ and $C_2 \in C(G - T_2)$ such that $C_1 \subseteq \overline{C}_2$. Since then $\overline{C}_2 \subseteq \overline{C}_1$, this concept is symmetric. It is easy to see that $T_1, T_2 \in T(G)$ are parallel, iff there is at most one $C \in C(G - T_1)$ with $C \cap T_2 \neq \emptyset$. Therefore, if $T_1, T_2 \in T(G)$ are parallel and different, there is exactly one $C \in C(G - T_1)$ containing vertices of T_2 and all other components of $G - T_1$ are contained in the same component of $G - T_2$. If C_i denotes this component of $G - T_i$ containing vertices of T_{i+1} ($i \mod 2$), then all elements of $C(G - T_i) - \{C_i\}$ are contained in C_{i+1}, hence the elements of $(C(G - T_1) \cup C(G - T_2)) - \{C_1, C_2\}$ are disjoint.

Lemma 4 *If $Cr(G)$ is uncountable, then there is a system of $|Cr(G)|$ disjoint fragments.*

Proof: By Zorn's lemma, there is a maximal $\mathcal{P} \subseteq T(G)$ such that every two elements of $\mathcal{P}$ are parallel. For every $T \in T(G)$, we can find a finite $V_T \subseteq V(G)$ such that $\mu(T; G(V_T)) \geq \mu(G) := n$. Suppose, there is a $T_0 \in T(G)$ with $T_0 \not\subseteq \bigcup_{T \in \mathcal{P}} V_T$. Then for every $T \in \mathcal{P}$, there is at most one $C \in C(G - T)$ with $C \cap T_0 \neq \emptyset$, because $|V_T \cap T_0| < n \leq \mu(T; G(V_T))$. Hence T_0 is parallel to all $T \in \mathcal{P}$, contradicting the maximality of $\mathcal{P}$. So for every $T_0 \in T(G)$, $T_0 \subseteq \bigcup_{T \in \mathcal{P}} V_T$ and hence $Cr(G) \subseteq \bigcup_{T \in \mathcal{P}} V_T$ holds. This implies

$$|Cr(G)| \leq |\bigcup_{T \in \mathcal{P}} V_T| = |\mathcal{P}| \leq |Cr(G)|.$$

□

Choose any $T_0 \in \mathcal{P}$. For $T \in \mathcal{P} - \{T_0\}$, C_T denotes the uniquely determined $C \in C(G - T)$ with $C \cap T_0 \neq \emptyset$, and let C_{T_0} be the empty graph. We define an order $\leq$ for $\mathcal{P}$ in the following way: for $T_1, T_2 \in \mathcal{P}$, $T_1 \leq T_2$ iff $C_{T_1} \subseteq C_{T_2}$. This order has the least element T_0.

Consider any interval $[T_0, T_1] := \{T \in \mathcal{P} : T_0 \leq T \leq T_1\}$. There are n disjoint T_0, T_1-paths $P_1, \ldots, P_n$ in G. For $T \in [T_0, T_1]$, $C_T \subseteq C_{T_1}$ and $T_0 \subseteq V(C_T) \cup T$ hold. So $|P_i \cap T| = 1$ for all $i = 1, \ldots, n$. Hence $T \subseteq \cup_{i=1}^{n} V(P_i)$ holds and $[T_0, T_1]$ is finite. For $T' \in [T_0, T_1] - \{T\}$, T' is contained completely in the T_0, T-subpaths of $P_1, \ldots, P_n$ or completely in the T, T_1-subpaths of $P_1, \ldots, P_n$, since there is only one component of $G - T$ containing vertices of T'. But then $C_{T'} \subseteq C_T$ or $C_T \subseteq C_{T'}$. Hence $[T_0, T_1]$ is a chain. If we define $E := \{(T_1, T_2) : |[T_1, T_2]| = 2\}$, then $B := (\mathcal{P}, E)$ is a T_0-branching, giving the order $\leq$.

If T_1, T_2 are incomparable in $\leq$, then $C_{T_1} \not\subseteq C_{T_2}$ and $C_{T_2} \not\subseteq C_{T_1}$. Since T_1 and T_2 are parallel, this implies $C_{T_1} \cap T_2 \neq \emptyset$ and $C_{T_2} \cap T_1 \neq \emptyset$. Then $C_1 \cap C_2 = \emptyset$ for all $C_i \in \mathcal{C}(G - T_i) - \{C_{T_i}\}$. Therefore, an antichain $\mathcal{A}$ of $(\mathcal{P}, \leq)$ provides a system of $|\mathcal{A}|$ disjoint fragments. But for $|Cr(G)| > \aleph_0$, $(\mathcal{P}, \leq)$ has an antichain $\mathcal{A}$ of cardinality $|Cr(G)|$. For this, consider $\mathcal{S}_i := \{T \in \mathcal{P} : \text{the } T_0, T\text{-path in } B \text{ has length } i\}$ for $i \in N$. Since $\mathcal{S}_i$ is an antichain in $(\mathcal{P}, \leq)$, we may assume $|\mathcal{S}_i| < |Cr(G)| = |\mathcal{P}|$ for all $i \in N$. Then there is a strictly increasing sequence $i_k \in N$ such that $|\mathcal{S}_{i_k}| > |\mathcal{S}_j|$ for all $j < i_k$ and $|\mathcal{S}_{i_0}| \geq \aleph_0$, since $|\mathcal{P}| > \aleph_0$. Then $|\mathcal{S}_{i_0}| < |\mathcal{S}_{i_1}| < \ldots$ and $\sup_{k \in N} |\mathcal{S}_{i_k}| = |\mathcal{P}|$. For every $k \in N$, there is an $S_k \in \mathcal{S}_{i_{k+1}-1}$ with $d^+(S_k; B) > |\mathcal{S}_{i_k}|$. By Ramsey's theorem, we get an infinite subsequence $S_{k_0}, S_{k_1}, \ldots$ of $S_k (k \in N)$ so that $S_{k_0}, S_{k_1}, \ldots$ is an antichain or $S_{k_0} \leq S_{k_1} \leq \ldots$. In the former case, $\cup_{\lambda=0}^{\infty} N^+(S_{k_\lambda}; B)$ is an antichain of cardinality $|Cr(G)|$. In the latter case, the same holds for $\cup_{\lambda=0}^{\infty} (N^+(S_{k_\lambda}; B) - \{V_{k_\lambda}\})$, where V_{k_λ} is the second vertex on the path from S_{k_λ} to $S_{k_{\lambda+1}}$ in B.

Note that Lemma 4 is not true for denumerable $Cr(G)$. So we need additional considerations for this case.

Lemma 5 *If there is an infinite $U \subseteq Cr(G)$ which is not separable in G, then there is an infinite number of disjoint fragments in G.*

Proof: For $u \in U$, choose a $T_u \in \mathcal{T}(G)$ with $u \in T_u$ such that $|T_u \cap U| = \max\{|T \cap U| : T \in \mathcal{T}(G) \wedge u \in T\}$. Since U is not separated by T_u, there is a $C_u \in \mathcal{C}(G - T_u)$ with $C_u \cap U = \emptyset$. Suppose $C_u \cap C_{u'} \neq \emptyset$, say, $x \in C_u \cap C_{u'}$. But $\overline{C}_u \cap \overline{C}_{u'} \neq \emptyset$ holds, as well, since $U \subseteq (\overline{C}_u \cup T_u) \cap (\overline{C}_{u'} \cup T_{u'})$ and U is infinite; say, $u_0 \in \overline{C}_u \cap \overline{C}_{u'} \cap U$. Then $T_0 := \{t \in T_u \cup T_{u'} : \text{there is a } u_0, t\text{-path } P \text{ with } V(P) \cap (T_u \cup T_{u'}) = \{t\}\}$ belongs to $\mathcal{T}(G)$ and separates x and u_0 (see [3], or Lemma 0 in [6]). Since U is not separable, $U \cap (T_u \cup T_{u'}) \subseteq T_0$ holds. Therefore, by the choice of T_u and $T_{u'}$, $T_u \cap U = T_0 \cap U = T_{u'} \cap U$. Hence $T_u \cap U \neq T_{u'} \cap U$ implies $C_u \cap C_{u'} = \emptyset$, proving the lemma. □

Lemma 6 *If there is an infinite $U \subseteq V(G) - (A \cup V(B))$ such that every infinite subset of U is separable in $G - A$, then there is an infinite, independent $S \subseteq U$ which is n-preserving in G.*

(The properties of G, A, B are described in the paragraph before Lemma 3.)

Proof: We construct a sequence $s_0, s_1, \ldots$ of distinct elements of U which is independent and n-preserving in G. □

Since U is separable in $G - A$, there is a $T_0 \in \mathcal{T}(G - A)$ separating U. Hence there is a $C_0 \in \mathcal{C}(G - (A \cup T_0))$ such that $C_0 \cap U \neq \emptyset$ and $U_0' := \overline{C}_0^{G-A} \cap U$ is infinite. Choose any

$s_0 \in C_0 \cap U$. As in the proof of Lemma 3, there is a finite $H_0 \supseteq T_0$ in $G - s_0$ such that $\mu(T_0 \cup A; G(H_0 \cup A \cup B)) \geq n$. Then $U_0 := U_0' - H_0$ is infinite.

Suppose, T_k, C_k, s_k and an infinite $U_k \subseteq U_{k-1} \subseteq U$ are defined. Then there are a $T_{k+1} \in \mathcal{T}(G-A)$ separating U_k and a $C_{k+1} \in \mathcal{C}(G-(A \cup T_{k+1}))$ such that $C_{k+1} \cap U_k \neq \emptyset$ and $U_{k+1}' := \overline{C}_{k+1} \cap U_k$ is infinite. Choose $s_{k+1} \in C_{k+1} \cap U_k$. Again, there is a finite $H_{k+1} \supseteq T_{k+1}$ in $G - s_{k+1}$ such that $\mu(T_{k+1} \cup A; G(H_{k+1} \cup A \cup B)) \geq n$. Then $U_{k+1} := U_{k+1}' - H_{k+1}$ is infinite.

Set $S := \{s_i : i \in N\}$. By construction, $s_k \in C_k$ and $\{s_i : i > k\} \subseteq V(\overline{C}_k)$. So s_k is separated from every s_i with $i > k$ by T_k in $G - A$. In particular, $S \subseteq U$ is infinite and independent. We have still to show that S is n-preserving in G. Suppose, there is an $S' \subseteq S$ such that $\mu(G - S') < n$. There is a $T' \in \mathcal{T}(G - S')$. By Lemma 1, there are a $T \subseteq T' \cup S'$ and distinct $C_1, C_2 \in \mathcal{C}(G - T)$ with $N(C_i; G) = T$ for $i = 1, 2$. Since $|T| > n$, $|T \cap S'| \geq 2$ holds. Consider

$$k_0 := \min\{k \in N : s_k \in T \cap S'\}.$$

Since $T_{k_0} \cup A$ separates s_{k_0} from $T \cap S' - \{s_{k_0}\} \neq \emptyset$ in G, we get $C_i \cap (T_{k_0} \cup A) \neq \emptyset$ for $i = 1, 2$. Hence T separates $T_{k_0} \cup A$ in G and so

$$|T \cap (A \cup B \cup H_{k_0})| \geq n,$$

since $\mu(T_{k_0} \cup A; G(H_{k_0} \cup A \cup B)) \geq n$. But, by construction, $\{s_k : k \geq k_0\} \cap H_{k_0} = \emptyset$ holds. So $T \cap (A \cup B \cup H_{k_0}) \subseteq T'$, contradicting $|T'| < n$.

Applying Lemmas 2 to 6, it is easy to prove Theorems 1 and 2.

Proof of Theorem 1: We may assume $\mu(G) = n + 1$. We apply Lemmas 3 and 6 for $A = V(B) = \emptyset$. If $Cr(G)$ is uncountable, then by Lemma 4, there is a system of $|Cr(G)|$ disjoint fragments of G, which, by Lemma 3, provides an n-preserving, independent set S in G with $|S| = |Cr(G)|$. So let us assume $|Cr(G)| = \aleph_0$. If there is an infinite $U \subseteq Cr(G)$ which is not separable in G, then by Lemmas 5 and 3, we get the desired infinite set S. So we may assume that every infinite $U \subseteq Cr(G)$ is separable in G. But then Lemma 6 completes the proof. □

Proof of Theorem 2: First, let us suppose that there is a finite $X \subseteq V(G)$ such that $\mu(G - X) > n$ and $|Cr(G - X)| \geq \aleph_0$. Then by Theorem 1, there is an independent, infinite $S \subseteq V(G - X)$ which is n-preserving in $G - X$, and this S, by Lemma 2, contains an infinite subset S_0 which is n-preserving in G and, of course, independent. □

So we may assume that for all finite $X \subseteq V(G)$ with $\mu(G - X) > n$, $Cr(G - X)$ is finite. Let A_0 be an infinite subset of $V(G)$, which we choose independent, if this is possible. Now we apply the procedure described after Lemma 2 and find an infinite $A \subseteq A_0$ and an induced subgraph $B \subseteq G - A$ with the properties stated there.

If $\mu(G - A) \geq n$, then A has the properties desired. So we may assume $\mu(G - A) < n$. If $G - A$ has an infinite number of disjoint fragments, then we get an infinite, independent, n-preserving S in G by Lemma 3, because only a finite number of these disjoint fragments F can have $F \cap B \neq \emptyset$, since $\mu(B) > n$. So we assume that there is no infinite system of disjoint fragments in $G - A$. Then for every $T \in \mathcal{T}(G - A)$, $\mathcal{C}(G - (A \cup T))$ is finite. Let us first consider the case that $Cr(G - A)$ is infinite. Then by Lemma 5, every infinite $U \subseteq Cr(G - A)$ is separable in $G - A$. But then $Cr(G - A) \cap B$ is finite, since B is not separable in $G - A$, and Lemma 6 implies the existence of an independent, infinite,

n-preserving S in G. So only the case remains that $Cr(G - A)$ is finite, which means that $T(G - A)$ is finite. Consider any $T \in T(G - A)$. Since $G - T$ is connected, for every $C \in C(G - (A \cup T))$, there is an $a_c \in N(C; G) \cap A$. Since $C(G - (A \cup T))$ is finite, $A_T := \{a_c : C \in C(G - (A \cup T))\}$ is finite, as well. Since $\mu(G(A_T \cup B)) \geq n + 1$ and $|T| < n$, $G - ((A - A_T) \cup T)$ is connected. Hence $\mu(G - A_1) > \mu(G - A)$, where $A_1 := A - \underset{T \in T(G-A)}{\overset{\cup A_T}{}}$ is infinite.

Considering A_1 instead of A, we can apply the same arguments as in the preceding paragraph. Continuing in this way, in at most $n - \mu(G - A)$ many steps, we get an infinite, n-preserving subset of A or an infinite, independent, n-preserving set.

As remarked before, it is not known, if an infinite, $(n + 1)$-connected graph G must contain an n-preserving S with $|S| = |G|$. But this is true, for instance, for locally finite graphs. Following the lines of the proof of Lemma 3, it is possible to obtain a more general result: *Let G be an infinite, $(n + 1)$-connected graph. Assume that there is a system of $|G|$ disjoint subgraphs C_λ of G such that for regular $|G|$, $|N(C_\lambda; G)| < |G|$ for all λ and for singular $|G|$, there is an $\aleph_\alpha < |G|$ such that $|N(C_\lambda; G)| \leq \aleph_\alpha$ holds for all λ. Then there is an n-preserving S in G with $|S| = |G|$.*

It is also true for $n = 1$.

Proposition 1 *Every infinite, 2-connected graph G has a 1-preserving set S with $|S| = |G|$.*

Proof: Choose any $z \in V(G)$ and consider $R_i := \{x \in V(G) : \rho(x, z) = i\}$ for $i \in N$, where $\rho(x, y)$ denotes the distance of x and y in G. We distinguish two cases.

1. *There is an $i_0 \in N$ with $|R_{i_0}| = |G|$.*
Hence $i_0 > 0$ and $C_0 := G(R_0 \cup \ldots \cup R_{i_0-1})$ is not empty.
(a) $|C(G - R_{i_0})| < |R_{i_0}|$.
For every component $C \neq C_0$ of $G - R_{i_0}$, choose an $r_c \in N(C; G)$. Then $S := R_{i_0} - \{r_c : C \in C(G - R_{i_0}) \wedge C \neq C_0\}$ is 1-preserving in G with $|S| = |G|$.
(b) $|C(G - R_{i_0})| = |R_{i_0}|$.
For every component $C \neq C_0$ of $G - R_{i_0}$, choose an $s_c \in C \cap R_{i_0+1} \neq \emptyset$. Then $S := \{s_c : C \in C(G - R_{i_0}) \wedge C \neq C_0\}$ is 1-preserving in G with $|S| = |G|$.

2. *For all $i \in N$, $|R_i| < |G|$.*
By Theorem 2, we may assume $|G| > \aleph_0$. Then there is a sequence $i_0 < i_1 < \ldots$ so that for all $k \in N$, $|R_{i_k}| > |R_i|$ holds for all $i < i_k$. Hence $\underset{k \in N}{\sup} |R_{i_k}| = |G|$. In addition, we may assume $i_{k+1} \geq i_k + 3$ for all $k \in N$ and $|R_{i_0}| \geq \aleph_0$. We define, by induction, 1-preserving sets S_k with $|S_k| \geq |R_{i_k}|$ such that also $\cup S_k$ is 1-preserving in G. For the definition of S_0, we distinguish again two cases.
(a) $|C(G - R_{i_0})| < |R_{i_0}|$.
Define S_0 as S in 1(a). For $x \in R_{i_0+1}$, there is an x, R_{i_0}-path P_x in $G - S_0$. For $V_0 := \underset{x \in R_{i_0+1}}{\cup} V(P_x)$, $|V_0| \leq \max\{\aleph_0, |R_{i_0+1}|\} < |R_{i_1}|$ holds.
(b) $|C(G - R_{i_0})| \geq |R_{i_0}|$.
Define S_0 as S in 1(b). For $x \in R_{i_0+2}$, there is an x, R_{i_0+1}-path P_x in $G - S_0$. For $V_0 := \underset{x \in R_{i_0+2}}{\cup} V(P_x)$, $|V_0| \leq \max\{\aleph_0, |R_{i_0+2}|\} < |R_{i_1}|$ holds.
Suppose now that for any $k \geq 1$, $S_\kappa \subseteq R_{i_\kappa} \cup R_{i_\kappa+1}$ for all $\kappa < k$ and $V_0 \subseteq V_1 \subseteq \ldots \subseteq V_{k-1} \subseteq V(G)$ with $|V_{k-1}| < |R_{i_k}|$ are defined.
(a) $|C(G - R_{i_k})| < |R_{i_k}|$.

Define S'_k as S in 1(a) and set $S_k := S'_k - V_{k-1}$. Then $|S_k| = |R_{i_k}|$ holds. For $x \in R_{i_k+1}$, there is an x, R_{i_k}-path P_x in $G - S_k$. For $V_k := V_{k-1} \cup \underset{x \in R_{i_k+1}}{\cup} V(P_x)$, $|V_k| \le |R_{i_k}| + |R_{i_k+1}| < |R_{i_k+1}|$ holds.

(b) $|\mathcal{C}(G - R_{i_k})| \ge |R_{i_k}|$.

Define S'_k as S in 1(b) and set $S_k := S'_k - V_{k-1}$. Then $|S_k| \ge |R_{i_k}|$ holds. For $x \in R_{i_k+2}$, there is an x, R_{i_k+1}-path P_x in $G - S_k$. For $V_k := V_{k-1} \cup \underset{x \in R_{i_k+2}}{\cup} V(P_x)$, $|V_k| \le |R_{i_k}| + |R_{i_k+2}| < |R_{i_k+1}|$ holds.

It is easy to see that $S := \underset{k \in N}{\cup} S_k$ is 1-preserving in G with $|S| = |G|$. □

Of course, there are analogous problems for infinite connectivity. *Does every $\aleph_\alpha$-connected graph G contain an $\aleph_\alpha$-preserving set S of cardinality $\aleph_\alpha$ or perhaps even of cardinality $|G|$?* For $|G| = \mu(G)$ this is true.

Proposition 2 *If $|G| = \mu(G) = \aleph_\alpha$, then there is an $\aleph_\alpha$-preserving S in G of cardinality $|G|$.*

Proof: We order $V(G)$ in an ω_α-sequence v_λ ($\lambda < \omega_\alpha$). For all $\lambda < \omega_\alpha$, we define by transfinite induction an $s_\lambda \in V(G)$, an $H_\lambda \subseteq V(G)$ with $|H_\lambda| \le |\lambda|$ or $|H_\lambda| < \aleph_0$ and for all $p := \{x,y\} \in \binom{H_\lambda}{2}$, an x, y-path P_p^λ. In addition, the ω_α-sequence H_λ is increasing and $H_\lambda \cap \{s_{\lambda'} : \lambda' \le \lambda\} = \emptyset$ holds. Set $s_0 := v_0$ and $H_0 := \{v_{i_0}\}$ for $i_0 := \min\{i : v_i \in N(v_0; G)\}$. Consider λ_0 with $0 < \lambda_0 < \omega_\alpha$ and suppose that for all $\lambda < \lambda_0$, s_λ, H_λ, and P_p^λ for all $p \in \binom{H_\lambda}{2}$ are defined. Set

$$S_{\lambda_0} := \{s_\lambda : \lambda < \lambda_0\}$$

and $H'_{\lambda_0} := \underset{\lambda < \lambda_0}{\cup} (H_\lambda \cup \underset{p \in \binom{H_\lambda}{2}}{\cup} V(P_p^\lambda))$. Then $|S_{\lambda_0}| \le |\lambda_0| < \aleph_\alpha$ and $|H'_{\lambda_0}| \le |\lambda_0| < \aleph_\alpha$ or $|H'_{\lambda_0}| < \aleph_0$ hold. Hence $R := V(G) - (S_{\lambda_0} \cup H'_{\lambda_0}) \ne \emptyset$ and we set $s_{\lambda_0} := v_{i_{\lambda_0}}$ for $i_{\lambda_0} := \min\{i : v_i \in R\}$. Since $d(s_\lambda; G) = \aleph_\alpha$, there is a (first) $b_\lambda \in N(s_\lambda; G) \cap R - \{s_{\lambda_0}\}$ for all $\lambda \le \lambda_0$. Then $H_{\lambda_0} := H'_{\lambda_0} \cup \{b_\lambda : \lambda \le \lambda_0\}$ is finite or of cardinality at most $|\lambda_0|$. For every $p := \{x,y\} \in \binom{H_{\lambda_0}}{2}$, set $\lambda_p := \min\{\lambda \le \lambda_0 : p \subseteq H_\lambda\}$. Since

$$T_p := S_{\lambda_0} \cup \{s_{\lambda_0}\} \cup \underset{\lambda_p \le \lambda < \lambda_0}{\cup} V(P_p^\lambda) - \{x,y\}$$

is of cardinality less than $\aleph_\alpha$ and G is $\aleph_\alpha$-connected, there is an x, y-path $P_p^{\lambda_0}$ in $G - T_p$ with $|P_p^{\lambda_0}| \ge 3$. □

So we have defined by transfinite induction a partition of $V(G)$ into $S := \{s_\lambda : \lambda < \omega_\alpha\}$ and $H := \underset{\lambda < \omega_\alpha}{\cup} H_\lambda$, where $|S| = \aleph_\alpha$. Consider $p := \{x,y\} \in \binom{H}{2}$ and $\lambda_p := \min\{\lambda : p \subseteq H_\lambda\}$. Then $G(H)$ contains the system of openly disjoint x, y-paths P_p^λ ($\lambda_p \le \lambda < \omega_\alpha$). Hence $\mu(G(H)) = \aleph_\alpha$. Since $N(s_{\lambda_0}; G) \cap (H_\lambda - \underset{\lambda' < \lambda}{\cup} H_{\lambda'}) \ne \emptyset$ for all $\lambda_0 \le \lambda < \omega_\alpha$, $d(s_{\lambda_0}, H; G) = \aleph_\alpha$ holds, and S is $\aleph_\alpha$-preserving in G.

I do not know, if in the case $|G| > \mu(G) = \aleph_\alpha$, G has an $\aleph_\alpha$-preserving subset of cardinality $\aleph_\alpha$. This is equivalent to the following conjecture:

C_α: *There does not exist an $\aleph_\alpha$-connected graph G such that for all $S \subseteq V(G)$ of cardinality $\aleph_\alpha$, $\mu(G - S) < \aleph_\alpha$ holds.*

The non-trivial implication of this equivalence, namely that C_α implies the existence of an $\aleph_\alpha$-preserving set of cardinality $\aleph_\alpha$, is shown by transfinite induction in a similar way as in the preceding proof. (If $B := G - V_0$ is $\aleph_\alpha$-connected for a $V_0 = \{v_\lambda : \lambda < \omega_\alpha\}$, then define b_λ only in the case $d(s_\lambda, B; G) < \aleph_\alpha$ and consider an x, B-path P_x^λ for x in H_λ instead of P_p^λ.)

For a proof of C_α, it would be enough to show that such a "critically $\aleph_\alpha$-connected" graph G has a system of disjoint subgraphs C_λ ($\lambda < \omega_{\alpha+1}$) with $|N(C_\lambda; G)| = \aleph_\alpha$ (this sufficiency follows as in Lemma 3). Perhaps it is possible to show the existence of such a system similar to Lemma 4.

It is somewhat surprising that for the directed case, no results corresponding to Theorems 1 and 2 are valid. Even worse, for every positive integer n and $\aleph_\beta$, n-critical[1] digraphs of connectivity number n and of order $\aleph_\beta$ exist, as the following example shows.

Let $V_0, V_1, \ldots$ be a sequence of disjoint sets of the cardinality $\aleph_\beta$. Since $\left|\binom{V_0}{n}\right| = |V_1|$, there is a bijection $f_1 : \binom{V_0}{n} \to V_1$. Define the directed graph $G_1 = (V_0 \cup V_1, E_1)$ by

$$E_1 := \{(x,y) : x \in V_1 \wedge y \in V_0\} \cup \bigcup_{M \in \binom{V_0}{n}} \{(x, f_1(M)) : x \in M\}.$$

Suppose that $G_i = (V_0 \cup V_1 \cup \ldots \cup V_i, E_i)$ is defined. Using a bijection $f_{i+1} : \binom{V_0 \cup \ldots \cup V_i}{n} \to V_{i+1}$, define $G_{i+1} = (V_0 \cup \ldots \cup V_i \cup V_{i+1}, E_{i+1})$ by

$$E_{i+1} := E_i \cup \{(x,y) : x \in V_{i+1} \wedge y \in V_0\} \cup \bigcup_{M \in \binom{V(G_i)}{n}} \{(x, f_{i+1}(M)) : x \in M\}.$$

It is easy to check that the digraph $G := \bigcup_{i=1}^{\infty} G_i$ is n-connected, but the deletion of any n vertices of G produces a digraph containing a vertex of indegree 0.

The same construction provides a counterexample for the conjecture C_α in the directed case. Let $\aleph_\beta$ be a cardinal such that $\aleph_\beta^{\aleph_\alpha} = \aleph_\beta$ (hence $\aleph_\beta \geq 2^{\aleph_\alpha} > \aleph_\alpha$). For example, $\aleph_\gamma := 2^{\aleph_\alpha}$ and then, by Hausdorff's formula (see, for instance, formula (6.18) in [9]), $\aleph_{\gamma+n}$ for all $n \in N$ have this property. Let cf $\aleph_\beta$ denote the cofinality of $\aleph_\beta$. Then $\aleph_\beta^{\aleph_\alpha} = \aleph_\beta$ implies cf $\aleph_\beta > \aleph_\alpha$, since $\aleph_\beta^{\mathrm{cf}\,\aleph_\beta} > \aleph_\beta$ (see, for instance, Theorem 9 in [9]). Let us consider an ω_β-sequence of disjoint sets $V_0, V_1, \ldots$, each of cardinality $\aleph_\beta$, and define digraphs G_λ ($0 < \lambda < \omega_\beta$) as above. This is possible, since for every such λ, there is a bijection $f_\lambda : \binom{\bigcup_{\kappa < \lambda} V_\kappa}{\aleph_\alpha} \to V_\lambda$. Then the digraph $G := \bigcup_{0 < \lambda < \omega_\beta} G_\lambda$ of order $\aleph_\beta$ is $\aleph_\alpha$-connected, but for every $S \subseteq V(G)$ of cardinality $\aleph_\alpha$, $G - S$ has a vertex of indegree 0, since cf $\aleph_\beta > \aleph_\alpha$.

Opposite to this construction, a digraph G of order $\aleph_\alpha$ and of connectivity $\aleph_\alpha$ must contain an "$\aleph_\alpha$-preserving" set S of cardinality $\aleph_\alpha$. This assertion is proved in a similar manner as Proposition 2. So there cannot be a "critically $\aleph_\alpha$-connected" digraph of order $\aleph_\alpha$. Whereas, in general, conjecture C_α is not true in the directed case, the problem remains to determine the cardinals $\aleph_\beta$ which can occur as the order of a "critically $\aleph_\alpha$-connected" digraph. One could conjecture the following statement:

$\vec{C}_\alpha$: There is an $\aleph_\alpha$-connected digraph G of order $\aleph_\beta$ such that for every $S \subseteq V(G)$ of cardinality $\aleph_\alpha$, the connectivity number of $G - S$ is less than $\aleph_\alpha$, iff $\aleph_\beta^{\aleph_\alpha} = \aleph_\beta$ holds.

Added in proof. In the meantime, C. Thomassen proved Conjecture C_α. But it remains open, if an $\aleph_\alpha$-connected graph G of order exceeding $\aleph_\alpha$ has an $\aleph_\alpha$-preserving set S such that $|S| > \aleph_\alpha$ or perhaps even $|S| = |G|$ holds.

---

[1] A digraph is called n-critical, if the deletion of any n vertices decreases the connectivity number by n.

References

[1] B. Bollobás: *Combinatorics*. Cambridge University Press, Cambridge 1986.

[2] R. L. Graham, B. L. Rothschild, J. H. Spencer: *Ramsey Theory*. John Wiley and Sons, New York 1980.

[3] R. Halin: Über trennende Eckenmengen in Graphen und den Mengerschen Satz. *Math. Ann.* **157**(1964), 34–41.

[4] R. Halin: *Graphentheorie I*. Wissenschaftliche Buchgesellschaft, Darmstadt 1980.

[5] W. Mader: Endlichkeitssätze für k-kritische Graphen. *Math. Ann.* **229**(1977), 143–153.

[6] W. Mader: Zur Struktur minimal n-fach zusammenhängender Graphen. *Abh. Math. Sem. Uni. Hamburg* **49**(1979), 49–69.

[7] W. Mader: On k-critically n-connected graphs. In Progress in Graph Theory, edited by J. A. Bondy and U. S. R. Murty, 1984 (Academic Press) 389–398.

[8] N. H. Williams: *Combinatorial Set Theory*. North-Holland Publishing Company, Amsterdam 1977.

[9] Th. Jech: *Set Theory*. Academic Press, New York 1978.

ORDERED GRAPHS WITHOUT INFINITE PATHS

E. C. MILNER*
Department of Mathematics and Statistics
University of Calgary
Calgary, Alberta, T2N 1N4
Canada

Abstract

Let $G = (V, E)$ be a graph on the well ordered set V which has no infinite path. If the order type of V, tpV, is a limit ordinal less than w_1^{w+2}, then there is a subset $V' \subseteq V$ having the order type (under the induced ordering) which is an independent set. If $tpV = w_1^{w+2}$ then this statement is independent of the axioms of set theory.

1 Introduction

In this paper, unless stated otherwise, lower case greek letters always denote ordinal numbers and, in particular, κ, μ denote infinite cardinals, i.e. initial ordinals or terms of the sequence $\omega, \omega_1, \omega_2, \ldots$. The symbol α^β when α is infinite will always denote ordinal exponentiation but, as usual, 2^ω denotes the cardinal of the continuum. If X, Y are subsets of an ordered set, then $X < Y$ means that $x < y$ holds for all $x \in X$ and $y \in Y$. For a set A and a cardinal λ, we define $[A]^\lambda = \{X \subseteq A : |X| = \lambda\}$, and $[A]^{<\lambda}$ and $[A]^{\leq\lambda}$ are similarly defined.

The partition symbol

$$(1.1) \qquad \alpha \to (\beta_0, \beta_1, \ldots, \beta_\rho, \ldots)^r_{\rho<\gamma},$$

where r is a positive integer, means that the following statement is true: *whenever A is a well ordered set of order type $tp(A) = a$ and $f : [A]^r \to \gamma$, then there are $\rho < \gamma$ and $B \subseteq A$ such that $tp(B) = \beta_\rho$ and $f(X) = \rho$ for all $X \in [B]^r$.* The negation of (1.1) is indicated by replacing the arrow $\to$ by a non-arrow $\not\to$.

Partition relations of the form (1.1) play an important role in infinitary combinatorics, and the book [4] gives a comprehensive survey for the case when the α and β_p are cardinal numbers. The most familiar relation of this kind is Ramsey's theorem [15] which states that $\omega \to (\omega)^r_k$ for finite r and k (this is an abbreviation for $\omega \to (\omega, \omega, \ldots, \omega)^r_k$). It is fairly well known that Ramsey's theorem does not immediately extend to higher cardinals. For example, Sierpinski [17] showed that $\omega_1 \not\to (\omega_1)^2_2$ (or, more precisely $2^\omega \not\to (\omega_1)^2_2$). There is, however, a very useful extension of Ramsey's theorem (for the case $r = 2$) due to Erdös, Dushnik and Miller [2] which states that, for any infinite cardinal κ,

$$\kappa \to (\kappa, \omega)^2.$$

*Research supported by NSERC grant #A5198.

G. Hahn et al. (eds.), Cycles and Rays, 161–180.
© 1990 by Kluwer Academic Publishers. Printed in the Netherlands.

In graph-theoretic language this means that for any graph on a set of size κ either there is an infinite complete subgraph or there is a "large" independent set, where "large" here means "of cardinality κ". The use of the term "large" is rather subjective. For example, in an ordered set of order type κ^2, a subset with order type κ may, in some sense, be considered to be rather small. So a natural question to ask is whether the stronger relation $\kappa^2 \rightarrow (\kappa^2, \omega)^2$ also holds It is false. In fact, the stronger negative relation $\kappa^2 \not\rightarrow (\kappa + 1, \omega)^2$ holds as is easily seen by considering the graph on the set $\{(\alpha, \beta) : \alpha, \beta < \kappa_1\}$ ordered alphabetically in which two points (α, β) and (α', β') are joined by an edge if and only if $\alpha < \alpha'$ and $\beta > \beta'$ (or $\alpha > \alpha'$ and $\beta < \beta'$). A similar argument shows that $\alpha \not\rightarrow (\kappa + 1, \omega)^2$ for any $\alpha < \kappa^+$, where κ^+ denotes the successor cardinal of κ.

The above example shows that for an ordered graph to have a "large" independent set (in the order type sense), it is not enough that there should be no infinite complete subgraphs. In [5] we considered this question for a class of much thinner graphs — those not containing an infinite path. We write

(1.2) $\alpha \rightarrow (\alpha, \text{ infinite path})^2$

to mean that: *whenever G is a graph on an ordered set of type α, then either G contains an infinite path or there is an independent set of type α.* In [5] we proved:

Theorem 1.1 $\alpha \rightarrow (\alpha, \text{ infinite path})^2$ *holds for limit* $\alpha < \omega_1^{\omega+2}$.

Obviously (1.2) is false if $\alpha = \beta + 1$ is a successor ordinal (consider the graph in which the last point is joined to all the rest), but the condition $\alpha < \omega_1^{\omega+2}$ seems to be rather curious. The method that we used to prove Theorem 1.1 depended upon a certain set-mapping result (Theorem 2.3), and this result applies only to ordered sets whose order type is strictly less than $\omega_1^{\omega+2}$. Despite this, in [5] we speculated that Theorem 1.1 might be true without this restriction on the size of α. Recently, much to our surprise, Larson [10] and Baumgartner and Larson [1] have shown that the relation

$$\omega_1^{\omega+2} \rightarrow (\omega_1^{\omega+2}, \text{ infinite path})^2$$

is independent of the axioms of set theory (see §7).

2   Set-Mappings

A set-mapping on a set S is a function $f : S \rightarrow P(S)$ such that $x \notin f(x)$ for every $x \in S$. A subset $A \subseteq S$ is f-*free* if $A \cap f[A] = \emptyset$. A very useful combinatorial tool is the following theorem of Hajnal [7].

Theorem 2.1 *If κ is an infinite cardinal and $\mu < \kappa$, and if f is a set-mapping on κ such that $|f(x)| < \mu(x \in \kappa)$, then there is an f-free subset of cardinality κ.*

Theorem 2.1 was conjectured by Ruziewicz [16] in 1935. Partial solutions were given by Lazar [9] (κ regular) and Piccard [14] ($\text{cf}(\kappa) = \omega$), and Erdös [3] proved the conjecture under the additional hypothesis that the generalized continuum hypothesis holds, finally Hajnal [7] proved it without this assumption. It is very easy to see that the condition $|f(x)| < \mu < \kappa$ cannot be replaced by the weaker condition $|f(x)| < \kappa$. For consider the set-mapping in which $f(x) = \{\alpha : \alpha < x\}$ $(x \in \kappa)$; in this case there is not even a free set consisting of two elements. However, in this example, although the cardinalities of the $f(x)$ are all less than κ, the order types of the $f(x)$ are not bounded below κ. Erdös and Specker [6] noticed the following easy extension of Theorem 2.1.

Theorem 2.2 *If f is a set mapping on the cardinal κ such that $tp(f(x)) < \alpha$, where $\alpha < \kappa$, then there is an f-free set of cardinality κ.*

Theorem 2.2 follows immediately from Theorem 2.1 in the case when κ is a singular cardinal since the hypothesis implies $|f(x)| < |\alpha|^{+} < \kappa$. Since we use it, we shall give the proof of Theorem 2.2 for the simple case when κ is a regular cardinal — this is essentially the same as Lazar's proof of Theorem 2.1 for the regular case.

Proof: Assume κ is regular. Suppose for a contradiction that there is no f-free subset of size κ. Recursively define subsets F_ν $(\nu < \kappa)$ so that F_ν is a maximal f-free subset of $S_\nu = \kappa \setminus \cup \{F_\xi \cup f[F_\xi] : \xi < \nu\}$. Since, by assumption, $|F_\xi \cup f[F_\xi]| < \kappa (\xi < \nu)$, it follows by the regularity of κ that $\cup\{F_\xi \cup f[F_\xi] : \xi < \nu\}$ is not cofinal in κ and so $F_\nu \neq \emptyset$. Also, by the construction $F_0 < F_1 < \dots$. Let $x \in F_\alpha$. Since F_ν is a maximal f-free subset in S_ν and $x \in S_\nu$, it follows that $f(x) \cap F_\nu \neq \emptyset$ $(\nu < \alpha)$. It follows that the order type $tp(f(x)) \geq \alpha$, and this is a contradiction. $\qquad\square$

In [5] we considered set-mappings on ordered sets when the order type is not necessarily a cardinal. For ordinals α, λ denote by $\mathrm{SM}(\alpha, \lambda)$ the following assertion: *If f is a set-mapping of order α on λ, i.e. if $tp(f(x)) < \alpha$ $(x \in \lambda)$, then there is an f-free subset of the full type λ.* Thus Theorem 2.2 says that $\mathrm{SM}(\alpha, \kappa)$ holds for κ an infinite cardinal and $\alpha < \kappa$. In [5] it is determined for which α, λ, $\mathrm{SM}(\alpha, \lambda)$ holds in the range $\alpha < \omega_1$ and $\lambda < \omega_2$. In §5 we will give the proof of the following result of [5] which will be needed for the proof of Theorem 1.1.

Theorem 2.3 $\mathrm{SM}(\omega, \omega_1, \gamma)$ *holds for* $\gamma < \omega_1^{\omega+2}$.

It is easy to see that $\mathrm{SM}(\omega, \lambda)$ is false if λ is not a multiple of ω_1, but the condition on the size of λ is also needed for $\mathrm{SM}(\omega, \lambda)$ to hold.

Theorem 2.4 $\mathrm{SM}(\omega, \lambda)$ *is false if* $\omega_1^{\omega+2} \leq \lambda < \omega_2$.

3 Some remarks on ordered sets

We shall first describe the structure of an ordered set in terms of certain subintervals (I is an *interval* if $(\forall x, y, z)(x \leq y \leq z \ \& \ x, z \in I \Rightarrow y \in I)$) which will provide us with a systematic method for constructing well ordered sets having a prescribed order type of the form $\omega_\alpha \gamma$ with $\gamma < \omega_{\alpha+1}$.

To begin, note that any ordinal γ, $1 < \gamma < \omega_{\alpha+1}$, has a representation as a sum

$$(3.1) \qquad \gamma = \Sigma(\xi < \chi)\omega_\alpha^{\rho_\xi},$$

in which $\chi \leq \omega_\alpha$ and each term $\omega_\alpha^{\rho_\xi}$ is strictly less than γ. We denote by $\chi(\gamma)$ the smallest value of χ for which there is such a representation for γ, and we call this *a standard representation* for γ.

Let S be a well ordered set of type $tp(S) = \omega_\alpha \gamma < \omega_{\alpha+1}$. We will define a sequence of subintervals of S, $I(S) = \langle I_\nu : \nu < \omega_\alpha \rangle$, and a regressive function $\varphi = \varphi(S) : \omega_\alpha \backslash \{0\} \to \omega_\alpha$ (i.e. $\varphi(\nu) < \nu$) so that the following conditions are satisfied:

$$\text{(3.2)} \qquad\qquad\qquad\qquad I_0 = S;$$

$$\text{(3.3)} \qquad\qquad\qquad\qquad tp(I_\nu) = \omega_\alpha^{\sigma_\nu}(1 \leq \nu < \omega_\alpha);$$

$$\text{(3.4)} \qquad\qquad\qquad\qquad S = \cup\{I_\nu : |I_\nu| = 1\};$$

$$\text{(3.5)} \qquad\qquad\qquad I_\nu \subset I_{\varphi(\nu)} \text{ and } \sigma_\nu < \sigma_{\varphi(\nu)} \quad (1 \leq \nu < \omega_\alpha);$$

$$\text{(3.6)} \qquad\qquad\qquad I_\nu \cap I_{\hat\rho} = \emptyset \text{ for } \varphi(\nu) < \rho < \nu;$$

$$\text{(3.7)} \qquad\qquad\qquad I_{\hat\rho} < I_\nu \text{ if } \rho < \nu \text{ and } \varphi(\rho) = \varphi(\nu);$$

$$\text{(3.8)} \qquad\qquad tp\{\rho : \varphi(\nu) < \rho < \nu; \ \varphi(\rho) = \varphi(\nu)\} < \chi(tp(I_{\varphi(\nu)})).$$

If $\gamma = 1$ this is obvious. Simply put $I_{1+\nu} = \{\nu\}$ and $\varphi(1 + \nu) = 0$ for $\nu < \omega_\alpha$. Now assume that $\gamma > 1$ and use induction. If (3.1) is a standard representation for γ, then we may write $S = \cup\{S_\xi : \xi < \chi\}$, where $tp(S_\xi) = \omega_\alpha^{1+\rho_\xi}$, and $S_0 < S_1 < \ldots < S_\xi < \ldots$. By the induction hypothesis, there are sequences of intervals $I(S_\xi) = \langle J_{\xi,\nu} : \nu < \omega_\alpha \rangle$ and regressive functions $\varphi_\xi = \varphi(S_\xi)$ for $\xi < \chi$ such that the above conditions are satisfied. Let

$$g : \chi \times \omega_\alpha \to \omega_\alpha \backslash \{0\}$$

be any bijective map satisfying

$$\text{(3.9)} \qquad\qquad\qquad g(\xi, 0) < g(\xi', 0) \qquad (\xi < \xi'),$$

$$\text{(3.10)} \qquad\qquad\qquad g(\xi, \nu) < g(\xi, \nu') \qquad (\nu < \nu').$$

Put $I_0 = S$, $I_\nu = J_{g^{-1}(\nu)}$ $(1 \leq \nu < \omega_\alpha)$. Also, define a function $\varphi = \varphi(S) : \omega_\alpha \backslash \{0\} \to \omega_\alpha$ as follows. For $1 \leq \rho < \omega_\alpha$ and $g^{-1}(\rho) = (\xi, \nu)$, put $\varphi(\rho) = 0$ if $\nu = 0$, and put $\varphi(\rho) = g(\xi, \varphi_\xi(\nu))$ if $\nu \neq 0$. The condition (3.10) ensures that φ is regressive. We have to check that (3.3)–(3.8) all hold.

Since g is bijective, it follows from the definition that $I(S)$ contains each $I(S_\xi)$ $(\xi < \chi)$ as a proper subsequence, and since the S_ξ are pairwise disjoint, the conditions (3.3)–(3.6) immediately follow from the corresponding conditions for the $I(S_\xi)$. Likewise, (3.7) and (3.8) hold if $\varphi(\nu) \neq 0$. But (3.7) and (3.8) also hold when $\varphi(\nu) = 0$ since (3.9) hold and $S_0 < S_1 < \ldots$.

Note that, since $\varphi = \varphi(S)$ is regressive, for each $\nu < \omega_\alpha$ there are ordinals $\nu_0 = \nu > \nu_1 > \ldots > \nu_k = 0$, where $k = k(\nu)$ is finite and $\nu_{i+1} = \varphi(\nu_i)$ $(i < k)$. By (3.5) it follows that

$$I_\nu = I_{\nu_0} \subset I_{\nu_1} \subset \ldots \subset I_{\nu_k} = I_0.$$

If $\nu_{i-1} > \rho > \nu_i$, then by (3.6) $I_\rho \cap I_{\nu_{i-1}} = \emptyset$ and hence $I_\rho \cap I_\nu = \emptyset$. It follows from this and (3.7) that whenever $\rho < \nu < \omega_1$, then either $I_\nu < I_\rho$ or $I_\nu > I_\rho$ or $I_\nu \subset I_\rho$. In fact, if $\rho \neq 0$ (or if $\rho = 0$ and $tp(S)$ is indecomposable) and $I_\nu \subset I_\rho$, then in fact

$$\text{(3.11)} \qquad\qquad I_\nu \subset_N I_\rho$$

holds, where $\subset_N$ indicates that I_ν is a *non-cofinal subset* of I_ρ. This is because, by (3.5), I_ν is a subinterval of I_ρ of smaller type and so cannot be cofinal in I_ρ since $tp(I_\rho)$ is indecomposable.

The way that the above interval description of $\omega_\alpha \gamma$ is used to construct a set of type $\omega_\alpha \gamma$ is simply by imitating this structure. If we construct somehow certain sets Z_ν so that $tp(Z_\nu) = tp(I_\nu)$ $(\nu < \omega_\alpha)$ and so that

$$Z_\nu \Delta Z_\rho \Leftrightarrow I_\nu \Delta I_\rho$$

holds for $\rho < \nu < \omega_\alpha$, where Δ indicates any one of the binary relations $<$, $>$, $\subset$ or $\subset_N$, then the set $Z = \bigcup\{Z_\nu : \nu < \omega_\alpha$ and $|Z_\nu| = 1\}$ has the same order type as $S = \bigcup\{I_\nu : \nu < \omega_\alpha$ and $|I_\nu| = 1\}$, i.e. $\omega_\alpha \gamma$.

4   A decomposition theorem

In order to understand why the set-mapping result does not work beyond $\omega_1^{\omega+2}$, we need a certain decomposition theorem (Theorem 4.2) for ordered sets.

Note first that, by an easy induction argument on n, the partition relation

$$\text{(4.1)} \qquad\qquad \omega_1^n \to (\omega_1^n)_\omega^1 \quad (n < \omega)$$

holds. This says that ω_1^n is not a union of countably many sets of smaller type. Of course, ω_1^ω is a union of countably many "small" sets, $\omega_1^\omega = \bigcup\{S_n : n < \omega\}$ with $tp(S_n) = \omega_1^n$ $(n < \omega)$, but it is a little surprising that the same is true for any ordinal $\alpha < \omega_2$. This is the so-called Milner–Rado paradox

$$\text{(4.2)} \qquad\qquad \alpha \not\to (\omega_1^n)_{n<\omega}^1 \quad (\alpha < \omega_2),$$

which is easily proved by induction on α (e.g. see [13]).

We shall need the following lemma from [5].

Lemma 4.1 *If $\rho < \omega + 2$ and X_ξ $(\xi < \omega_1)$ are subsets of ω_1^ρ having order type strictly smaller than ω_1^ρ, then there is a countable set $D \subseteq \omega_1$ such that $\bigcup\{X_\xi : \xi \in D\} \neq \omega_1^\rho$.*

Proof: For $\rho < \omega$ we can take for D any countable subset of ω_1 in view of (4.1). Suppose $\rho = \omega$. Then there are $n < \omega$ and an uncountable set $C \subseteq \omega_1$ such that $tp(X_\xi) < \omega^n$ for $\xi \in C$, and the lemma holds for any $D \in [C]^\omega$. Finally, suppose that $\rho = \omega + 1$. We may write $\omega_1^{\omega+1} = \bigcup\{A_\nu : \nu < \omega_1\}$, where $tp(A_\nu) = \omega_1^\omega$ $(\nu < \omega_1)$ and $A_0 < A_1 < \dots$. For each $\xi < \omega_1$ there are $n(\xi) < \omega$ and $\alpha(\xi) < \omega_1$ such that $tp(X_\xi \cap A_\alpha) < \omega_1^{n(\xi)}$ for every $\alpha > \alpha(\xi)$. There is an uncountable set $C \subseteq \omega_1$ such that $n(\xi) = n$ for all $\xi \in C$. Choose $D \in [C]^\omega$ and $\alpha < \omega_1$ such that $\alpha > \alpha(\xi)$ for all $\xi \in D$, then $A_\alpha \backslash \bigcup \{X_\xi : \xi \in D\} \neq \emptyset$. $\square$

Lemma 4.1 fails for $\omega_1 + 2 \leq \rho < \omega_2$, for we have the following seemingly paradoxical decomposition theorem of [5].

Theorem 4.2 *Let A be a well ordered set, $tp(A) < \omega_2$. Then there are subsets $X_\xi \subseteq A$ $(\xi < \omega_1)$ such that $tp(X_\xi) < \omega_1^{\omega+2}$ and $\cup\{X_\xi : \xi \in D\} = A$ for every $D \in [\omega_1]^\omega$.*

Proof: Without loss of generality, we may assume that $tp(A) = \omega_1^\rho < \omega_2$. We prove the theorem by induction on ρ. For $\rho \le \omega + 1$ the result is obvious (put $X_\xi = A$). Suppose $\rho \ge \omega + 2$.

Case 1. $cf(\omega_1^\rho) = \omega$. Then we may write $A = \cup\{A_n : n < \omega\}$, where $A_0 < A_1 < \ldots$, and $tp(A_n) = \omega_1^{p(n)} < \omega_1^\rho$. By the induction hypothesis there are sets $X_{n\xi} \subseteq A_n$ $(\xi < \omega_1)$ such that $tp(X_{n\xi}) < \omega_1^{\omega+2}$ and the union of any ω of the $X_{n\xi}$ (n fixed) cover A_n. Put $X_\xi = \cup\{X_{n\xi} : n \in \omega\}$ $(\xi < \omega_1)$.

Case 2. $cf(\omega_1^\rho) = \omega_1$. In this case we may write $A = \cup\{A_\nu : \nu < \omega_1\}$, where $A_0 < A_1 < \ldots$ and $tp(A_\nu) = \omega_1^{p(\nu)} < \omega_1^\rho$. By the induction hypothesis there are sets $X_{\nu\xi} \subseteq A_\nu$ $(\xi < \omega_1)$ such that $tp(X_{\nu\xi}) < \omega_1^{\omega+2}$ and the union of any ω of these cover A_ν. Also, by (4.2), there is a partition of each A_ν into subsets $A_{\nu n}$ $(n < \omega)$ such that $tp(A_{\nu n}) < \omega_1^n$. For each $\nu < \omega_1$, let $f(v, .) : \nu \to \omega$ by any injective map, and define

$$X_\xi = \cup\{X_{\nu\xi} : \nu \le \xi\} \cup \cup\{A_{\nu n} : \xi < \nu < \omega_1, n \le f(\nu, \xi)\} \qquad (\xi < \omega_1).$$

It is easy to see that

$$tp(X_\xi) \le \cup\{tp(X_{\nu\xi}) : \nu \le \xi\} + \omega_1^\omega \cdot \omega_1 < \omega_1^{\omega+2} \qquad (\xi < \omega_1),$$

since $X_{0\xi} < X_{1\xi} < \ldots < X_{\xi\xi}$ and $tp(X_{\nu\xi}) < \omega_1^{\omega+2}$ $(\nu \le \xi)$. Now consider any $D \in [\omega_1]^\omega$. We want to show that $A_\nu \subseteq \cup\{X_\xi : \xi \in D\} = D^*$ for every $\nu < \omega_1$. D contains an increasing sequence of ordinals ξ_n $(n < \omega)$ with limit, say $\xi < \omega_1$. If $\nu < \xi$, then $\nu < \xi_n$ for some integer n and

$$D^* \supseteq \cup\{X_{\nu\xi_m} : n \le m < \omega\} = A_\nu.$$

On the other hand, if $\nu \ge \xi$, then the $f(\nu, \xi_n)$ $(n < \omega)$ are distinct integers and so

$$D^* \supseteq \cup\{A_{\nu n} : n < \omega\} = A_\nu. \qquad \square$$

5   Proof of Theorems 1.3 and 1.4

First we prove Theorem 2.4 that $SM(\omega, \lambda)$ is false for $\omega_1^{\omega+2} \le \lambda < \omega_2$. It is an immediate consequence of Theorem 4.2. By that result there are subsets X_ξ $(\xi < \omega_1)$ of λ such that $tp(X_\xi) < \omega_1^{\omega+2}$ and the union of any ω of them covers λ. Let $g : \omega_1 \to \lambda$ be any bijective map and consider the set mapping f on λ defined by $f(x) = \{g(\xi) : \xi < \omega_1, g(\xi) \ne x, x \notin X_\xi\}$. Clearly $f(x)$ is finite since the union of any infinite number of the X_ξ covers λ. Also, if A is any subset of λ such that $tp(A) \ge \omega_1^{\omega+2}$ and $y \in A$, then there is $x \in A \backslash (X_{g^{-1}(y)} \cup \{y\})$, and so $y \in f(x)$ and A is not f-free. $\qquad \square$

Proof of Theorem 2.3: Let $tp(S) = \omega_1\gamma < \omega_1^{\omega+2}$ and let f be a finite set mapping on S. If $\gamma = 1$ the result follows from Theorem 2.2, and so we assume that $\gamma > 1$. We shall use the construction described in §3 in order to construct an f-free subset $Z \subseteq S$ of order type $\omega_1\gamma$.

Note first that *if $A \subseteq S$, then there is $C(A) \in [S]^{\le\omega}$ such that*

$$(5.1) \qquad tp(A \backslash f^{-1}[D]) = tp(A) \text{ whenever } D \in [S \backslash C(A)]^{\le\omega}.$$

We prove this by induction on $\alpha = tp(A)$. If A is countable then (5.1) holds with $C(A) = f[A]$. If α is decomposable, then $A = A_0 \cup A_1$, where $A_0 < A_1$ and $tp(A_i) < \alpha$ $(i < 2)$, and (5.1) holds with $C(A) = C(A_0) \cup C(A_1)$. A similar argument applies if $cf(\alpha) = \omega$. Therefore, we may assume that $\alpha = \omega_1^\sigma$ for some $\sigma < \omega + 2$.

If there is no countable set $C(A)$ satisfying (5.1), then there are pairwise disjoint countable sets D_ν $(\nu < \omega_1)$ such that $tp(A \backslash f^{-1}[D_\nu]) < tp(A)$. Now by Lemma 4.1 there are $N \in [\omega_1]^\omega$ and $x \in A \backslash \cup \{A \backslash f^{-1}[D_\nu] : \nu \in N\}$. Thus $f(x) \cap D_\nu \neq \emptyset$ for $\nu \in N$. Therefore $f(x)$ is infinite, and this is a contradiction.

Let $I(S) = \langle I_\nu : \nu < \omega_1 \rangle$, $\varphi = \varphi(S)$ be as defined in §3. We shall define by recursion subsets $Z_\nu \subseteq S$ $(\nu < \omega_1)$ so that the following conditions are satisfied:

$$(5.2) \qquad\qquad tp(Z_\nu) = tp(I_\nu);$$

$$(5.3) \qquad\qquad Z_\nu \Delta Z_\rho \Leftrightarrow I_\nu \Delta I_\rho (\rho < \nu; \Delta \in \{<, >, \mathsf{C}, \mathsf{C}_N\});$$

$$(5.4) \qquad\qquad Z_\nu \cap C(Z_\rho) = \emptyset \quad (\rho < \nu);$$

$$(5.5) \qquad\qquad Z_\nu \cap (f[Z_\rho] \cup f^{-1}[Z_\rho]) = \emptyset \quad (\rho < \nu, \ |Z_\rho| = 1);$$

$$(5.6) \qquad\qquad Z_\nu \subseteq I_\nu \text{ if } \varphi(\nu) = 0 \text{ and } \chi(\gamma) < \omega.$$

The condition (5.6) is rather special and is introduced to take care of the case when $tp(S) = \omega_1 \gamma$ is decomposable.

Put $Z_0 = S$. First, we ensure that the special condition (5.6) is satisfied. Put $\chi = \chi(\gamma)$ if $\chi(\gamma)$ is finite and $\chi = 0$ otherwise. In the case when $\chi(\gamma)$ is finite, we may assume that

$$S = I_1 \cup I_2 \cup \ldots \cup I_\chi,$$

where $I_1 < I_2 < \ldots < I_\chi$ and $tp(I_n) = \omega_1^{1+\sigma_n}$ $(1 \leq n \leq \chi)$. Put $Z_\nu = I_\nu \cup \{C(I_n) : n \leq \chi\}$ $(1 \leq \nu \leq \chi)$. Then (5.2)–(5.6) hold trivially for $\nu \leq \chi$. Now suppose that $\tau > \chi$ and that the Z_ν have been defined for $\nu < \tau$.

We want to define Z_τ. Note that, in view of the definition of χ, we may assume that either $\varphi(\tau) > 0$, or that $\varphi(\tau) = 0$ and $tp(S)$ is indecomposable; in either case, $tp(Z_{\varphi(\tau)}) = tp(I_{\varphi(\tau)})$ is indecomposable.

Put $A = \cup\{C(Z_\nu) : 0 \leq \nu < \tau\}$, $B_0 = \cup\{Z_\nu : \nu < \varphi(\tau), |Z_\nu| = 1\}$, $B_1 = \cup\{Z_\nu : \varphi(\tau) < \nu < \tau, |Z_\nu| = 1\}$, $B = B_0 \cup B_1$, $K = \{\nu < \tau : \varphi(\nu) = \varphi(\tau)\}$.

By (3.7) and (3.11) we have that $I_\nu < I_\tau \mathsf{C}_N I_{\varphi(\tau)}$ for $\nu \in K$, and so $Z_\nu \mathsf{C}_N Z_{\varphi(\tau)}$ $(\nu \in K)$ by (5.3). Also by (3.8) $tp(K) < \chi(tp(I_{\varphi(\tau)}))$, and so

$$(5.7) \qquad\qquad Z' = \cup\{Z_\nu : \nu \in K\} \mathsf{C}_N Z_{\varphi(\tau)}.$$

By (5.5) we have that

$$(5.8) \qquad\qquad Z_{\varphi(\tau)} \cap f^{-1}[B_0] = \emptyset.$$

Also, $B_1 \cap C(Z_{\varphi(\tau)}) = \emptyset$ by (5.4), and therefore, since B_1 is countable,

$$(5.9) \qquad\qquad tp(Z_{\varphi(\tau)} \backslash f^{-1}[B_1]) = tp(Z_{\varphi(\tau)}),$$

by (5.1). Since $A \cup f[B]$ is countable and $tp(Z_{\varphi(\tau)}) = tp(I_{\varphi(\tau)}) = \omega_1^{1+\sigma}$ (where $\sigma = p_{\varphi(\tau)}$), and since (5.7)–(5.9) hold, it follows that there is a subset Z_τ of $Z_{\varphi(\tau)}$ such that

$$Z' < Z_\tau \subset Z_{\varphi(\tau)} \backslash (A \cup f[B] \cup f^{-1}[B]),$$

and $tp(Z_\tau) = tp(I_\tau) < tp(I_{\varphi(\tau)})$. Moreover, Z_τ can be chosen to be non-cofinal in $Z_{\varphi(\tau)}$ if I_τ is non-cofinal in $I_{\varphi(\tau)}$. It is easy to see that (5.2)-(5.5) now hold for $\nu = \tau$ with this choice of Z_τ. This defines Z_ν for all $\nu < \omega_1$ such that (5.2)-(5.6) hold.

By (5.2) and (5.3) it follows that $Z = \cup\{Z_\nu : \nu < \omega_1, |Z_\nu| = 1\}$ has the same order type as S, i.e. $tp(Z) = \omega_1\gamma$. Also Z is f-free by (5.5).   □

6   Graphs without infinite paths — Proof of Theorem 1.1

Let S be an ordered set of order type $tp(S) = \omega\beta < \omega_1^{\omega+2}$, and let $G = (S, E)$ be a graph on S with no infinite path. We want to show that there is an independent set $Z \subseteq S$ having the same order type $\omega\beta$. For $x \in S$, let $E(x) = \{y \in S : \{x, y\} \in E\}$, and for $X \subseteq S$, let $E[X] = \cup\{E(x) : x \in X\}$.

Case 1. $\beta < \omega_1$. We begin with the observation that, whenever $T \subseteq S$ is a subset such that $tp(T) = \omega\gamma$, then there is a finite set $F(T) \subset S$ such that

(6.1) $tp(T \backslash E[X]) = tp(T)$ $(\forall X \in [S \backslash F(T)]^{<\omega}).$

To see this, suppose first that $\gamma = \omega^\sigma$ is indecomposable. If there is no finite set $F(T)$ satisfying (6.1), then there are infinitely many pairwise disjoint finite sets X_n $(n < \omega)$ such that $tp(T \backslash E(X_n)) < \omega\gamma$. Since

$$\omega^{1+\sigma} \to (\omega^{1+\sigma})_k^1$$

holds for finite k, it follows that for each n there is $x_n \in X_n$ such that

$$tp(T \backslash E(x_n)) < tp(T).$$

Thus, if $Y_n = \cap\{E(x_i) : i \leq n\}$ $(n < \omega)$, then each Y_n is infinite and $Y_0 \supseteq Y_1 \supseteq \dots$. Hence we can choose distinct elements $y_n \in Y_n$ so that $y_0, x_0, y_1, x_1, \dots$ is an infinite path in G. This is a contradiction. In the general case, we may write $T = T_0 \cup T_1 \cup \dots \cup T_k$, where $k < \omega$, $tp(T_i) = \omega^{1+\sigma(i)}$ $(i \leq k)$ and $T_0 < T_1 < \dots < T_k$.

Then (6.1) holds with $F(T) = \cup\{F(T_i) : i \leq k\}$.

Now let $I(S) = \langle I_n : n < \omega \rangle$, $\varphi(S) = \varphi$ be as defined in §3. We shall inductively define subsets Z_n $(n < \omega)$ so that $Z = \cup\{Z_n : n < \omega, |Z_n| = 1\}$ is independent and has the same order type as S. We will choose the Z_n so that the following conditions are satisfied:

(6.2) $tp(Z_n) = tp(I_n);$

(6.3) $Z_n \triangle Z_i \Leftrightarrow I_n \triangle I_i$ $(i < n, \triangle \in \{<, >, \subset, \subset_N\};$

(6.4) $Z_n \cap F(Z_i) = \emptyset$ $(i < n, |Z_i| \neq 1);$

(6.5) $Z_n \cap E[Z_i] = \emptyset$ $(i < n, |Z_i| = 1);$

(6.6) $Z_n \subseteq I_n$ if $\varphi(n) = 0$ and $tp(S) = \omega\gamma$ is decomposable.

Put $Z_0 = I_0 \ (= S)$. First we make sure that the special condition (6.6) is satisfied. Put $\chi = \chi(\gamma)$ if this is finite; otherwise, put $\chi = 0$. If $\chi(\gamma)$ is finite then we may assume that

$$S = I_1 \cup I_2 \cup \ldots \cup I_\chi,$$

where $I_1 < I_2 < \ldots < I_\chi$. Now put $Z_n = I_n \backslash \cup \{F(I_i) : i \leq \chi\} \ (1 \leq n \leq \chi)$. Clearly (6.2)–(6.6) hold for $n \leq \chi$. Now assume that $m > \chi$ and that the Z_n have been defined for $n < m$. We want to define Z_m. Since $m > \chi$, it follows that $tp(Z_{\varphi(m)}) = tp(I_{\varphi(m)})$ is indecomposable.

Put $A = \cup\{F(Z_i) : i < m, |Z_i| \neq 1\}$, $B_1 = \cup\{Z_i : i < \varphi(m), |Z_i| = 1\}$, $B_2 = \cup\{Z_i : \varphi(m) < i < m, |Z_i| = 1\}$, $K = \{i < m : \varphi(i) = \varphi(m)\}$.

By (6.5), $Z_{\varphi(m)} \cap E[B_1] = \emptyset$. Also, by (6.4) $B_2 \cap F(Z_{\varphi(m)}) = \emptyset$, and so

$$tp(Z_{\varphi(m)} \backslash E[B_2]) = tp(Z_{\varphi(m)}),$$

by (6.1). By (6.3), and the fact that K is finite and $tp(Z_{\varphi(m)})$ is indecomposable, it follows that $Z' = \cup\{Z_i : i \in K\} \subset_N Z_{\varphi(m)}$. Therefore, since A is finite, we can choose a subset Z_m of $Z_{\varphi(m)}$ so that

$$Z' < Z_m \subset_N Z_{\varphi(m)} \backslash (A \cup E[B_1 \cup B_2]).$$

It is clear that (6.2)–(6.6) now hold for $n = m$ with this choice of Z_m.

By (6.2) and (6.3), it follows that $Z = \cup\{Z_i : i < \omega, |Z_i| = 1\}$ has the same order type as S. Also Z is an independent set by (6.5).

Case 2. $\beta = \omega_1\gamma$. Note first that, for any set $T \subseteq S$, there is some element $x(T) \in T$ which has finite relative valency, i.e. $E(x) \cap T$ is finite. Otherwise, if $E(x) \cap T$ is infinite for each $x \in T$, then there is an infinite path in G. Now define a well ordering of the elements of S so that $S = \{x_\sigma : \sigma < \lambda\}$, where $x_\sigma = x(S \backslash \{x_\rho : p < \sigma\})$. In this well ordering an element x_σ is joined to only finitely many elements x_τ with $\tau > \sigma$. Put $f(x_\sigma) = \{x_\tau : \sigma < \tau < \lambda, x_\tau \in E(x_\sigma)\}$. Then f is a finite set mapping on the ordered set S of order type $\omega_1\gamma < \omega_1^{\omega+2}$ and so by Theorem 2.3 there is an f-free subset $Z \subseteq S$ with the same order type $\omega_1\gamma$. The set Z is also an independent set for the graph G for, if $x_\sigma, x_\tau \in Z$ and $\sigma < \tau$, then $x_\tau \notin E(x_\sigma)$ since $x_\tau \notin f(x_\sigma)$.

Case 3. $\omega\beta = \omega_1\gamma + \omega\delta$, where $\delta < \omega_1$. In this case we may write $S = A \cup B$, where $A < B$, $tp(A) = \omega_1\gamma$ and $tp(B) = \omega\delta$. By Cases 1 and 2 we can assume that A and B are independent sets. We want to show that there are subsets $A' \subseteq A$ and $B' \subseteq B$ such that $tp(A') = \omega_1\gamma$, $tp(B') = \omega\delta$ and $A' \cup B'$ is independent.

Assume first that γ is indecomposable. Consider a new graph G' on the vertex set B in which two vertices b, b' are joined by an edge if and only if $E(b) \cap E(b') \ (\subseteq A)$ is infinite. It is easy to see that G' contains no infinite path (otherwise G does), and so by Case 1, there is a subset $B' \subseteq B$ which has order type $\omega\delta$ and contains no edge of the graph G'. It follows that $E(b) \cap E(b')$ is finite for distinct elements b, b' in B', and so $A_1 = \{a \in A : |E(a) \cap B'| \geq 2\}$ is countable. Thus $A' = A \backslash A_1$ also has order type $\omega_1\gamma$ and each point of A' is joined to at most one point of B'. Since $2 \cdot \omega = \omega$ and $tp(B') = \omega\delta$, it follows that $B' = B_0 \cup B_1$, where $B_0 \cap B_1 = \emptyset$ and $tp(B_i) = \omega\delta \ (i < 2)$. Put $A_0 = \{a \in A' : E(a) \cap B_0 = \emptyset\}$, $A_1 = A' \backslash A_0$. Then $A' = A_0 \cup A_1$ and since γ is indecomposable, it follows that there is $i < 2$ such that $tp(A_i) = \omega_1\gamma$. Then $A_i \cup B_i$ is an independent set with order type $\omega_1\gamma + \omega\delta$.

Finally, suppose that γ is decomposable. Then we may write $A = A_0 \cup \ldots \cup A_k$, where $k < \omega$, $tp(A_i) = \omega_1 \gamma_i$ $(i \leq k)$ and γ_i is indecomposable. Using the above argument $k+1$ times, we obtain $A' \subseteq A$, $B' \subseteq B$ such that $tp(A') = \omega_1 \gamma$, $tp(B') = \omega\delta$ and $A' \cup B'$ is independent.   □

7   Independence results

Our proof that

(7.1) $\alpha \to (\alpha, \text{infinite path})^2$ for $\alpha = \omega\beta < \omega_1^{\omega+2}$,

made use of the fact that $SM(\omega, \alpha)$ holds. Since $SM(\omega, \omega_1^{\omega+2})$ is false, we cannot use this method to establish (7.1) for the case $\alpha = \omega_1^{\omega+2}$. At the time of writing [5] we thought that (7.1) was probably true without this limit on the size of α, and this was left as an open question. The recent papers by Larson [10], [11], [12] and Baumgartner and Larson [1] make considerable progress with this problem.

In [10], Larson showed that it is consistent that

(7.2) $\omega_1^{\omega+2} \to (\omega_1^{\omega+2}, \text{infinite path})^2$

holds. More recently she showed [11] that Martin's axiom (MA) (and $2^\omega > \omega_1$) implies that (7.1) holds for any limit ordinal $\alpha < 2^\omega$. On the other hand, Baumgartner and Larson [1] have shown that Jensen's diamond principle (for a description of this see e.g. Kunen [8]) implies that

(7.3) $\alpha \not\to (\alpha, \text{infinite path})^2$ $(\omega_1^{\omega+2} \leq \alpha < \omega_2)$.

Thus, in particular (7.2) is independent of the usual axioms of set theory. It is not known exactly what axioms do ensure (7.2) or (7.3). For example, it is not known if (7.3) holds under the continuum hypothesis (which is weaker than diamond) although Larson [12] has shown that in this case (7.3) does at least hold for a cofinal set of limit ordinals $\alpha < \omega_2$. In the remaining sections we shall only describe Larson's clever proof of the consistency of (7.2) since this will not involve us in any sophisticated set-theory and is in the same spirit as our proof of (7.1).

8   Some preliminaries

In order to prove the consistency of (7.2) we will use the following axiom

(∗) $(\forall \mathcal{F} \in [{}^\omega\omega]^{\leq \omega_1})(\exists g \in {}^\omega\omega)(\forall f \in \mathcal{F})(f \prec g)$,

where $f \prec g$ means that $\{n : g(n) \leq f(n)\}$ is finite. (Larson [10] actually uses the slightly weaker axiom that there is no ω_1-scale, but this causes no essential change to the argument.)

Note that the consistency of (∗) follows from the following well known consequence of MA known as Solovay's lemma (see e.g. Kunen [8]) SL_ω : *If* $A, B \in [P(\omega)]^{\leq \kappa}$, $\kappa < 2^\omega$, *and if* $|B \setminus \cup A'| = \omega$ *whenever* $B \in \mathcal{B}$ *and* $A' \in [\mathcal{A}]^{<\omega}$, *then there is* $C \subseteq \omega$ *such that* $|C \cap A| < \omega$ $(A \in \mathcal{A})$ *and* $|C \cap B| = \omega$ $(B \in \mathcal{B})$.

Proof that SL_ω $(\& 2^\omega > \omega_1) \Rightarrow (*)$: Let $\mathcal{F} \in [{}^\omega\omega]^{\omega_1}$. For $f \in \mathcal{F}$ put $A_f = \{(m,n) \in \omega \times \omega : n \leq f(m)\}$, and for $m \in \omega$ put $B_m = \{(m,n) : n < \omega\}$. By SL_ω there is a set $C \subseteq \omega \times \omega$ such that $|C \cap A_f| < \omega$ $(f \in \mathcal{F})$ and $|C \cap B_m| = \omega$ $(m \in \omega)$. Put $g(m) = \min\{n : (m,n) \in B_m \cap C\}$ $(m < \omega)$. $\square$

Let $\mathcal{C}$ denote the set of all functions $f \in {}^\omega\omega$ which are non-decreasing and unbounded. Although the inverse, f^{-1}, of a function $f \in \mathcal{C}$ is not in general a function, we can nevertheless define two operations on $\mathcal{C}$ which closely correspond to "inverse". Let $f \in \mathcal{C}$. Then there are integers m_i, n_i $(i < \omega)$ such that $0 = m_0 < m_1 < \dots$, $0 \leq n_0 < n_1 < \dots$ and $f(m) = n_i$ $(m_i \leq m < m_{i+1})$. Put $n_{-1} = -1$ and define

$$f^{(-1)}(n) = m_i \ (n_{i-1} < n \leq n_i; i < \omega);$$

$$f^{[-1]}(n) = m_{i+1} - 1 \ (n_i \leq n < n_{i+1}; i < \omega); f^{[-1]}(n) = 0 \ (n < n_0).$$

We leave the reader to check that $f^{(-1)}, f^{[-1]} \in \mathcal{C}$ and that

(8.1) $$(f^{(-1)})^{[-1]} = f = (f^{[-1]})^{(-1)},$$

(8.2) $$(\forall f, g \in \mathcal{C})(f \prec g \Leftrightarrow g^{(-1)} \prec f^{(-1)} \Leftrightarrow g^{[-1]} \prec f^{[-1]}).$$

It follows from (8.1) and (8.2) that $(*)$ is equivalent to the dual statement

(**) $$(\forall \mathcal{F} \in [\mathcal{C}]^{\leq \omega_1})(\exists g \in \mathcal{C})(\forall f \in \mathcal{F})(g \prec f).$$

Larson's construction of subsets of $\omega_1^{\omega+2}$ having the same order type is based upon the following. For $\alpha, \beta < \omega_1$ and $m < \omega$ define

$$W(\alpha, \beta, m) = \{(\alpha, \beta, m, \beta_1, \dots, \beta_m) : \beta_i < \omega_1\},$$

$$W(\alpha, m) = \cup\{W(\alpha, \beta, m) : \beta < \omega_1\},$$

$$W(\alpha) = \cup\{W(\alpha, m) : m < \omega\},$$

$$W = \cup\{W(\alpha) : \alpha < \omega_1\}.$$

Ordered alphabetically, $tp(W(\alpha, \beta, m)) = \omega_1^m$, $tp(W(\alpha, m)) = \omega_1^{m+1}$, $tp(W(\alpha)) = \omega_1^{\omega+1}$ and $tp(W) = \omega_1^{\omega+2}$. We will write $\xi < \eta$ to indicate that the finite sequences of ordinals ξ, η are ordered alphabetically. For any subset $Z \subseteq W$, define $Z(\alpha, \beta, m) = Z \cap W(\alpha, \beta, m)$ etc. The word "large" will be used in the following different senses, but the intended meaning should always be clear from the context.

Definition 1 A subset A of an ordered set of type ω_1^{m+1} $(m < \omega)$ is *large* if $tp(A) = \omega_1^{m+1}$. A set $J \subseteq \omega_1 \times \omega$ is *large* if, for uncountably many α, $J_\alpha = \{m : (\alpha, m) \in J\}$ is infinite. A subset $Z \subseteq W$ is *large* if $Z = \cup\{Z(\alpha, m) : (\alpha, m) \in J\}$, where J is large and each $Z(\alpha, m)$ $((\alpha, m) \in J)$ is large (we call J the *index set* of Z and $Z(\alpha, m)$ the $(\alpha, m) -$ th *block* of Z).

We first show that a large subset $Z \subseteq W$ really is large, i.e. has order type $\omega_1^{\omega+2}$. It is a consequence of the following easy lemma.

Lemma 8.1 *Let $M \in [\omega]^\omega$ and suppose that, for $m \in M$, $U(m)$ is a cofinal subset of $\omega_1^{\omega+1}$ of order type ω_1^{1+m}. Then $U = \cup\{U(m) : m \in M\}$ has order type $\omega_1^{\omega+1}$.*

Proof: Let $M = \{m(n) : n < \omega\}$, $m(0) < m(1) < \ldots$. We may write $\omega_1^{\omega+1} = \cup\{S_\nu : \nu < \omega_1\}$, where $S_0 < S_1 < \ldots$ and $tp(S_\nu) = \omega_1^\omega$ $(\nu < \omega_1)$. For $\xi < \omega_1$ write $\xi = \omega\nu_\xi + n_\xi$ $(\nu_\xi < \omega_1, n_\xi < \omega)$ and inductively choose ordinals $\alpha(\xi) < \omega_1$ so that $\alpha(0) < \alpha(1) < \ldots$ and $tp(U(m(n_\xi))) \cap S_{\alpha(\xi)} = \omega_1^{m(n_\xi)}$. Then $tp(U) \geq tp(U \cap \{S_{\alpha(\xi)} : \xi < \omega_1\}) = \omega_1^{\omega+1}$. $\qquad\square$

Corollary 1 *If $Z \subseteq W$ is large, then $tp(Z) = \omega_1^{\omega+2}$.*

Proof: We have that $Z = \cup\{Z(\alpha, m) : (\alpha, m) \in J\}$, where J is large and each $Z((\alpha, m)$ is large $((\alpha, m) \in J)$. Thus, for uncountably many α, there are infinitely many integers m such that $Z(\alpha, m)$ is cofinal subset of $W(\alpha)$. Hence, by the lemma $tp\, Z(\alpha) = \omega_1^{\omega+1}$, and so $tp(Z) = \omega_1^{\omega+2}$. $\qquad\square$

The next lemma will show how the hypothesis (*) will be used.

Lemma 8.2 *Assume (*) holds. Let $J \subseteq \omega_1 \times \omega$ be large and let $A = \{\alpha : J_\alpha$ is infinite$\}$. Let $f : J \to \omega$ be such that $f(\alpha, .)$ is non-decreasing and unbounded for each $\alpha \in A$. Then there are a large set $K \subseteq J$ and functions h, $H \in C$ such that*

$$h(m) < f(\alpha, m) < H(m)$$

for all $(\alpha, m) \in K$.

Proof: For $(\alpha, n) \in A \times \omega$, let $m = m(\alpha, n)$ be the least integer such that $m \geq n$ and $(\alpha, m) \in J$. Define $t : A \times \omega \to \omega$ so that $t(\alpha, n) = f(\alpha, m(\alpha, n))$. Then $t(\alpha, .) \in C(\alpha \in A)$. It follows by (*) and (**) that there are h, $H \in C$ such that $h \prec t(\alpha, .) \prec H$ for all $\alpha \in A$. It follows that, for each $\alpha \in A$, there is $n_\alpha < \omega$ such that $h(n) < t(\alpha, n) < H(n)$ for all $n \geq n_\alpha$. The lemma holds with $K = \{(\alpha, m) \in J : \alpha \in A, m \geq n_\alpha\}$. $\qquad\square$

We shall identify ω_1^m with the set of all m-tuples $(\xi_0, \ldots, \xi_{m-1})$ $(\xi_i < \omega_1)$ ordered alphabetically, and if $\xi = (\xi_0, \ldots, \xi_{m-1}) \in \omega_1^m$ and $\xi_m \in \omega_1$, then we write $(\xi, , \xi_m)$ for $(\xi_0, \ldots, \xi_m)$. If S is a large subset of ω_1^{m+1}, we define

$$S^\square = \{\xi \in \omega_1^m : (\xi, \sigma) \in S \text{ for uncountably many } \sigma\}.$$

Thus $S^\square$ is a large subset of ω_1^m. Also, for $\xi \in S^\square$, let $S^\square(\xi) = \{\sigma : (\xi, \sigma) \in S\}$.

Lemma 8.3 *Let $m \in \omega$, $|A| \leq \omega_1$ and, for $a \in A$, let f_a be a 1–1 function on some large subset S_a of ω_1^{m+1}. Then there are large subsets $T_a \subseteq S_a$ such that the sets $f_a[T_a]$ $(a \in A)$ are pairwise disjoint.*

Proof: If $m = 0$ this is clear since there are pairwise disjoint uncountable sets $V_a \subseteq f_a[S_a]$ $(a \in A)$, and the lemma holds with $T_a = f_a^{-1}[V_a]$. Assume $m > 0$ are use induction.

For $\xi \in S_a^\square$, let $S_a^\square(\xi) = \{\sigma_{a,\nu}(\xi) : \nu < \omega_1\}$, where $\sigma_{a,0}(\xi) < \sigma_{a,1}(\xi) < \ldots$. For $(a, \nu) \in A \times \omega_1$, define a 1–1 function $h_{a,\nu}$ on $S_a^\square$ by

$$h_{a,\nu}(\xi) = f_a(\xi, \sigma_{a,\nu}(\xi)).$$

By the induction hypothesis, there are large subsets $T_{a,\nu} \subseteq S_a^\square$ such that the sets $h_{a,\nu}[T_{a,\nu}]$ $((a, \nu) \in A \times \omega_1)$ are pairwise disjoint. The lemma holds with

$$T_a = \{(\xi, \sigma_{a,\nu}(\xi)) : \nu < \omega_1, \xi \in T_{a,\nu}\} \quad (a \in A). \qquad\square$$

If $\xi = (\xi_0, \ldots, \xi_m)$, $\eta = (\eta_0, \ldots, \eta_m) \in \omega_1^{m+1}$, and $n \leq m+1$, we write

$$\xi \equiv_n \eta \Leftrightarrow (\forall i < n)(\xi_i = \eta_i).$$

A function f defined on a large subset S of ω_1^{m+1} is said to be *n-steady* on S if

$$(\forall \xi, \eta \in S)(f(\xi) = f(\eta) \Leftrightarrow \xi \equiv_n \eta),$$

and f is *steady* if it is n-steady for some $n \leq m+1$. Note that f is constant if it is 0-steady, and 1-1 if it is $(m+1)$-steady.

Lemma 8.4 *If $m \in \omega$ and f is a function defined on a large subset $S \subseteq \omega_1^{m+1}$, then f is steady on some large set $T \subseteq S$.*

Proof: This is clear in the case $m = 0$ since f is either constant or 1-1 on some uncountable set $T \subseteq S$. Assume $m > 0$ and use induction.

For $\xi \in S^{\square}$, let $S^{\square}(\xi) = \{\sigma_\nu(\xi) : \nu < \omega_1\}$, where $\sigma_0(\xi) < \sigma_1(\xi) < \ldots$. For $\nu < \omega_1$ define f_ν on $S^{\square}$ by $f_\nu(\xi) = f(\xi, \sigma_\nu(\xi))$. By the induction hypothesis, there are a large set $T_\nu \subseteq S^{\square}$ and $n_\nu \leq m$ such that f_ν is n_ν-steady on T_ν $(\nu < \omega_1)$. There is a large set $A \subseteq \omega_1$ such that $n_\nu = n$ for $\nu \in A$.

Suppose $n > 0$. For $\xi = (\xi_0, \ldots, \xi_{m-1}) \in \omega_1^m$, let $\xi|n = (\xi_0, \ldots \xi_{n-1})$. For fixed $\nu \in A$ and any $\xi \in T_\nu$, the value $f_\nu(\xi)$ is determined by $\xi|n$, say $f_\nu(\xi) = g_\nu(\xi|n)$. Since g_ν $(\nu \in A)$ is 1-1, it follows by Lemma 8.5 that there is a large set $U_\nu \subseteq T_\nu|n = \{\xi|n : \xi \in T_n\}$ such that the sets $g_\nu[U_\nu]$ $(\nu \in A)$ are pairwise disjoint. Then f is n-steady on the large set

$$T = \{(\xi, \sigma_\nu(\xi)) : \nu \in A, \xi \in T_\nu, \xi|n \in U_\nu\}.$$

If $n = 0$, then each f_ν assumes some constant value on T_ν $(\nu \in A)$ and an easy argument shows that f is either constant or $(m+1)$-steady on some large set $T \subseteq S$. $\quad\square$

We conclude this section with another easy technical lemma that will be needed later.

Lemma 8.5 *For $n < \omega$, let $k(n) \leq \omega$ and let $(\alpha_i(n) : i < k(n))$ be an increasing sequence of ordinals. Then there is an infinite set $M \subseteq \omega$ and an increasing function $q : \omega \to \omega$ such that $A = \{\alpha_i(m) : i < \omega, m \in M, m \geq q(i)\}$ has order type $tp(A) \leq \omega^2$.*

Proof: Successively choose infinite sets $N_i = \{n_{i,j} : i \leq j < \omega)\} \subseteq \omega$ $(i < \omega)$ so that $n_{i,i} < n_{i,i+1} < \ldots, N_{i+1} \subseteq N_i \backslash \{n_{i,i}\}$ and so that, for each i, either $i \geq k(n)$ for all $n \in N_i$, or $i < k(n)$ for all $n \in N_i$ and α_i is non-decreasing on N_i. The lemma holds with $M = \{n_{i,i} : i < \omega\}$ and $q(i) = n_{i,i}$.

To see this observe that (i) if $\alpha_i(m)$ and $\alpha_i(n)$ both belong to A and $m \leq n$, then $\alpha_i(m) \leq \alpha_i(n)$ and (ii) if $\alpha_i(m)$ and $\alpha_j(n)$ belong to A and $i \leq j$ and $m \leq n$, then $\alpha_i(m) \leq \alpha_i(n) \leq \alpha_j(n)$. Since $q(j)$ is increasing, it follows from (ii) that, if $\alpha_i(m) \in A$, then there are only finitely many pairs (j, n) with $j > i$ and $n < \omega$ such that $\alpha_j(n) \in A$ and $\alpha_j(n) < \alpha_i(m)$. By this and (i) it follows that the set of elements of A which are less than $\alpha_i(m)$ has order type strictly less than $\omega(i+1)$. Hence $tp(A) \leq \omega^2$. $\quad\square$

9   The main result modulo a lemma

Let $G = (W, E)$ be a graph on W. We want to show, assuming (*), that either G has an infinite path or there is a large independent set.

Let G^* be the enriched graph on W in which distinct points u, v are joined by an edge if, and only if, either $\{u, v\} \in E$ or if u and v are infinitely connected, i.e. they are joined by infinitely many paths which are pairwise disjoint apart from u and v. It is easy to see that G has no infinite path if and only if G^* has no infinite path. Since G^* has more edges than G, it will be enough if we show that there is a large independent set for the graph G^*. So we may assume that $G = G^*$, i.e.

(9.1)   if u, v are infinitely connected, then $\{u, v\} \in E$.

Call an edge $\{u, v\}$ of G a *local edge* if $u, v \in W(\alpha)$ for some α. In view of Theorem 1.1, we may also assume, without loss of generality, that

(9.2) G has no local edges.

We now define a certain set mapping on W associated with the graph G which will be used to reduce the problem of finding a large independent set to that of finding a large free set.

Definition 2 For large $U \subseteq W$ and $s \in U$, denote by $B1_U(s)$ the block $U(\alpha, m)$ of U which contains s. The *block link map* induced by U, $link_U$, is defined by

$$link_U(s) = \{t \in U : E(t) \supseteq B1_U(s)\}.$$

The main idea of Larson's proof is contained in the following lemma.

Lemma 9.1 (Sargasso Lemma (sic!)[10]) *Assume (*) holds. Suppose that G is a graph on W with no infinite path and that (9.1) and (9.2) hold. Then there are a large subset $U \subseteq W$ and a function $LOW \in C$ such that*

(9.3) $link_U$ *is a finite set mapping on U;*

(9.4) $(\forall s, t \in U)(\{s, t\} \in E \Leftrightarrow s \in link_U(t)$ *or* $t \in link_U(s))$;

(9.5)   *if $s \in U(\beta, n)$, $t \in U(\alpha, m)$ and $s \in link_U(t)$, then $\beta < \alpha$ and $LOW(m) < n$.*

A proof of the lemma will be given in the next section and we conclude this section by showing how it implies (7.2).

Proof of (7.2) (Assuming the Sargasso Lemma): Let G be a graph on W having no infinite path. As we observed, we may assume that (9.1) and (9.2) hold. Let U and LOW be the large set and function given by the Sargasso Lemma so that (9.3)–(9.5) hold. Let J be the index set of U. For $(\alpha, m) \in J$ let $Link_U(\alpha, m)$ denote the constant finite set $link_U(t)$ for $t \in U(\alpha, m)$.

For each $(\alpha, m) \in J$, partition $U(\alpha, m)$ into pairwise disjoint large subsets $U_i(\alpha, m)$ $(i < \omega)$, and for $t \in U(\alpha, m)$ let $i(t)$ denote the index i such that $t \in U_i(\alpha, m)$. By (9.3) $\text{Link}_U(\alpha, m)$ is finite for $(\alpha, m) \in J$ and so we can choose an integer $I(\alpha, m)$ such that

$$(9.6) \qquad I(\alpha, m) > i(s) \text{ for every } s \in \text{Link}_U(\alpha, m).$$

We may assume that $I(\alpha, .) \in C$ whenever J_α is infinite and so, by Lemma 8.4, there are a large set $K \subseteq J$ and a function $H \in C$ such that

$$(9.7) \qquad H(m) > I(\alpha, m) \text{ whenever } (\alpha, m) \in K.$$

Let $g \in C$ be any function so that

$$(9.8) \qquad m > g(n) \Rightarrow \text{LOW}(m) > n.$$

Now put $V(\beta, n) = \cup\{U_j(\beta, n) : j > H(g(n))\}$ for $(\beta, n) \in K$.

Clearly the set $V = \cup\{V(\beta, n) : (\beta, n) \in K\}$ is a large subset of U. We claim that V is an independent set for the graph G. To see this suppose for a contradiction that $s \in V(\beta, n)$ and $t \in V(\alpha, m)$ are joined by an edge.

By (9.2) $\beta \neq \alpha$, say $\beta < \alpha$. Then $s \in \text{Link}_U(\alpha, m)$ by (9.4) and (9.5), and so by (9.7), (9.6) and the definition of $V(\beta, n)$,

$$H(m) > I(\alpha, m) > i(s) > H(g(n)).$$

Since $H \in C$, it follows that $m > g(n)$ and hence $\text{LOW}(m) > n$ by (9.8). But this contradicts (9.5) and the proof is complete. $\qquad\square$

10 A proof of the Sargasso Lemma

We need some preliminary preparation and some new notation.

Let $s = (\alpha, \beta_0, m, \beta_1, \ldots, \beta_m) \in W$. If ξ is any one of the sequences

$$(\alpha, m), (\alpha, \beta_0, m), (\alpha, \beta_0, m, \beta_1), \ldots, (\alpha, \beta_0, m, \beta_1, \ldots, \beta_m),$$

then the *initial segments* of ξ are all those sequences up to and including ξ in the list. In particular, we shall denote by $\hat{\xi}$ the term immediately preceding ξ in the list (if there is one), and the term following ξ in the list is called a *one-point extension* of ξ. For any set $V \subseteq W$ we shall denote by V^* the set of all initial segments of members of V, and by $\hat{V}$ the set $\{\hat{\eta} : \eta \in V\}$. Also, for any $\xi \in W^*$ we define $\text{Ext}(\xi) = \{s \in W : \xi \text{ is an initial segment of } s\}$. In particular, if $s \in W$, then $\text{Ext}(\hat{s})$ is the set of all $u \in W$ which agree with s except possibly in the last coordinate.

Lemma 10.1 *Let $G = (W, E)$ be a graph with no infinite path. Then there are a large set $V \subseteq W$ and a finite set mapping g on V such that, for $s, t \in V$,*

$$(10.1) \qquad t \in g(s) \Rightarrow \{s, t\} \in E,$$

$$(10.2) \qquad t \notin g(s) \Rightarrow |E(t) \cap \text{Ext}(\hat{s})| \leq \omega,$$

$$(10.3) \qquad |g(s)| \text{ is constant on each block of } V.$$

Proof: Let $u \in \hat{W}$. Define for as long as possible a sequence of distinct elements $t_0, t_1, t_2, \ldots$ of W so that $| \text{ Ext } (u) \cap E(t_0) \cap \cdots \cap E(t_i)| = \omega_1$. Since there is no infinite path in G, this must terminate after some finite number of steps, say k. Put $V_u = \text{Ext} (u) \cap E(t_0) \cap \cdots \cap E(t_{k-1})$, $g_u = \{t_i : i < k\}$. The set $V = \cup\{V_u : u \in \hat{W}\}$ is clearly a large subset of W and g, defined by $g(s) = g_u \cap V$ $(s \in V_u)$, is a finite set mapping on V which satisfies (10.1) and (10.2). By (4.1) we may replace each block of V by a large subset so that (10.3) also holds. $\square$

The next lemma provides the important first step in proving the Sargasso Lemma.

Lemma 10.2 (Set-mapping reduction lemma [10]) *Let $G = (W, E)$ be a graph with no infinite path. Then there is a large subset U of W such that $link_U$ is a finite set mapping and such that, for $s, t \in U$,*

$$(10.4) \qquad \qquad \{s, t\} \in E \Leftrightarrow s \in link_U(t) \text{ or } t \in link_U(s).$$

Proof: By Lemma 10.1 there are a large subset V of W and a finite set mapping g such that (10.1)–(10.3) hold. Let J be the index set of V.

Fix an arbitrary element $w^* \in W$. We can assume $w^* \notin V$. We shall define functions $g_i : V \to W$ for $i < \omega$ as follows. By (10.3), for $(\alpha, m) \in J$ there is an integer $k = k(\alpha, m)$ such that $|g(s)| = k$ for all $s \in V(\alpha, m)$. Define $g_i(s)$ for $s \in V(\alpha, m)$ so that $g(s) = \{g_i(s) : i < k\}$ and $g_i(s) = w^*$ for $i \geq k$. Applying Lemma 9.1 k times we find a large subset of $V(\alpha, m)$ on which every g_i is steady. Thus, without loss of generality, we may assume that every g_i is steady on each block of V.

Note that, if $s, t \in V$ and $s \in \text{link}_V(t)$, then $s \in g(t)$ by (10.2) and so link_V is a finite set mapping. It also follows from (10.2) that $\text{link}_U = U \cap \text{link}_V$ for any large subset U of V, and so link_U is also a finite set mapping.

We want to construct a large subset U of V so that (10.4) holds. We shall do this by recursively defining certain sequences $\text{seq}(\nu, n) \in V^*$ for $(\nu, n) \in \omega_1 \times \omega$ so that $\text{SEQ} = \{\text{seq } (\nu, n) : (\nu, n) \in \omega_1 \times \omega\}$ is closed under initial segments, $\text{SEQ} \supseteq J$ and each $\xi \in \text{SEQ} \setminus V$ has ω_1 different one-point extensions in SEQ. This will ensure that the set $U = V \cap \text{SEQ}$ is large and has the same index set J.

Let $\nu < \omega_1$ and suppose that $\text{seq}(p, r)$ has been defined for all $p < \nu$ and $r < \omega$. We want to define $\text{seq}(\nu, n)$ for $n < \omega$. Put $X_\nu = \{\text{seq}(p, r) : p < \nu, r < \omega\}$.

Case 1. ν is even. In this case choose $\alpha < \omega_1$ to be the least ordinal such that $J_\alpha = \{m_0, m_1, \ldots\} \neq \emptyset$ and such that $(\alpha, m) \notin X_\nu$ for any $m \in \omega$. Define $\text{seq}(\nu, n) = (\alpha, m_n)$ $(n < \omega)$. (This will ensure that the index set of U is J).

Case 2. ν is odd. Let $\xi_0, \xi_1, \ldots, \xi_n, \ldots$ be a 1–1 enumeration of all those sequences in $X_\nu \setminus W$. For each $n < \omega$, $\text{seq}(\nu, n)$ will be chosen to be a one-point extension of ξ_n in V^*, but we need to avoid a certain countable set of extensions. Let $n < \omega$ and suppose that we have already chosen $\text{seq}(\nu, m)$ for $m < n$. Put $X_{\nu,n} = X_\nu \cup \{\text{seq}(\nu, m) : m < n\}$.

The bad extensions of ξ_n that we wish to avoid are members of the three sets A, B, C which are defined as follows. The set

$$A = \{g_l(s) : \xi \in X_\nu, \ l < \omega, \ g_l \text{ is constant on Ext}(\xi), \text{ and } s \in \text{Ext}(\xi)\}.$$

A is countable since X_ν is. Let L be the set of all $l < \omega$ such that ξ_n does not determine the value of g_l (i.e. there are extensions of ξ_n in V giving different values of g_l).

Then for $(l, u) \in L \times V$ there is at most one one-point extension of ξ_n (say ξ) which determines the value of g_l to be u (i.e. such that $g_l(s) = u$ for every extension s of ξ in V). This follows from the fact that g_l is steady on each block of V. Denote this unique ξ by $\xi_\ell(u)$ if it exists and put $\xi_\ell(u) = \xi_n$ if there is no such ξ. The set

$$B = \{\xi_\ell(u) : l \in L, u \in X_{\nu,n} \cap V\}$$

is countable. For $s \in V$ define $C_s = E(s) \cap \text{Ext}(\xi_n)$ if this set is countable, and define $C_s = \emptyset$ otherwise. Then the set

$$C = \cup\{C_s : s \in X_{\nu,n} \cap V\}$$

is countable.

Since there are ω_1 different one-point extensions of ξ_n which belong to V^*, it follows that we can choose such an extension $\text{seq}(\nu, n)$ which does not belong to $X_{\nu,n} \cup A \cup B \cup C$. This completes the inductive definition of SEQ. It is clear from the construction that $U = V \cap$ SEQ is a large subset of V with the same index set J. It remains to check that (10.4) holds.

Let $s, t \in U$ and suppose that $\{s, t\} \in E$. We can assume that s appears before t in the construction of SEQ. Suppose for definiteness that $t = \text{seq}(\nu, n) = (\alpha, \beta_0, m, \beta_1, \ldots, \beta_m)$. Thus, with the above notation, $\xi_n = \hat{t}$ and $s \in X_{\nu,n}$. Now $t \in$ SEQ and so $t \notin C_s$. Therefore, since $t \in E(s) \cap \text{Ext}(\xi_n)$, it follows that $E(s) \cap \text{Ext}(\xi_n)$ is uncountable. Thus $s = g_l(t)$ for some $l < \omega$ by (10.2). We claim that s is the constant value of g_l on the block $V(\alpha, m)$. Suppose not. Then there is some initial segment $\zeta \neq (\alpha, m)$ of t such that g_l is constant on all extensions of ζ (in V), but not constant on all extensions of $\hat{\zeta}$. Now s must appear before ζ in the construction of SEQ (since we avoid the appropriate set A when s is added) and so ζ could not have been chosen in SEQ (since at the appropriate stage we avoid the set B). This contradiction shows that $g_l(x) = s$ for all $x \in V(\alpha, m)$. It follows that $s \in \text{link}_U(t) = U \cap \text{link}_V(t)$. $\square$

We now prove the Sargasso Lemma, which for the convenience of the reader we restate.

Lemma 9.1 (Sargasso Lemma [10]) *Assume (*) holds. Suppose that G is a graph on W with no infinite path and that (9.1) and (9.2) hold. Then there are a large subset $U \subseteq W$ and a function $LOW \in C$ such that*

(9.3) link_U *is a finite set mapping on* U;

(9.4) $(\forall s, t \in U)(\{s, t\} \in E \Leftrightarrow s \in \text{link}_U(t) \text{ or } t \in \text{link}_U(s))$;

(9.5)   *if* $s \in U(\beta, n)$, $t \in U(\alpha, m)$ *and* $s \in \text{link}_U(t)$, *then* $\beta < \alpha$ *and* $LOW(m) < n$.

Proof: By Lemma 10.2 there is a large subset U of W such that (9.3) and (9.4) hold. We need to refine U to a large subset so that (9.5) also holds. We shall do this in several steps, but at the end of each stage we shall denote the resulting large subset by the same letter U. This will result in an economy of notation, and there is no harm in doing this since the properties that we shall successively claim for U are all hereditary. For example, because of (10.4), it follows that (9.3) and (9.4) hold if U is replaced by any large subset.

Let J be the index set for U, $A = \{\alpha : J_\alpha$ is infinite $\}$. For $(\alpha, m) \in J$, denote by $\text{Link}_U(\alpha, m)$ the constant value of $\text{link}_U(t)$ for $t \in U(\alpha, m)$. For $x \in U$ we define $\beta(x) = \beta$ and $n(x) = n$ if $x \in U(\beta, n)$.

For $\alpha \in A$, define $L(\alpha) = \{\beta : U(\beta) \cap \cup\{\text{Link}_U(\alpha, m) : m \in J_\alpha\} \neq \emptyset\}$. Then $L(\alpha)$ is countable. Hence there is a large subset A' of A such that $L(\alpha) \cap A' \subseteq \alpha$ for all $\alpha \in A'$. Thus, replacing U by $\cup\{U(\alpha, m) : \alpha \in A', m \in J_\alpha\}$, we may assume that U satisfies

(10.5) \qquad if $s \in U(\beta, n), t \in U(\alpha, m)$ and $s \in \text{link}_U(t)$, then $\beta < \alpha$,

which is one half of (9.5).

If $s, s' \in \text{link}_U(t)$, $s \neq s'$, then s and s' are infinitely connected in G and hence $\{s, s'\} \in E$ by the assumption (9.1). Therefore, since G has no local edges by (9.2), it follows that

(10.6) \qquad $|\text{link}_U(t) \cap U(\beta)| \leq 1$ for $t \in U$ and $\beta \in A$.

Since Link_U is a finite set mapping, for $(\alpha, m) \in J$ there is an integer $f(\alpha, m)$ such that $|\text{link}_U(\alpha, m)| < f(\alpha, m)$ and $n(x) < f(\alpha, m)$ whenever $x \in \text{Link}_U(\alpha, m)$. By Lemma 8.4 there are a large subset $K \subseteq J$ and a function $H : \omega \to \omega$ such that

$$f(\alpha, m) < H(m) \text{ holds for all } (\alpha, m) \in K.$$

Let $V = \cup\{U(\alpha, m) : (\alpha, m) \in K\}$. If $s, t \in V$ and $s \in \text{link}_V(t)$ then $s \in \text{link}_U(t)$ by (9.4). Thus, $\text{Link}_V(\alpha, m)$ is a subset of $\text{Link}_U(\alpha, m)$ and hence

$$|\text{Link}_V(\alpha, m)| < H(m) \text{ for } (\alpha, m) \in K.$$

Now, if $s, s' \in \text{link}_V(t)$ and $s' < s$, then as we just observed, $\{s, s'\} \in E$ and so $s' \in \text{link}_V(s)$ by (9.4) and (10.5). Thus, replacing U by V and J by K, we may assume that U has the additional property that

(10.7) \qquad if $s \in U(\beta, n) \cap \text{link}_U(t)$, then $|\text{link}_U(t) \cap \{s' : s' < s\}| < H(n)$.

For $(\alpha, m) \in J$ there is an integer $k(\alpha, m)$ $(< H(m))$ such that $\text{Link}_U(\alpha, m) = \{x_i(\alpha, m) : i < k(\alpha, m)\}$, where $x_0(\alpha, m) < x_1(\alpha, m) < \dots$. Put $\beta_i(\alpha, m) = \beta(x_i(\alpha, m))$ for $i < k(\alpha, m)$. Then $\beta_0(\alpha, m) < \beta_1(\alpha, m) < \dots$ by (10.6). By Lemma 10.1, for each $\alpha \in A$ there are an infinite subset J'_α of J_α and a function $q_\alpha : \omega \to \omega$ such that the set

$$S(\alpha) = \{\beta_i(\alpha, m) : i < \omega, m \in J'_\alpha, m \geq q_\alpha(i)\}$$

has order type at most ω^2. Thus S is a set mapping of order $\omega^2 + 1$ on A and so by Theorem 2.2 there is an S-free subset A' of cardinality ω_1. Let $K = \{(\alpha, m) : \alpha \in A', m \in J'_\alpha\}$, $V = \cup\{U(\alpha, m) : (\alpha, m) \in K\}$. Then V is a large subset of U.

Note that, if $(\alpha, m) \in K$, $i < \omega$, $m \geq q_\alpha(i)$ and if $\beta_i(\alpha, m)$ exists, then it belongs to $S(\alpha)$ and so does not belong to A', and consequently $x_i(\alpha, m)$ (if it exists) does not belong to V. Therefore, if $m \geq q_\alpha(i)$ for every $i \leq H(n)$, then $\text{Link}_V(\alpha, m)$ does not contain any of the first $H(n)$ elements of $\text{Link}_U(\alpha, m)$, and hence $x \notin \text{Link}_V(\alpha, m)$ if $n(x) = n$ by (10.7). In other words, $\text{Link}_V(\alpha, m)$ contains no element x with $n(x) \leq n$ provided that m is sufficiently large.

For $(\alpha, m) \in K$ define

$$\text{low}(\alpha, m) = \min(\{n(x) : x \in \text{link}_U(\alpha, m)\} \cup \{m\}).$$

From the above remark, it follows that $\text{low}(\alpha, .)$ is unbounded on $K_\alpha = J'_\alpha$ and therefore is increasing on some infinite subset K'_α of K_α. From Lemma 8.4 it follows that there are a large subset L of K and an unbounded non-decreasing function LOW such that

$$\text{LOW}(m) < \text{low}(\alpha, m) \text{ for all } (\alpha, m) \in L.$$

Then (9.5) holds if we replace U by the large subset $\cup \{U(\alpha, m) : (\alpha, m) \in L\}$. $\qquad \square$

References

[1]   J. Baumgartner and J. Larson, A diamond example of an ordinal graph with no infinite path, preprint, University of Florida (1984). To appear in Annals of Pure and Applies Logic.

[2]   B. Dushnik and E. W. Miller, Partially ordered sets, *Amer. J. Math.* **63**(1941), 605.

[3]   P. Erdös, Some remarks on set theory, *Proc. Amer. Math. Soc.* **1**(1950), 133–137.

[4]   P. Erdös, A. Hajnal, A. Maté and R. Rado, Combinatorial set theory: partition relations for cardinals, Akademiai Kiado, Budapest (1984).

[5]   P. Erdös, A. Hajnal and E. C. Milner, Set mappings and polarized partition relations, Coll. Math. Soc. Janos Bolyai, Combinatorial Theory and its Applications, Balatonfured (1969), 327–363.

[6]   P. Erdös, and E. Specker, On a theorem in the theory of relations and a solution of a problem of Knaster, *Colloq. Math.* **8**(1961), 19–21.

[7]   A. Hajnal, Proof of a conjecture of S. Ruziewicz, *Fund. Math.* **50**(1961), 123–128.

[8]   K. Kunen, Set Theory, An Introduction to Independence Proofs, North Holland, Amsterdam (1980).

[9]   D. Lazar, On a problem in the theory of aggregates, *Compositio Math.* **3**(1936), 304.

[10] J. Larson, A consequence of no short scale for ordinal graphs with no infinite paths, *J. London Math. Soc.* (2) **33**(1986), 193–202.

[11] J. Larson, Martin's Axiom and ordinal graphs: large independent sets or infinite paths, preprint, University of Florida (1985). To appear in Annals of Pure and Applied Logic.

[12] J. Larson, A GCH example of an ordinal graph with no infinite paths, *Trans. Amer. Math. Soc.* **303**(1987), 383-393.

[13] E. C. Milner and R. Rado, The pigeon-hole principle for ordinal numbers, *Proc. London Math. Soc.* (3) **15**(1965), 750–768.

[14] S. Picard, Sur un problème de M. Ruziewicz de la théorie des relations, pour les nombres cardinaux $m < \aleph_\Omega$, *Comptes Rendus Varsovie* **30**(1937), 12–18.

[15] F. P. Ramsey, On a problem of formal logic, *Proc. London Math. Soc.* (2) **30**(1930), 264–286.

[16] S. Ruziewicz, Une généralisation d'un théorème de M. Sierpinski, *Pub. Math. de l'Université de Belgrade* **5**(1936), 23–27.

[17] W. Sierpinski, Sur un problème de la théorie des relations, *Ann. Scuola Norm. Sup. Pisa* **2**(1933), 285–287.

ENDS OF INFINITE GRAPHS, POTENTIAL THEORY AND ELECTRICAL NETWORKS

M.A. PICARDELLO
Dipartimento di Matematica Pura e Applicata
Università dell'Aquila
I-67100 L'Aquila
Italy

W. WOESS
Institut für Mathematik und Angewandte Geometrie
Montanuniversität Leoben
A-8700 Leoben
Austria
and
Dipartimento di Matematica
Università di Milano
Via C. Saldini 50
I-20133 Milano
Italy

Abstract

We give an introductory survey on concepts and results concerning harmonic functions on infinite graphs with the goal of describing the interplay between graph sturcture and potential theory. A particular emphasis is on the connection between the Martin boundary for harmonic functions and the space of ends of the underlying graph. A variety of results is described.

1 Introduction

The purpose of this paper is to describe the interplay between potential theory on infinite graphs and their asymptotic structure. We consider infinite, connected locally finite graphs Γ. Except where mentioned otherwise, Γ will be unoriented and without multiple edges; loops however are allowed. We are interested in studying positive "harmonic" functions on Γ. The most natural definition of harmonicity is given in terms of the obvious mean value property: a function h on the set of vertices V of Γ is harmonic if its value at any vertex x is the average of its values at the neighbours $y \sim x$ (we shall write $y \sim x$ when the vertices y and x are adjacent).

We can rephrase this definition in a more algebraic fashion, which involves the adjacency matrix A of Γ, renormalized by dividing each row by its row sum. In other words, denote

G. Hahn et al. (eds.), Cycles and Rays, 181–196.

by $\deg(x)$ the number of edges connecting at the vertex x, and let

$$p(x,y) = \begin{cases} 1/\deg(x), & \text{if } x \sim y, \\ 0, & \text{otherwise.} \end{cases} \tag{1}$$

The matrix P acts on the space of functions on V by the rule

$$Pf(x) = \sum_y p(x,y)f(y). \tag{2}$$

Then the harmonic functions are the fixed points of the linear operator P. More generally, we can extend the definition of harmonicity by considering "non-isotropic averages", that is, operators P which are

$$\text{nearest-neighbour: } p(x,y) > 0 \text{ if and only if } x \sim y \tag{3a}$$

and

$$\text{stochastic: } \sum_y p(x,y) = 1 \text{ for every } x. \tag{3b}$$

From now on, the notion of harmonicity will be always understood in this more general sense. In the final section non-nearest-neighbour operators will also be considered. Our purpose is to describe all positive harmonic functions,

$$\mathcal{H}^+ = \{h : V \to \mathbf{R}^+ | Ph = h\}, \tag{4}$$

in terms of their asymptotic values "at infinity". For this, we need first to construct a boundary $\mathcal{M}$ of Γ which is large enough. The abstract construction of the "Martin boundary" $\mathcal{M}$, described in §2, relies on the behaviour at infinity of the random walk on Γ induced by the operator P, which, from now on, will be referred to as "transition operator". We will always assume that this random walk is transient (otherwise, positive harmonic functions are constant). In order to link the structure of the Martin boundary with the geometry of the graph, we need a less abstract realization of $\mathcal{M}$. A natural candidate is the space Ω of all ends of Γ introduced in [12] and [14] and described in §3. Ends are represented by infinite paths in Γ. Two paths give rise to the same end if they cannot be separated by finite sets of vertices.

Now the problem is to provide sufficient conditions for $\mathcal{M} = \Omega$ (in the sense that they give rise to homeomorphic compactifications of Γ). Equality does not hold for all graphs and transition operators, as is pointed out in §3 by looking at typical examples, including lattices and trees. Sufficient conditions for $\mathcal{M} = \Omega$ are given in §4, in terms of "diameter" and "conductance" of ends. These notions become more familiar if the graph is thought of as an electric network, following [22, 9]; see also [20, 33]. Then ends can be visualized as asymptotic branches of the network without bifurcations. However, the lack of bifurcations in an end is a condition which depends only on the structure of Γ, not on the transition operator. The size of the end may become larger and larger and the "electric resistance" between vertices moving to infinity along two different paths within the end may grow. Our conditions are expressed in terms of bounds on the rate of growth of the "diameter" and the "transversal conductance" of each end.

In the case of trees, the study of the Martin boundary was begun in [10] (in the special case of homogeneous trees and group-invariant nearest neighbour transition operators) and

[4]. In this environment the geometry is simpler, and ends are in one-to-one correspondence with rays originating from any reference vertex. It was shown in [6] and [25] that $\mathcal{M} = \Omega$ for a large class of transition operators on trees which are not nearest neighbour but of bounded range.

An application of our results to potential theory on trees is given in §5. Consider the normalized adjacency matrix P on a homogeneous tree T and its cone of positive harmonic functions $\mathcal{H}^+$. It is easy to see that harmonic functions satisfy the mean value property not only with respect to the neighbours of each vertex x but also with respect to the balls of any radius, that is the set of vertices at any given distance $r(x)$ from x. Then we can choose and fix a positive integer $r(x)$ for every vertex x, and consider the operator P_r defined at each vertex x by the average over the ball $B(x, r(x))$ with center x and radius $r(x)$. We have seen that its cone $\mathcal{H}_r^+$ of positive harmonic function contains $\mathcal{H}^+$.

Is it true that, if at each x a positive function h satisfies the mean value property with respect to $B(x, r(x))$, then h is harmonic in the usual sense? This amounts to asking whether $\mathcal{H}_r^+ = \mathcal{H}^+$. It turns out that equality holds if and only if the Martin boundaries $\mathcal{M}$ of P and $\mathcal{M}_r$ of P_r coincide. However, $\mathcal{M} = \Omega$. Therefore we consider the graph Γ_r with the same vertices as T, whose edges correspond to the jumps allowed by P_r and study its space of ends Ω_r. In general, $\Omega_r \subset \Omega$ strictly. In terms of a "Lipschitz condition" on the radius function $r(x)$, we provide sufficient conditions for the equalities $\Omega_r = \Omega$ and $\mathcal{M}_r = \Omega_r$.

We shall not deal extensively with the group-theoretical side of this subject. The interested reader is referred to the survey paper [37], and to the "classic" references [1, 19].

2 The Martin boundary

In this section we give an outline of the Martin boundary theory for discrete transition operators, due to Doob [7]. Except for the last section, P will always be assumed nearest neighbour, but this hypothesis is not relevant here. Instead of assuming (3a), we may consider a finite range transition operator P and Γ, that is, only jumps to finitely many vertices are allowed at each point. It is convenient to think of P as a table of transition probabilities: its entry $p(x, y)$, $x, y \in V$, is the probability of jumping from x to y. Then the entry $p^{(n)}(x, y)$ of the power P^n of P represents the probability of hitting y from x in n steps.

We introduce the *Green kernel*

$$G(x, y) = \sum_{n \geq 0} p^{(n)}(x, y). \tag{5}$$

Its probabilistic significance is the following: if X_n is the random vertex visited at time n, then $G(x, y)$ is the expected number of visits at vertex y starting at $X_0 = x$. Since Γ is connected, either $G(x, y) = \infty$ for all x, y, or $G(x, y) < \infty$ for every x, y. In the first case, we say that the random walk X_n is recurrent. For our purpose, this eventuality is not interesting, because it is easy to see that all positive harmonic functions are constant [18, Prop. 6.4]. Therefore we shall concentrate on the transient case and assume that

$$G(x, y) < \infty \text{ for all } x, y \in V. \tag{6}$$

We want to express positive harmonic functions on Γ in terms of some "boundary values". It may be useful to keep in mind the analogy with potential theory on the disc D. There, positive harmonic functions can be thought of as electrostatic potentials arising from distributions of electric charges on the boundary of the circle ∂D. All these potentials can be represented as "superpositions" (sums or integrals) of the elementary potentials determined by point charges in ∂D, that is, Poisson kernels. We want to give an abstract construction of Poisson kernels associated with P on Γ. In analogy with the disc, each Poisson kernel will ideally correspond to a "boundary point" of Γ.

Choose and fix a reference vertex o and define the *Martin kernel*

$$K(x,y) = G(x,y)/G(o,y). \tag{7}$$

(Here, we have to assume that $G(o,y) > 0$ for all y, which is always true in the nearest neighbour case, as Γ is connected.) It is easily seen that $K(.,y)$ is harmonic everywhere except at y. Thus, if we "move y to infinity" in such a way that $K(x,y)$ has a limit, this limit gives us a positive harmonic function on Γ (normalized so as to take the value 1 at the vertex o). How can we make precise this limit process? It is easy to proceed abstractly. For the moment, we do not think of the variable y as moving to infinity. On the other hand, we define the *Martin compactification* as the smallest metrizable compatification $\hat{V}$ of V such that all the Martin kernels $K(x,.)$ extend continuously to $\hat{V}$ and separate the points of $\hat{V}$ in the first variable.

The *Martin boundary* is then defined as $\mathcal{M} = \hat{V} - V$. The extensions $K(x,m)$, $m \in \mathcal{M}$, are the *Poisson kernels*. Although it is abstract, this construction solves the problem of reconstructing positive harmonic functions on Γ by their boundary data. Indeed, as for the disc, to reconstruct a positive harmonic function by its boundary behaviour means to express it as an integral on $\mathcal{M}$ of Poisson kernels, i.e., if $h \in \mathcal{H}^+$, then there is a positive Borel measure ν_h on $\mathcal{M}$, such that for every $x \in V$,

$$h(x) = \int_{\mathcal{M}} K(x,.)d\nu_h. \tag{8}$$

Up to dilations, we can limit attention to positive harmonic functions normalized to take the value 1 at o, that is, to the base $\mathcal{B}$ of the convex cone $\mathcal{H}^+$. When speaking of a convex set of functions, we can look at extreme points. The extreme points of $\mathcal{B}$ turn out to be minimal harmonic functions on Γ, and, indeed, Poisson kernels. Now the integral representation of a positive harmonic function $h \in \mathcal{B}$ boils down to expressing h as a convex combination (or integral) of extreme points. On the basis of a cone, this convex disintegration is always available by Choquet's theory (see [5, 24] or [18, Ch. 6] for details).

Moreover,

$$\text{if } K(.,m_1) \equiv K(.,m_2) \text{ then } m_1 = m_2 \text{ in } \mathcal{M}. \tag{9}$$

This allows us to think of the points of $\mathcal{M}$ as the Poisson kernels. In principle, not all Poisson kernels need to be minimal positive harmonic functions; that is, not all points of $\mathcal{M}$ need to be extreme points of $\mathcal{B}$. However, for the purposes of the convex disintegration we can, and always will, restrict attention to extreme points ("minimal boundary").

Let us now return to the "geometric" point of view, which suggests to construct Poisson kernels as limits of Martin kernels $K(x,y)$ as $y \to \infty$. What do we mean by $y \to \infty$? We should look at infinite paths π in Γ, and verify whether or not $K(x,y)$ converges to a limit as y tends to infinity along π. Two paths π_1, π_2 with this property should be considered equivalent if the limits $K(x, \pi_1)$ and $K(x, \pi_2)$ are the same for every $x \in V$. Then each class of equivalence which determines a *minimal* $K(., \pi)$ should correspond to a point in the Martin boundary. This is the so-called "Martin method". However, in order to obtain by this method a truly geometric construction of $\mathcal{M}$, we should restrict attention to simple paths in Γ (without repeated vertices) and should provide a geometric characterization of equivalence of paths and of the topology of $\mathcal{M}$. We shall proceed to this program in the next section.

Here, as a final remark, we remind the reader of the two interesting interpretations which arise naturally in our context. One is "electric": Poisson kernels are regarded as elementary electrostatic potentials. In other words, the graph Γ should be interpreted as an infinite electric network. Its edges $[x, y]$ are resistors with conductance $p(x, y)$ and positive harmonic functions are electric potentials. The other is the probabilistic one: the transition operator is the generator of a transient random walk $\{X_n, n = 0, 1, \ldots\}$. Then the "points at ∞", i.e., the admissible escape trajectories of the random walk, are always representable as points of $\mathcal{M}$.

3   The space of ends of an infinite graph

We follow here [12] and [14]. See also [17, 29]. Throughout this section, a *simple path* in Γ will be a sequence of subsequently adjacent distinct vertices.

Two paths are *equivalent* if they cannot be separated by finite sets in V. Classes of equivalence are called *ends* and the set of all ends is denoted by Ω. We now endow Ω with a compact Hausdorff topology. For every finite set $F \subset V$, consider the infinite connected components Γ_i of $\Gamma - F$ and denote by C_i the set of vertices of Γ_i. Adjoin to C_i the set of ends represented by paths lying in C_i and denote by $\overline{C}_i \subseteq V \cup \Omega$ the augmented set. For every $x \in (V - F) \cup \Omega$, denote by $C(F, x)$ the augmented connected component $\overline{C}_i$ which contains x. Then the family

$$\{C(F, x) | F \subset V \text{ finite}, x \in (V - F) \cup \Omega\} \tag{10}$$

is a basis of our topology. The family $\{C(F, x) | F \subset V \text{ finite}\}$ is a local basis at x. It is easy to see that V is discrete, open and dense and Ω is compact.

This approach leads naturally to the belief that there should be very strict connections between $\mathcal{M}$ and Ω, and the remainder of the paper is devoted to outlining them. However, there are cases where Ω reduces to a single point, while $\mathcal{M}$ does not (see the example of the Euclidean lattices below). Therefore, the most general result that we can hope for is the existence of a natural surjection from $\mathcal{M}$ to Ω, when $\mathcal{M}$ arises from a nearest neighbour transition operator. It is comforting to discover that this statement is rather easy to prove (see [26] in the present context, and [32] for a more general and abstract setting).

Theorem 1 *If P is nearest neighbour and transient, then there exists a continuous surjection $\tau : \hat{V} \to V \cup \Omega$, which is trivial on V and maps $\mathcal{M}$ onto Ω.*

In fact, to prove Theorem 1, the assumption (3a) that the edges of the graph Γ correspond to the transitions allowed by P can be partially relaxed. If x and y are neighbours, then we may assume that one of the numbers $p(x, y)$ and $p(y, x)$ is non-zero, but not necessarily both, and

$$p(x, y) > 0 \text{ implies } x \sim y. \tag{11a}$$

(Here, as elsewhere, it may be helpful to regard the edges of Γ as pairs of oriented edges with transition probabilities depending on the orientation. Now we are considering the case when jumping along one of the two edges in a pair may not be allowed, that is, that oriented edge is "missing".) However, now we have to assume that P is *uniformly irreducible*; that is, there is a positive integer L such that

$$y \sim x \text{ implies } p^{(k)}(y, x) > 0 \text{ for some } k \leq L. \tag{11b}$$

In other words, the minimal number of steps needed to connect any two neighbours must be bounded throughout Γ. Also this assumption is aimed at adapting the transition operator to the geometry of Γ. We assume that vertices which are close in Γ are at bounded distance under the action of P.

Now it is clear that the space of ends is a true geometric realization of the Martin boundary only when τ is a homeomorphism. By compactness and the previous result, τ is a homeomorphism if it is injective. We shall provide sufficient conditions in the next section; here we present some significant examples.

A. *Trees and tree-like graphs.*

Trees are the typical setting where the identity $\mathcal{M} = \Omega$ holds, that is, τ is a homeomorphism. In a tree T, simple paths are rays and two rays are equivalent if and only if they join after a finite number of edges. If we choose a reference vertex o and restrict attention to rays issuing from o, then any two distinct rays are inequivalent and each end coincides with a ray.

For nearest neighbour transition operators on a tree, the homeomorphism $\mathcal{M} \simeq \Omega$ was proved in [4]. In this case, every Poisson kernel turns out to be locally constant on Ω.

If the tree is homogeneous of degree q (that is, every vertex belongs to exactly q edges), then it is the Cayley graph of a free group (if q is even) or of an appropriate free product (see [BP] for details and references in an analytical setting). In these cases, it is natural to restrict attention to group-invariant transition operators. In the special case of free groups, the fact that $\mathcal{M} \simeq \Omega$ was originally proved in [10]. This result was extended in [6] to all *finitely supported* (not only nearest neighbour!) transition operators invariant under the free group. In this case, the Poisson kernel is no longer locally constant on Ω, and the argument proves existence of limits at infinity of Martin kernels by making use of a fixed point theorem.

Finally, all these results find a natural generalization to the setting of nearest neighbour transition operators on graphs Γ which admit *uniformly spanning trees* [25]. A uniformly spanning tree for Γ is a tree T with the same set of vertices of Γ and whose metric is equivalent to the metric of Γ. Groups whose Cayley graphs admit such trees are finite extensions of free groups [38]. The approach of [25] requires that the transition operator be very well adapted to the geometry of Γ and T. This requirement reduces to assuming a positive lower bound for the non-zero transition probabilities. The relevance of this condition (which may be weakened, see [26] and below) will become more clear in Example D. Basically, it guarantees that the probabilities to make steps along two different edges cannot differ significantly. Hence, from the probabilistic (or electric) viewpoint, each edge does in fact contribute to the connectivity properties of Γ.

If we restrict attention to Cayley graphs, further results in a tree-like spirit are known. In particular, consider two discrete groups G_1 and G_2, with Cayley graphs Γ_1 and Γ_2 and define the graph $\Gamma = \Gamma_1 * \Gamma_2$ as the corresponding Cayley graph of the free product $G_1 * G_2$. We refer the reader to [36] for a characterization of the Martin boundary of Γ in terms of those of Γ_1 and Γ_2.

B. *Euclidean lattices.*

For a lattice with dimension at least two, there is only one end. Let us look at the Martin boundary of the integer lattice $\mathbf{Z}^n$, described in [16, 23]. We assume that the transition operator is group-invariant and with finite range: $p(x, y) = p(o, y - x)$ for every $x, y \in \mathbf{Z}^n$ and $x \to p(o, x)$ is finitely supported. Consider the moment

$$\mu = \sum_{x \in \mathbf{Z}^n} x p(o, x). \tag{12}$$

If $\mu = 0$, then all positive harmonic functions are constant (see [31] for $n \geq 3$; for $n \leq 2$ we even have that P is recurrent). Therefore, in this case, also $\mathcal{M}$ is a singleton.

But, if $\mu \neq 0$ then P is transient, for every n. If $n = 1$, i.e., the lattice is $\mathbf{Z}$, it is immediately seen that $\mathcal{M}$ consists of two points ($\neq \infty$). If $n > 1$, $\mathcal{M}$ turns out to be homeomorphic with the unit sphere S^{n-1} in $\mathcal{R}^n$. More precisely, start by contracting $\mathbf{Z}^n$ into the unit ball in $\mathbf{R}^n$ by the map $x \to \frac{1}{1+\|x\|}x$. After this contraction is performed, it becomes clear that S^{n-1}, with the relative Euclidean topology, is a natural geometric boundary for $\mathbf{Z}^n$. We are left with the problem of studying the directions of convergence of the Martin kernel. It is known, and easy to see, that the minimal positive harmonic functions are exponentials, i.e., functions of the type $x \to e^{x \cdot u}$ for some $u \in \mathbf{R}^n$ [5, 8] (the dot denotes inner product). Observe that

$$Pe^{x \cdot u} = \phi(u)e^{x \cdot u} \text{ with } \phi(u) = \sum_y p(o, y)e^{y \cdot u}. \tag{13}$$

Therefore we must choose u such that $\phi(u) = 1$. Let $\|y_n\| \to \infty$ in the direction v; that is, let $y_n/\|y_n\| \to v \in S^{n-1}$. If $K(x, y_n)$ converges, then $\lim_n K(x, y) = e^{x \cdot u}$ for some $u = u(v)$. We want to find u.

The result proved in [23] is the following:

Theorem 2 [23] *If $y_n/\|y_n\| \to v$, then $K(x, y_n) \to e^{x \cdot u}$, where the exponent $u = u(v)$ is the unique solution of*

$$\phi(u) = 1 \text{ and } \frac{grad\ \phi(u)}{\|grad\ \phi(u)\|} = v.$$

In other words, the direction of convergence of the Martin kernel is the direction of the gradient of the eigenvalue map ϕ evaluated at the exponent u. A similar result can be proved for the cartesian product of two homogeneous trees; see the forthcoming paper [28].

C. *Hyperbolic lattices.*

These lattices are Cayley graphs of discrete groups of conformal automorphisms of the complex unit disc D ("Fuchsian groups"). The fundamental domain in D of a Fuchsian group F and its F-translates give rise to a hyperbolic tessellation of D, as in Figure 1. See [21] for more examples.

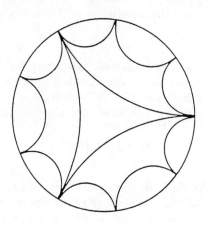

Figure 1

Choose a point in the fundamental domain (for instance, the origin o), and look and its F-orbit (the "centers" of the translates of the domain). The graph Γ is obtained by connecting two such points if they belong to adjacent domains. It turns out to be a graph with exponential growth and with certain "tree-like" properties, but it may have only one end. The transition operator is assumed to be finitely supported on Γ and F-invariant. Consider the set Λ of a accumulation points of the vertices of Γ in the circle ∂D. It was shown in [30] that Λ provides a geometric description of the Martin boundary of Γ for a large class of Fuchsian groups (see [30] for more details; the following result is slightly mis-stated in [37]).

Theorem 3 [30] *If F is a non-exceptional Fuchsian group, then there is a continuous surjection $M \to \Lambda$ which is injective except at most at countably many points, where it is two to one.*

We warn the reader that transition operators on Fuchsian groups, or trees, which allow jumps of unbounded length may give rise to a considerably different boundary theory. See [13] for an example.

D. *The ladder.*

   This final example is introduced with the aim of stressing the relationship between the geometry of the graph (the disposition of its edges) and the size of the transition probabilities (the likelihood of jumping along those edges). The relevance of this relationship, which already appeared in Example A, relies upon the following observation. Suppose that the transition probabilities associated with a given sequence of edges decay fast enough. Then we may approximate the transition probability with zero from a certain edge onward. For the purpose of studying harmonic functions, this amounts to removing those edges. This may change the end structure of the graph. Therefore the Martin boundary should depend not on the geometry of Γ, but on its geometry "weighted" with P.

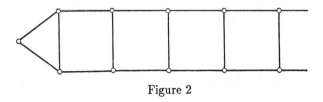

Figure 2

Consider the graph Γ given by the ladder in Figure 2. It has only one end, and if the transition probabilities along its edges are non-zero and bounded below, the results of [25] show that there is only one positive harmonic function, up to normalization.

However, if we remove all the vertical edges except a finite number, then there are two ends and a two-dimensional space of positive harmonic functions. In other words, $\mathcal{M}$ consists of two points. An easy computation shows that, if the transition probabilities along the vertical edges of the original graph decay at a fast (exponential) rate, then again $\mathcal{M}$ consists of two points, although Ω is a singleton. In this case, P is not sufficiently "well adapted" to Γ.

E. *Transversal conductance of some electric networks.*

Let us make use again of the interpretation of Γ as an electric network. At first the ladder of Figure 2 looks well connected, and we expect that there is only one elementary electrostatic potential (up to multiples). But if the transversal conductance decays very quickly, then the network behaves as if it was disconnected into two distinct lines of resistors; each line will give rise to one electrostatic potential.

More generally, what would happen for ladders with many layers? Consider the example of the network Γ in Figure 3.

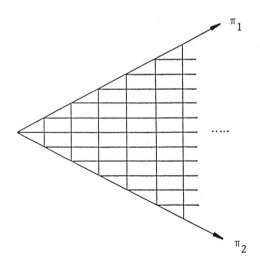

Figure 3

Again, Γ has only one end. But now we expect that the dimension of the cone of positive harmonic functions, i.e., the size of $\mathcal{M}$, will be larger than one if the vertical conductivities decay too quickly. Indeed, in this case the two external paths π_1 and π_2 (and also other intermediate paths) will behave as if they were electrically disconnected from a certain point onward. The sufficient conditions for $\mathcal{M} = \Omega$, that we are going to present in the next section, will need some assumptions on the rate of decay of transversal conductivities.

4 Diameter, minimal conductance and transversal current of an end

In this section we give sufficient conditions for $\mathcal{M} = \Omega$, in the sense that the surjection τ of Theorem 1 is bijective. If $F \subset V$ is finite, we denote by $diam(F)$ its diameter in the discrete metric of Γ and define its (geodesic) completion $\tilde{F}$ as the set of vertices which belong to a shortest path connecting two vertices of F. Regarding Γ again as an electric network, we introduce the *minimal conductance*

$$u(F) = \min\{p(x,y)|p(x,y) > 0, x, y \in \tilde{F}, \ x \neq y\}; \tag{14}$$

in the trivial case of a singleton, we set $u(F) = 1$. In addition, we define the *specific current* between two vertices x, y as

$$U(x,y) = G(x,y)/G(y,y) = \sum p(x,t_1)p(t_1,t_2)\cdots p(t_{n_1},y), \tag{15}$$

where the sum is taken over all finite paths $[x, t_1, \ldots t_{n-1}, y]$ of length ≥ 1 such that $t_j \neq y$ for all j. It is important to realize that from the probabilistic point of view, $U(x,y)$ is the hitting probability of y from x (see [9] and [4]), and that $K(x,y) = U(x,y)/U(o,y)$.

For every end ω, we can select sequences of finite sets $F_n \subset V$ such that for all n,

$$C(F_{n-1},\omega) \supseteq F_n \cup C(F_n,\omega),$$

$\{C(F_n,\omega)\}$ is a basis at ω of the topology, and $\qquad\qquad\qquad\qquad$ (16)

$$\tilde{F}_{n-1} \cap F_n = \emptyset.$$

Such a sequence is called *(strictly) contracting* towards ω (see Figure 4).

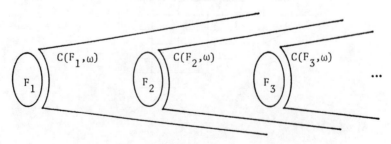

Figure 4

Now we define the *transversal current* between two vertices x, y in the nth set F_n of a sequence contracting towards ω as

$$U_{tr}(x, y) = \sum p(x, t_1)p(t_1, t_2)\cdots p(t_{n-1}, y), \tag{17}$$

where the sum is taken over all finite paths from x to y which visit y only at the end and do not ever visit F_{n+1}. Similarly, the transversal current of F_n is

$$U_{tr}(F_n) = \min\{U_{tr}(x, y) | x, y \in F_n, \ x \neq y\}. \tag{18}$$

If F_n is a singleton, one sets $U_{tr}(F_n) = 1$. Finally, the diameter, minimal conductance and transversal current of the end ω are defined by

$$diam(\omega) = \inf_{\{F_n\}} \liminf_n diam(F_n);$$

$$u(\omega) = \inf_{\{F_n\}} \liminf_n u(F_n); \tag{20}$$

$$U_{tr}(\omega) = \inf_{\{F_n\}} \liminf_n U_{tr}(F_n).$$

Observe that the transversal current is small when there are equivalent paths in ω with a poor "electric connection". This may not imply that the geometric connection is poor (that is, the diameter becomes large), or the minimal conductance is low (that is, the electric resistance of each resistor is high). On the other hand, one obviously has $U_{tr}(F_n) \geq u(F_n)^{diam(F_n)}$ (limit attention to geodesic paths in $\tilde{F}_n$) and $U_{tr}(\omega) \geq u(\omega)^{diam(\omega)}$.

Theorem 4 [26] *Suppose that there exists a sequence $\{F_n\}$ of finite sets, (strictly) contracting towards $\omega \in \Omega$, such that*

$$\sum_n U_{tr}(F_n) = \infty. \tag{21}$$

Then:
   (i) the surjection τ of Theorem 1 is injective at ω, i.e., $\tau^{-1}(\omega) = 1$;
   (ii) ω is an extreme point of $\mathcal{M}$.

In particular, if (21) holds for every ω, then $\mathcal{M} = \Omega$, and all points are extreme. Condition (21) holds, for instance, if (a) $\sum u(F_n)^{diam(F_n)} = \infty$, or if (b) $U_{tr}(\omega) > 0$, or if (c) $u(\omega) > 0$ and $diam(\omega) < \infty$.

5 The mean value property on homogeneous trees

In this section, we denote by P the normalized adjacency matrix of the homogeneous tree T of degree $q + 1$, $q \geq 2$. Here we shall also consider another transition operator on T, not nearest neighbour, defined as follows. Let $x \to r(x)$ be a positive integer-valued function on the set of vertices. Then define the transition operator P_r by

$$p_r(x, y) = \begin{cases} 1/\#B(x, r(x)) & \text{if } dist(x, y) \leq r(x), \\ 0, & \text{otherwise.} \end{cases} \tag{22}$$

Let us denote by $\mathcal{H}_r^+$ the cone of positive harmonic functions with respect to P_r. We have already observed in §1 that $\mathcal{H}_r^+ \supseteq \mathcal{H}^+$. Here we give sufficient conditions for $\mathcal{H}_r^+ = \mathcal{H}^+$, in terms of Martin boundaries and ends. The problem arises as a discrete analogue of the mean value property and its (restricted) converse in classical potential theory. For results of this type on the disc (and on more general domains in Euclidean space), see [11, 15, 34, 35, 2, 3]. All the results of this section are taken from [27].

As usual, we denote by $\mathcal{M}$ the Martin boundary of the operator P and by Ω the space of ends of the underlying tree. The corresponding spaces for P_r will be denoted by $\mathcal{M}_r$ and Ω_r. In fact, Ω_r is the space of ends of another graph, denoted Γ_r, which corresponds to P_r more naturally than the original tree. It has the same vertices as T, and two vertices x, y delimit an edge if and only if

$$dist(x, y) \le \max\{r(x), r(y)\}. \tag{23}$$

Observe that $p_r(x, y) > 0$ implies $x \sim y$ in Γ_r, but not conversely; compare with (11a). Since Γ_r has more edges than T, it is easy to see that there is a continuous surjection $\sigma : V \cup \Omega \to V \cup \Omega_r$, σ is trivial on V, and $\sigma(\Omega) = \Omega_r$. We know that $\mathcal{M} = \Omega$ (Example A of §3). It is not difficult to show the following:

Proposition 1 [27] *If* $\mathcal{M}_r \simeq \Omega$, *then* $\mathcal{H}_r^+ = \mathcal{H}^+$. *(The converse is also true if the Martin boundary is understood in the minimal sense.)*

The proof consists of showing that the Poisson kernels for P and P_r are the same. We use two tools. The first is the fact that every continuous function on Ω has a unique continuous extension on $T \cup \Omega$, harmonic on T (the "Dirichlet problem", see [6]). The second tool is the "probabilistic Fatou theorem" [18, Theorem 10.43]: if $\psi \in L^\infty(\Omega)$ and h is its Poisson integral on T, then, almost surely, $h(X_n) \to \psi(X_\infty)$, where X_∞ is the limit random variable on Ω of the random walk X_n induced by P_r.

Thus the converse of the mean value property, i.e., the question whether $\mathcal{H}_r^+ = \mathcal{H}^+$, is in fact a problem related to the geometric characterization of the Martin boundary of the transition operator P_r. However, in contrast with the typical examples for trees (Example A, §3), the operator P_r is not of bounded range, unless the function r is bounded on T.

On the other hand, one already has the following result:

Corollary 1 [25] *If* r *is bounded, then* $\mathcal{H}_r^+ = \mathcal{H}^+$.

Indeed, under this hypothesis, $\mathcal{M}_r \simeq \Omega$ (see Example A in §3), and Proposition 1 applies. This leads us to study under which (more general) conditions $\mathcal{M}_r \simeq \Omega$. Observe that now we will have two continuous surjections, namely $\tau : \mathcal{M}_r \to \Omega_r$ (Theorem 1) and $\sigma : \Omega \to \Omega_r$ (remarks before Proposition 1). We shall give conditions under which σ and τ are both injective. In the spirit of [35], the conditions will be stated in terms of a Lipschitz inequality concerning the function r.

In the sequel, we shall assume that, for some $\Phi : \mathcal{N}_0 \to \mathcal{N}_0$, and for every $x, y \in V$,

$$|r(x) - r(y)| \le \Phi(dist(x, y)). \tag{24}$$

The first question is: how large can Φ be chosen if σ has to be injective?

Proposition 2 [27] *Suppose that* $\limsup_n \Phi(n) - n < \infty$ *(i.e., that there exists* $R > 0$ *such that* $|r(x) - r(y)| \le dist(x, y) + R$). *Then the continuous surjection* $\sigma : \Omega \to \Omega_r$ *is a homeomorphism.*

The proof of the proposition shows that σ is one-to-one, by the following argument. Suppose that ω_1 and ω_2 are two different ends in Ω, represented by two infinite geodesics (rays) π_1 and π_2 of T starting from the reference vertex o. Then π_1 and π_2 also represent two different ends of Γ_r, i.e., outside of a ball $B(o, \rho)$ (defined with respect to the tree-metric), for sufficiently large ρ there is no path in Γ_r which connects π_1 and π_2.

As we have already observed, $y \sim x$ in Γ_r does not necessarily imply $p_r(y, x) > 0$. Hence, to be able to apply Theorem 1, we have to prove uniform irreducibility in the sense of (11b) with respect to the structure of Γ_r.

Proposition 3 [27] *Suppose that* $\limsup_n \Phi(n)/n < 1$. *Then* P_r *is uniformly irreducible.*

The idea of the proof is the following. We can reach y from x in one step via P_r. How many steps are necessary to come back? Since the slope of the function r is less than one, if $r(x)$ is large, $r(w)$ will be large for w between x and y, and it will not take too long to come back. In fact, if $|r(x) - r(y)| \le A\ dist(x, y) + B$, where $0 \le A < 1$ and $B > 0$, then the number $L = \left\lceil \frac{1+B}{1-A} \right\rceil$ meets the requirements of condition (11b).

Observe that this argument fails if $A = 1$. Suppose for instance, that $r(x) = dist(x, o) + 1$. Then the largest distance we can cover in k steps moving out of the origin is $\sum_{j=1}^{k} j \simeq k^2/2$, so the time needed to move from o to a point at distance $n > 0$ grows as $\sqrt{2n}$, but we can come back to o in just one step.

Finally, to apply Theorem 4 we need good control over the transversal current of a sequence of sets contracting in Γ_r towards an end ω. For this, we consider "barriers", i.e., finite sets that the trajectories of P_r must intersect in the process of moving to infinity along ω. As the length of the admissible jumps may increase while moving out, these will only be one-way barriers: they are necessarily hit while moving *out* from the origin, but not conversely. We can choose the barriers far enough apart as to be included in disjoint balls. If Φ grows slwoly enough, inside each ball B we can find "good" upper and lower bounds for the radius function. This allows us to minimize P_r inside B by a *substochastic* transition operator P'_ρ with *constant* radius ρ (depending on B).

Now, for operators with constant radius it is not too difficult to estimate the transversal current of a ball, and we obtain the following result (which shows how slowly r should grow).

Theorem 5 [27] *If* T *is the homogeneous tree of degree* $q + 1$ *and*

$$\limsup_n \frac{\Phi(n)}{\log_q n} < \frac{2}{3},$$

then τ *is injective. Thus* $\mathcal{M}_r = \Omega$ *and* $\mathcal{H}_r^+ = \mathcal{H}^+$.

The constant $\frac{2}{3}$ is the best constant which we could establish by this method. Thus, the method applies only to a "nice" radius function bounded by $\frac{2}{3} \log_q(dist(x, o))$ plus a constant.

What happens if the growth is larger, for instance linear? With the assumption of a Lipschitz condition, all our attempts to construct a function h in $\mathcal{H}_r^+ - \mathcal{H}^+$ have led to unbounded oscillations of h, giving rise to negative values.

On the other hand, if we set $r(o) = 2$ and

$$r(x) = 3 \cdot 2^{[\log_2 d(x,o)]} \text{ for } x \neq o, \tag{25}$$

then we can give a counterexample, that is, a function in $\mathcal{H}_r^+$ (in fact, a bounded one) which is not harmonic. However, in this example, the Lipschitz condition fails for the radius function r under consideration: it has unbounded jumps, although it grows at most linearly with respect to the distance from o. The reader is referred to [27] for all details.

References

[1] R. Azencott and P. Cartier: Martin boundaries of random walks on locally compact groups. *Proc. 6th Berkeley Symposium on Math. Statistics and Probability* **3**(1972), 87–129.

[2] J. R. Baxter: Restricted mean values and harmonic functions. *Trans. Amer. Math. Soc.* **167**(1972), 451–463.

[3] J. R. Baxter: Harmonic functions and mass cancellation. *Trans. Amer. Math. Soc.* **245**(1978), 375–384.

[4] P. Cartier: Fonction harmoniques sur un arbre. *Symposia Math.* **9**(1972), 203–270.

[5] G. Choquet and J. Deny: Sur l'équation de convolution $\mu = \mu * \sigma$. *C.R. Acad. Sci. Paris* **250**(1960), 799–801.

[6] Y. Derriennic: Marche aléatoire sur le groupe libre et frontière de Martin. *Zeitschr. Wahrscheinlichkeitstheorie Verw. Geb.* **32**(1975), 261–276.

[7] J. L. Doob: Discrete potential theory and boundaries. *J. Math. Mech.* **8**(1959), 433–458.

[8] J. L. Doob, J. L. Snell and R. E. Williamson: Application of boundary theory to sums of independent random variables. Contributions to Probability and Statistics, Stanford Univ. Press (1960), 182–197.

[9] P. Doyle and J. L. Snell: Random Walks and Electric Networks. Carus Math. Monographs, Math. Assoc. Amer., 1984.

[10] E. B. Dynkin and M. B. Malyutov: Random walks on groups with a finite number of generators. *Soviet Math. Doklady* **2**(1961), 399–402.

[11] W. Feller: Boundaries induced by nonnegative matrics. *Trans. Amer. Math. Soc.* **83**(1956), 19–54.

[12] H. Freudenthal: Über die Enden diskreter Räume und Gruppen. *Comment. Math. Helv.* **17**(1944), 1–38.

[13] H. Furstenberg: Random walks and discrete subgroups of Lie groups. Advances in Probability and Related Topics, Vol. 1 (P. Ney, ed.), Dekker, New York (1971), 1–63.

[14] R. Halin: Über unendliche Wege in Graphen. Math. Annalen 157(1964), 125–137.

[15] D. Heath: Functions possessing restricted mean value properties. Proc. Amer. Math. Soc. 41(1973), 588–595.

[16] P. L. Hennequin: Processus de Markoff en cascade. Ann. Inst. H. Poincaré 18(1963), 109–196.

[17] H. A. Jung: Connectivity in infinite graphs. Studies in Pure Math. (L. Mirsky, ed.), Academic Press, New York (1971), 137–143.

[18] J. G. Kemeny, J. L. Snell and A. W. Knapp: Denumerable Markov Chains. 2nd ed., Springer, New York — Heidelberg — Berlin, 1976.

[19] V. A. Kaimanovich and A. M. Vershik: Random walks on discrete groups: boundary and entropy. Annals Probab. 11(1983), 457–490.

[20] T. Lyons: A simple criterion for transience of a reversible Markov chain. Annals Probab. 11(1983), 393–402.

[21] W. Magnus: Noneuclidean Tesselations and their Groups. Academic Press, New York, 1974.

[22] C. St. J. A. Nash-Williams: Random walks and electric currents in networks. Proc. Cambridge Phil. Soc. 55(1959), 181–194.

[23] P. Ney and F. Spitzer: The Martin boundary for random walk. Trans. Amer. Math. Soc. 121(1966), 116–132.

[24] R. R. Phelps: Lectures on Choquet's Theorem. Van Nostrand Math. Studies, Vol. 7, New York, 1966.

[25] M. A. Picardello and W. Woess: Martin boundaries of random walks: ends of trees and groups. Trans. Amer. Math. Soc. 302(1987), 185–205.

[26] M. A. Picardello and W. Woess: Harmonic functions and ends of graphs. Proc. Edinburgh Math. Soc. 31(1988), 457–461.

[27] M. A. Picardello and W. Woess: A converse to the mean value property on homogeneous trees. Trans. Amer. Math. Soc., in print.

[28] M. A. Picardello and P. Sjögren: Preprint, Univ. Maryland (1988).

[29] N. Polat: Aspects topologiques de la séparation dans les graphes infinis. Math. Zeitschr. 165(1979), 73–100.

[30] C. Series: Martin boundaries of random walks on Fuchsian groups. Israel J. Math. 44(1983), 221–242.

[31] F. Spitzer: Principles of Random Walk. Van Nostrand, Princeton, 1974.

[32] J. C. Taylor: The Martin boundaries of equivalent sheaves. *Ann. Inst. Fourier* **20**(1970), 433–456.

[33] C. Thomassen: Resistances and currents in infinite electrical networks. *J. Combin. Theory Ser. B* in print.

[34] W. Veech: A zero-one law for a class of random walks and a converse to Gauss' mean value theorem. *Annals Math.* **97**(1973), 189–216.

[35] W. Veech: A converse to the mean value theorem for harmonic functions. *Amer. J. Math.* **97**(1975), 1007–1027.

[36] W. Woess: A description of the Martin boundary for nearest neighbour random walks on free products. "Probability Measures on Groups", *Springer Lect. Notes in Math.* **1210**(1986), 203–215.

[37] W. Woess: Harmonic functions on infinite graphs. *Rendiconti Sem. Mat. Fis. Milano* **56**(1986), 53–63.

[38] W. Woess: Graphs and groups with tree-like properties. *J. Combin. Theory Ser. B* in print.

TOPOLOGICAL ASPECTS OF INFINITE GRAPHS

N. POLAT
Département de Mathématiques
Bâtiment 101
Université Claude Bernard (Lyon I)
43, Boulevard du 11 Novembre 1918
69622 Villeurbanne Cedex
France

Abstract

A uniform structure is defined on the vertex set of an infinite graph in such a way that its completion and its compactification are related to the two basic infiniteness properties of the graph: "height" and "width". These different uniformities with their associated proximity relations, and the concept of terminal expansion of an infinite graph — a sequence of particular subgraphs canonically associated with an increasing sequence of closed sets of the topological space of its ends — are used to study different combinatorial problems: an extension of Menger's theorem to graphs with ideal points, and some characterizations of the graphs which have special spanning trees.

Introduction

This paper is divided into two principal parts. In the first, the fundamental aspects of infinite graphs, "height" and "width", are studied by means of methods which are essentially topological. We define a uniform structure S_G on the vertex set of a graph G in such a way that the associated topological space is compact if and only if G is finite. More precisely, this uniform space is complete if and only if G is rayless, and totally bounded if and only if for each finite subset S of vertices, $G - S$ has only finitely many components. In Section 1.3, the completion of this space is obtained by canonically extending S_G to the set of all vertices and ends of G; while, in Section 1.4, the Smirnov compactification of the associated proximity space is connected to the usual relations of separation in graphs. The non-principal clusters of this proximity space are used to characterize two particularly important classes of sets of vertices, called *concentrated* and *fragmented*, which are respectively associated with Cauchy clusters and non-Cauchy clusters, corresponding once again to the two basic aspects of infiniteness of a graph (Section 1.5). Menger's theorem (already generalized by Zelinka and Halin to graphs with ideal points) is once more extended by introducing the non-principal clusters as new points (Section 1.6).

In order to study certain problems involving the ends of infinite graphs, we define, in the second part of this paper, the notion of *terminal expansion* of an infinite graph into particular subgraphs, called *multi-endings*, having only free ends.

197

G. Hahn et al. (eds.), Cycles and Rays, 197–220.
© *1990 by Kluwer Academic Publishers. Printed in the Netherlands.*

This concept is canonically associated with an expansion of the topological space of its ends into an increasing sequence of closed sets (Sections 2.6 and 2.7). We prove that every connected graph has a terminal expansion; that the minimal length of such expansions (the *terminal degree*) is less than or equal to ω_1 (it is less than or equal to $\omega + 1$ in particular for trees and countable graphs); and that, for any ordinal $\alpha \leq \omega_1$, there is a graph whose terminal degree is α (Section 2.8).

All these concepts have been used to study various combinatorial problems. Among them we only mention, in Sections 2.10 and 2.11, the characterizations of those graphs which have particular spanning trees, such as: coterminal trees, rayless trees, or trees whose associated uniform spaces are subspaces of those of the given graphs.

The proofs of all results, but of a few of them which are new, and which all are in the second part, can be found in [10], [11], [12], [13] and [14].

Notation

The terminology and notation will be for the most part that used in the above-mentioned papers.

Each ordinal α is defined as the set of ordinals less than α. If X is a set we denote by $|X|$ its cardinality, and by $\mathbf{P}(X)$ its power set. Furthermore, if χ is a cardinal (i.e., an initial ordinal) we set $[X]^\chi := \{A \in \mathbf{P}(X) : |A| = \chi\}$. The sets $[X]^{<\chi}$, $[X]^{\leq\chi}$, $[X]^{>\chi}$, $[X]^{\geq\chi}$ are defined similarly.

A *graph* G is a set $V(G)$ *(vertex set)* together with a set $E(G) \subseteq [V(G)]^2$ *(edge set)*. For the sake of simplicity, *throughout this paper we will write* $[V(G)]$ *for* $[V(G)]^{<\omega}$. We denote by $\langle x \rangle$ the graph whose vertex set is $\{x\}$. Similarly, if $e = \{x, y\}$ is an edge, $\langle e \rangle$ or $\langle x, y \rangle$ will denote the graph consisting of e and its end points x, y. For $x \in V(G)$ the set

$$V(x; G) := \{y \in V(G) : \{x, y\} \in E(G)\}$$

is the *neighbourhood* of x, and its cardinality is the *degree* of x. A graph is *locally finite* if all its vertices have finite degrees.

H is a *subgraph* of G ($H \subseteq G$) if $V(H) \subseteq V(G)$ and $E(H) \subseteq E(G)$. H is a *restriction* or an *induced subgraph* ($H \leq G$) if H is a subgraph such that $E(H) = [V(H)]^2 \cap E(G)$. For $A \subseteq V(G)$ we denote by $G|A$ the restriction of G to the set A. If B is any set, and H any graph, we define $G - B := G|(V(G) - B)$ and $G - H := G - V(H)$.

The *union* of a family $(G_i)_{i \in I}$ of graphs is the graph $\bigcup G_i$ given by $V(\bigcup_{i \in I} G_i) = \bigcup_{i \in I} V(G_i)$ and $E(\bigcup_{i \in I} G_i) = \bigcup_{i \in I} E(G_i)$. The intersection is defined similarly. If $(G_i)_{i \in I}$ is a family of subgraphs of a graph G, the restriction of G to the union of this family will be denoted by $\bigvee_{i \in I} G_i$.

If H is a subgraph of G, and X a subgraph of $G - H$, the *boundary of H with X* is the set

$$\mathcal{B}(H, X) := \{x \in V(H) : V(x; G) \cap V(X) \neq \emptyset\}.$$

In particular, $\mathcal{B}(H) := \mathcal{B}(H, G - H)$ is the *boundary* of H.

The set of components of G is denoted by $\mathcal{C}_G$, and if x is a vertex, then $\mathcal{C}_G(x)$ is the component of G containing x.

A *path* $W = \langle x_0, \ldots, x_n \rangle$ (resp. a *ray* or *one-way infinite path* $R = \langle x_0, x_1, \ldots \rangle$, a *double ray* or *two-way infinite path* $D = \langle \ldots, x_{-1}, x_0, x_1, \ldots \rangle$) is the graph with $V(W) = \{x_0, \ldots, x_n\}$ (resp. $V(R) = \{x_n : n \in \mathbf{N}\}$, $V(D) = \{x_n : n \in \mathbf{Z}\}$), $x_i \neq x_j$ if $i \neq j$ and $E(W) = \{\{x_i, x_{i+1}\} : 0 \leq i < n\}$ (resp. $E(R) = \{\{x_n, x_{n+1}\} : n \in \mathbf{N}\}$, $E(D) = \{\{x_n, x_{n+1}\} : n \in \mathbf{Z}\}$). A ray $\langle x_0, x_1, \ldots \rangle$ is said to *originate* at x_0. A path $\langle x_0, \ldots, x_n \rangle$ is called an $x_0 x_n$-*path*. For $A, B \subseteq V(G)$, an AB-*path* of G is an xy-path of G with $x \in A$ and $y \in B$, and such that x and y are the only vertices in $A \cup B$. Two paths are *disjoint* (resp. *internally disjoint*) if they have no vertices (resp. at most their extremities) in common. If A, B and C are subsets of $V(G)$, C *separates* A *from* B *in* G, or C is an AB-*separator* of G, if all AB-paths of G have vertices in C.

1 Basic Properties of Infinite Graphs

1.1 Infiniteness properties

Among the different ways of characterizing an infinite graph the following is fundamental to this study.

A *graph G is infinite if and only if it has a ray or a cofinite restriction having infinitely many components* (i.e., there is a finite subset S of vertices of G such that the subgraph $G - S$ has infinitely many components).

These two conditions will be called the *basic infiniteness properties* of a graph. The first part of this paper will be devoted to reformulations of these conditions, most of them topological, and with various consequences of their being satisfied or not.

1.2 Uniformity $\mathbf{S}_G$

Let G be a graph. To every $S \in [V(G)]$ we associate a relation $\tilde{S}$ on $V(G)$ defined by

$$(x, y) \in \tilde{S} \text{ if and only if } x = y \text{ or } \mathcal{C}_{G-S}(x) = \mathcal{C}_{G-S}(y).$$

$\tilde{S}$ is clearly an equivalence relation. Moreover $(S_1 \cup S_2)^\sim \subseteq \tilde{S}_1 \cup \tilde{S}_2$ for all $S_1, S_2 \in [V(G)]$. Then obviously:

The family $(\tilde{S})_{S \in [V(G)]}$ is a base of an ultrauniformity on $V(G)$, denoted by $\mathbf{S}_G$, whose induced topology is discrete.

We recall that an *ultrauniformity* is a uniformity generated by a family of equivalence relations; and that the induced topology is the topology on $V(G)$ such that, for every vertex x, the family $(\tilde{S}(x))_{S \in [V(G)]}$ is a local basis at the point x. All the uniform structures defined in this paper will be Hausdorff ultrauniformities, hence their induced topologies will all be zero-dimensional.

The interest of this uniform structure results from the fact that:

Theorem 1.1 *G is finite if and only if the space* $(V(G), \mathbf{S}_G)$ *is compact. More precisely,* $(V(G), \mathbf{S}_G)$ *is complete (resp. precompact, i.e., totally bounded) if and only if G is rayless (resp. every cofinite restriction of G has finitely many components).*

Notice that the completeness and the precompactness of this space are related respectively to the first and the second basic infiniteness property of the graph. It is this remark which motivates the study of this space and in particular of its completion and compactification.

1.3   Completion

1.3.1 We first recall the definition of an *end* (or *terminal class*) of a graph. This concept was introduced by Freudenthal [1] and independently by Halin [2].

The *ends* of a graph G are the classes of the equivalence relation $\sim_G$ defined on the set of all rays of G by: $R \sim_G R'$ if and only if there is a ray R'' whose intersections with R and R' are infinite; or equivalently if and only if $\mathcal{C}_{G-S}(R) = \mathcal{C}_{G-S}(R')$ for any $S \in [V(G)]$ (where $\mathcal{C}_{G-S}(R)$ denotes the component of $G - S$ containing a subray of R).

We will use the following notations:

— $[R]_G$: class of a ray R modulo $\sim_G$;

— $\mathbf{T}(G)$: set of all ends of G;

— $\hat{V}(G) := V(G) \cup \mathbf{T}(G)$.

1.3.2 For every $S \in [V(G)]$ we denote by $\hat{S}$ the natural extension of $\tilde{S}$ to $\hat{V}(G)$ defined by

$$(x, y) \in \hat{S} \text{ if and only if } x = y \text{ or } \mathcal{C}_{G-S}(x) = \mathcal{C}_{G-S}(y)$$

(where, if x is an end, $\mathcal{C}_{G-S}(x)$ is the component of any ray belonging to x).

As before:

The family $(\hat{S})_{S \in [V(G)]}$ *is a base of a Hausdorff ultrauniformity on* $\hat{V}(G)$*, denoted by* $\hat{\mathbf{S}}_G$.

We can then prove that:

Theorem 1.2 $(\hat{V}(G), \hat{\mathbf{S}}_G)$ *is the completion of* $(V(G), \mathbf{S}_G)$.

1.3.3 In the following we will denote by $\mathbf{S}_G^*$ the uniformity induced on $\mathbf{T}(G)$ by $\hat{\mathbf{S}}_G$.

$(\mathbf{T}(G), \mathbf{S}_G^*)$ *is a complete ultrauniform space which is compact if and only if every cofinite restriction of G has finitely many components containing a ray.*

An end which is isolated for the induced topology $\text{top}(\mathbf{S}_G^*)$ is said to be *free*.

1.3.4 Examples

1.3.4.1 All ends of the graph in Figure 1 have been represented by stars drawn in the prolongation of the corresponding rays. This representation anticipates the interpretation of ends as ideal points (Section 1.6). In this graph all ends but τ_ω are free.

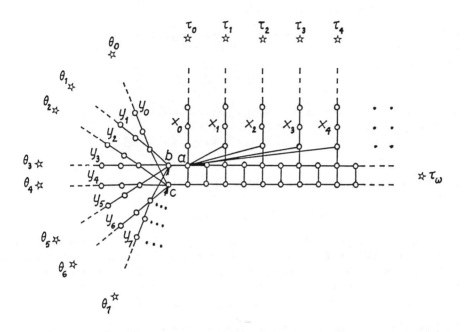

Figure 1

1.3.4.2 Let (T, a) be the dyadic (i.e., binary) tree rooted at a. Let φ be the natural bijection from 2^ω onto the set of all rays of T originating at the root a (Figure 2). This bijection clearly induces a homeomorphism from the boolean space 2^ω onto the space $(\mathbf{T}(T), \text{top}(\mathbf{S}_T^*))$.

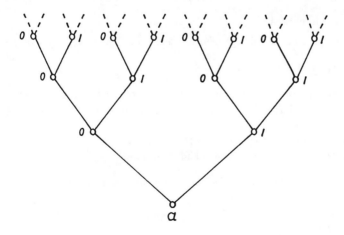

Figure 2

1.3.5 A special extension of a graph

For a graph G we choose once and for all a family $(R_x)_{x \in V(G)}$ of pairwise disjoint rays such that, for any vertex x, R_x originates at x and has only this vertex in common with G. We then define

$$G^+ := G \cup (\bigcup_{x \in V(G)} R_x)$$

and we denote by ρ_G the bijection from $\hat{V}(G)$ onto $\mathbf{T}(G^+)$ such that $\rho_G(x) = [R_x]_{G^+}$ if $x \in V(G)$, and $\rho_G([R]_G) = [R]_{G^+}$ if R is a ray of G.

This bijection ρ_G will prove useful in the second part of this paper. In particular we get immediately:

Proposition 1.1 $(\hat{V}(G), \hat{\mathbf{S}}_G) \cong (\mathbf{T}(G^+), \mathbf{S}_{G^+}^*)$.

1.4   Associated proximity. Compactification

To define a compactification of the topological space $(V(G), \mathrm{top}(\mathbf{S}_G))$, it seems natural — since we are working with graphs — to consider at first the proximity relation associated with the uniformity $\mathbf{S}_G$, as this is one of the separation relations usually defined in graph theory. We will recall the definitions of a proximity and of a cluster; all other definitions and properties related to proximity spaces can be found in [8].

Definition 1.1 A relation σ on the power set of a set X is an *(Efremovič) proximity* on X if and only if it satisfies the following properties for all $A, B, C \in \mathbf{P}(X)$:

P1. $A\sigma B$ implies $B\sigma A$;

P2. $(A \cup B)\sigma C$ if and only if $A\sigma C$ or $B\sigma C$;

P3. $A\sigma B$ implies $A \neq \emptyset$ and $B \neq \emptyset$;

P4. $A \not{\sigma} B$ implies the existence of a subset E such that $A \not{\sigma} E$ and $(X - E) \not{\sigma} B$;

P5. $A \cap B \neq \emptyset$ implies $A\sigma B$.

The pair (X, σ) is called a *proximity space*, and it is *Hausdorff* if σ also satisfies:

P6. For all $x, y \in X$, $x\sigma y$ implies $x = y$.

If $(X, \mathbf{U})$ is a uniform space then the relation σ on $\mathbf{P}(X)$ such that $A\sigma B$ if and only if $\mathcal{U} \cap (A \times B) \neq \emptyset$ for all $\mathcal{U} \in \mathbf{U}$, is a proximity on X which is said to be *associated* with $\mathbf{U}$. If $\mathbf{U}$ is Hausdorff then so is σ.

1.4.1 For a graph G the proximity on $V(G)$, denoted by σ_G, which is associated with $\mathbf{S}_G$, can be characterized by different equivalent properties.

Theorem 1.3 *Let $A, B \subseteq V(G)$. The following conditions are equivalent:*
   (i) $A\sigma_G B$;
   (ii) $\tilde{S} \cap (A \times B) \neq \emptyset$ for all $S \in [V(G)]$ (i.e., A and B cannot be separated by a finite subset of vertices);
   (iii) there exist $A' \in [A]^\omega$ and $B' \in [B]^\omega$ such that $A'\sigma_G B'$;
   (iv) there is an infinite family of pairwise disjoint AB-paths.

Definition 1.2 A *cluster* of a proximity space (X, σ) is a subset γ of $\mathbf{P}(X)$ satisfying:

C1. if A and B belong to γ, then $A\sigma B$;

C2. if $A\sigma C$ for every $C \in \gamma$, then $A \in \gamma$;

C3. if $A \cup B \in \gamma$, then $A \in \gamma$ or $B \in \gamma$.

1.4.2 For a graph G, the set of all clusters of the proximity space $(V(G), \sigma_G)$ will be denoted by $\mathbf{C}(G)$.

For $x \in \hat{V}(G)$ we denote by γ_x the *Cauchy cluster* generated by the Cauchy ultrafilter $\mathbf{U}_x$ converging to x, i.e.,

$$\gamma_x := \{A \subseteq V(G) : A\sigma_G U \text{ for all } U \in \mathbf{U}_x\}.$$

If x is a vertex, then $\gamma_x = \{A \subseteq V(G) : x \in A\}$, i.e., γ_x is the principal ultrafilter generated by x; it is called a *point cluster*.

The map which associates to each vertex or end x the Cauchy cluster γ_x is a bijection from $\hat{V}(G)$ to the set of Cauchy clusters of $(V(G), \sigma_G)$. The restriction of this map to $V(G)$ is a bijection onto the set of point clusters. This bijection enables us to identify a vertex with a point cluster, and an end with a Cauchy cluster which is not a point cluster.

1.4.3 The relation $\sigma_G^\#$ on $\mathbf{P}(\mathbf{C}(G))$ such that

$$\mathcal{A}\sigma_G^\#\mathcal{B} \text{ if and only if } A\sigma_G B \text{ for all } (A, B) \in (\cap\mathcal{A}) \times (\cap\mathcal{B})$$

is a proximity on $\mathbf{C}(G)$ whose associated space $(\mathbf{C}(G), \sigma_G^\#)$ is compact (i.e., every cluster of this space is a point cluster). This proximity space is called the *Smirnov compactification* of $(V(G), \sigma_G)$.

1.4.4 We will denote by $\hat{\sigma}_G$ (resp. σ_G^*) the proximity on $\hat{V}(G)$ (resp. $\mathbf{T}(G)$) which is associated with the uniformity $\hat{\mathbf{S}}_G$ (resp. $\mathbf{S}_G^*$). *The canonical mapping $x \to \gamma_x$ from $\hat{V}(G)$ into $\mathbf{C}(G)$, is a proximity isomorphism from $(\hat{V}(G), \hat{\sigma}_G)$ onto the proximity subspace of all Cauchy clusters.* This implies that, if $\mathbf{A}$ and $\mathbf{B}$ are sets of Cauchy clusters, then $\mathbf{A}\sigma_G^\#\mathbf{B}$ if and only if $\{x \in \hat{V}(G) : \gamma_x \in \mathbf{A}\}\hat{\sigma}_G\{x \in \hat{V}(G) : \gamma_x \in \mathbf{B}\}$.

1.5 Concentrated sets. Fragmented sets

We denote by $\mathbf{C}_\infty(G)$ the set of all clusters of $(V(G), \sigma_G)$ which are not point clusters, and by $\mathbf{C}_C(G)$ (resp. $\mathbf{C}_F(G)$) the subset of elements of this set which are (resp. which are not) Cauchy clusters. These two kinds of clusters will be characterized in terms of particular sets of vertices. By P2 we already know that any member of such a cluster is an infinite set.

Definition 1.3 Let $A \in [V(G)]^{\geq\omega}$.

(a) A is *concentrated* if there is an end τ such that $A - \hat{S}(\tau)$ is finite for any $S \in [V(G)]$ (A is said to be *"concentrated in τ"*).

(b) A is *fragmented* if $\tilde{S} \cap (A \times A) = id_A$ for some $S \in [V(G)]$ (i.e., the elements of A are pairwise separated by S).

For example, in the graph of Figure 1, the set $\{x_n : n < \omega\}$ is concentrated in τ_ω, while the sets $\{y_{2n} : n < \omega\}$ and $\{y_{2n+1} : n < \omega\}$ are fragmented — it suffices to delete the vertices a and b to pairwise separate their elements. Notice that $\{y_n : n < \omega\}$ is not fragmented.

We will give several characterizations of concentrated and fragmented sets of vertices. For $A \in [V(G)]^{\geq\omega}$ we let

$$\gamma_A := \cap\{\gamma \in \mathbf{C}_\infty(G) : A \in \gamma\} \quad (= \{B \subseteq V(G) : B\sigma_G C \text{ for all } C \in [A]^{\geq\omega}\}).$$

Theorem 1.4 *Let $A \in [V(G)]^{\geq\omega}$. The following conditions are equivalent:*
   (i) A is concentrated;
   (ii) every countably infinite subset of A is concentrated;
   (iii) $B\sigma_G C$ for all $B, C \in [A]^{\geq\omega}$;
   (iv) $\gamma_A \in \mathbf{C}_\infty(G)$;
   (v) $\gamma_A \in \mathbf{C}_C(G)$;
   (vi) if $A \in \gamma \in \mathbf{C}_\infty(G)$ then $\gamma = \gamma_A$;
   (vii) $\gamma_A \cap \mathbf{P}(A) = [A]^{\geq\omega}$.

Theorem 1.5 *Let $A \in [V(G)]^{\geq\omega}$. The following conditions are equivalent:*
   (i) A is fragmented;
   (ii) every countably infinite subset of A is fragmented;
   (iii) $B \not\sigma_G C$ for all disjoint subsets B and C of A;
   (iv) if $A \in \gamma \in \mathbf{C}(G)$ then $\gamma \cap \mathbf{P}(A)$ is an ultrafilter on A.

The following two theorems show the importance of these sets of vertices for infinite graphs.

Theorem 1.6 *Every infinite subset of vertices of a graph has a concentrated or a fragmented subset.*

Theorem 1.7 *Let $\gamma \in \mathbf{C}_\infty(G)$. Then γ is (resp. is not) a Cauchy cluster if and only if there is $A \in \gamma$ which is concentrated (resp. fragmented).*

In the next result we state in parallel several equivalent properties related to the two basic infiniteness properties of graphs; some of them, characterizing rayless graphs, were already given by Halin [3].

Theorem 1.8 *Let G be an infinite graph. The following are equivalent:*
   (i) G is rayless (resp. every cofinite restriction of G has finitely many components);
   (ii) $(V(G), \mathbf{S}_G)$ is complete (resp. precompact);
   (iii) $\mathbf{C}_C(G) = \emptyset$ (resp. $\mathbf{C}_F(G) = \emptyset$);
   (iv) $V(G)$ has no concentrated (resp. fragmented) subset;
   (v) every infinite subset of $V(G)$ has a fragmented (resp. concentrated) subset.

1.6   Extension of Menger's theorem

Since the vertices of an infinite graph can be identified with the point clusters of its associated proximity space, it seems quite reasonable to try to extend one of the most famous
theorems of graph theory, that is, Menger's theorem, to infinite graphs by admitting as
ideal points also those clusters which are not point clusters. For locally finite graphs this
has already been done, first by Zelinka [15] who considered only the free ends (i.e., the
free Cauchy clusters) as new points, and then by Halin [6] who considered all the ends as
new points. Halin also pointed out that his result did not hold for graphs with vertices of
infinite degree.

1.6.1 We will extend these results to arbitrary infinite graphs using all the clusters which
are not point clusters as new points. It is then necessary to define an appropriate separation
relation, and what is meant by a "path" between two sets of clusters. To this end we
introduce two relations of *preproximity*, δ_G and $\delta_G^\#$, which are associated with σ_G and
$\sigma_G^\#$, respectively. Preproximities are a generalization of proximities, and each preproximity
canonically determines a proximity (see [10, Section 2.2]).

δ_G is the relation on $\mathbf{P}(V(G))$ such that $A\delta_G B$ if and only if $S \cap (A \times B) \neq \emptyset$ for any
$S \in [V(G) - A \cup B]^{<\omega}$ (i.e., A and B cannot be separated by a finite subset of vertices
disjoint from A and B).

$\delta_G^\#$ is the relation on $\mathbf{P}(C(G))$ given by $\mathcal{A}\delta_G^\#\mathcal{B}$ if and only if $A\delta_G B$ for any $(A, B) \in$
$(\cap\mathcal{A}) \times (\cap\mathcal{B})$. $\delta_G^\#$ will be our separation relation. Now for the paths:

1.6.2 For $\mathcal{A}, \mathcal{B} \subseteq C(G)$ we denote by $\mathcal{W}_{\mathcal{A}\mathcal{B}}$ the set of all finite or infinite paths (i.e., rays
or double rays) W of G such that $V(W)^\#\delta_G^\#\mathcal{A}$ and $V(W)^\#\delta_G^\#\mathcal{B}$, where $V(W)^\# := \{\gamma \in$
$C(G) : V(W) \in \gamma\}$. Then $\mathcal{W}_{\mathcal{A}\mathcal{B}}^{\min}$ will be the set of minimal elements (with respect to
inclusion) of $\mathcal{W}_{\mathcal{A}\mathcal{B}}$.

1.6.3 Example

(Figure 3): For the sake of simplicity each vertex or end a has been identified with the
cluster γ_a.

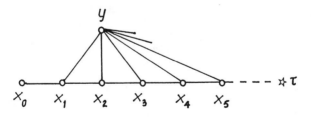

Figure 3

We have for example:

$$\mathcal{W}_{x_0\tau} = \{\langle x_0, x_1, x_2, \ldots\rangle, \langle x_1, x_2, \ldots\rangle\} \cup \{W : \langle x_1, y\rangle \subseteq W\},$$

$$\mathcal{W}_{x_0}^{\min} = \{\langle x_1, x_2, \ldots\rangle, \langle x_1, y\rangle\};$$

and

$$\mathcal{W}_{x_0x_2} = \{W : x_1 \in V(W)\}, \quad \mathcal{W}_{x_0x_2}^{\min} = \{\langle x_1\rangle\}.$$

1.6.4 The set $\mathcal{W}_{AB}^{\min}$ may be empty, as is shown by the example of a ray R; if τ is the end containing R, then clearly $\mathcal{W}_{V(R),\{\tau\}}^{\min} = \emptyset$. But *if $A \not\!\beta_G^\# B$ then the poset $(\mathcal{W}_{AB}, \supseteq)$ is inductive, and thus $\mathcal{W}_{AB}^{\min} \neq \emptyset$.*

1.6.5 Let $A, B \subseteq \mathbf{C}(G)$ be such that $A \not\!\beta_G^\# B$. Then we have $A \not\!\beta_G^\# B$, and thus $\mathcal{W}_{AB}^{\min} \neq \emptyset$. The elements of this set will be the AB-paths. We introduce a few more notations:

$\mathrm{SEP}(A,B) := \{S \in [V(G)] :$ there is an $(A, B) \in \cap A \times \cap B$ such that $\tilde{S} \cap (A \times B) = \emptyset$ and
$\qquad\qquad S \cap (A \cup B) = \emptyset\}$

$\mathrm{sep}(A,B) := \inf\{|S| : S \in \mathrm{SEP}(A,B)\}$

$\mathrm{DISJ}(A,B) := \{\Delta \in \mathrm{P}(\mathcal{W}_{AB}^{\min}) :$ for all $W, W' \in \Delta$, $W \neq W'$ implies $W \cap W' = \emptyset\}$

$\mathrm{disj}(A,B) := \sup\{|\Delta| : \Delta \in \mathrm{DISJ}(A,B)\}$.

We then have the following extension of Menger's theorem:

Theorem 1.9 *Let $A, B \subseteq \mathbf{C}(G)$ be such that $A \not\!\beta_G^\# B$. Then*

$$sep(A,B) = disj(A,B).$$

1.6.6 Example

Let us take the example given by Halin at the end of his paper [6] to show that his result does not hold for non-locally finite graphs. Let G be the graph which consists of a double ray W and of a vertex $x \notin V(W)$ adjacent to all vertices of W (Figure 4). If τ and τ' are the two ends of G, then $\mathrm{disj}(\tau, \tau') = 2 = \mathrm{sep}(\tau, \tau')$ since $\{W, \langle x \rangle\} \in \mathrm{DISJ}(\tau, \tau')$, and $\{x, y\} \in \mathrm{SEP}(\tau, \tau')$ for every $y \in V(W)$.

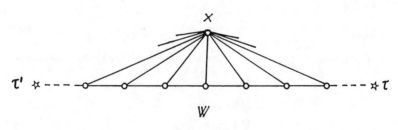

Figure 4

2   Terminal Expansions of Infinite Graphs

2.1   Introduction

From now on, we will assume that the end set of a graph G is endowed with the topology induced by $\mathbf{S}_G^*$ (cf. 1.3.3).

For the proofs of certain results pertaining to the ends of a graph G which proceed by induction, it will be convenient to have a kind of decomposition of the end set of G. One way to do this is to define a classification of the ends of G according to rank. This has been done by Jung [7]. His classification is actually the one we get by considering the iterated Cantor–Bendixon derivatives of the space $\mathbf{T}(G)$ [13, Section 6]. Unfortunately this process is in most cases completely ineffective. For example, if T is the dyadic tree, then $\mathbf{T}(T)$ has no isolated points, and thus is equal to its Cantor–Bendixon derivative. Any two ends of the dyadic tree have exactly the same properties, they cannot be differentiated from one another.

Another procedure is to decompose, as far as possible, the end space of G into discrete subspaces, since several properties which hold for every end, generally also hold for any discrete subspace of ends. For example, let (T, a) be the rooted dyadic tree, and let φ be the natural bijection defined in 1.3.4.2. For any $n < \omega$ let

$$T_n := \cup\{\varphi(s) : s \in 2^\omega \text{ and } |s^{-1}(1)| \leq n\}$$

and let $T_\omega = T$. Then the subspace $\mathbf{T}_{T_{n+1}}(T) - \mathbf{T}_{T_n}(T)$ is discrete, and $\mathbf{T}(T)$ is the closure of $\bigcup_{n<\omega} \mathbf{T}_{T_n}(T)$, where $\mathbf{T}_{T_n}(T) = \{[R]_T : R \text{ is a ray of } T_n\}$.

The sequence $(\mathbf{T}_{T_n}(T))_{n \leq \omega}$ (resp. $(T_n)_{n \leq \omega}$) will be called an *expansion* of the space $\mathbf{T}(T)$ (resp. a *terminal expansion* of the tree T).

If b is any vertex of T different from a, and if, for any end τ of T, R_τ is the unique ray in τ originating at b, then $(T'_n)_{n \leq \omega}$, where

$$T'_n := \begin{cases} \cup\{R_\tau : \tau \in \mathbf{T}_{T_n}(T)\} & \text{if } n < \omega \\ T & \text{if } n = \omega \end{cases}$$

is another terminal expansion of T which is associated with the expansion $(\mathbf{T}_{T_n}(T))_{n \leq \omega}$ in the sense that, for each $n \leq \omega$, $\mathbf{T}_{T'_n}(T) = \mathbf{T}_{T_n}(T)$.

It will be very easy to define the concept of expansion of a topological space; the case is different, however, with that of a terminal expansion of an arbitrary graph. For a rooted tree G, the method, introduced above for the dyadic tree, can always be used to associate a terminal expansion of G to any given expansion of $\mathbf{T}(G)$: at each step one has to consider the new ends and to add to the subtree already obtained the corresponding rays which start at the root. But for an arbitrary connected graph, we will first have to define what sort of subgraph will play the role of the rays in the case of a tree. These special subgraphs, which will be the terms of a terminal expansion, will be called *endings* and more generally *multi-endings*.

Before defining these concepts we will describe some problems for which they proved to be useful. These are problems about the existence of three different special types of trees in infinite graphs.

2.2   Special trees

Definition 2.1 A subgraph H of G is *terminally faithful* (resp. *coterminal*) if the map $\epsilon : \mathbf{T}(H) \to \mathbf{T}(G)$ given by $\epsilon([R]_H) = [R]_G$ for every ray R of H, is injective (resp. bijective).

We denote by $\mathbf{T}_H(G)$ the image of ϵ, i.e., the set of ends of G having rays of H as elements.

2.2.1   In [2] Halin asked if any connected graph has a *coterminal spanning tree* (CST for short), and he proved that it holds true for countable one-ended graphs. We will give only partial answers to this problem, stronger than Halin's but nevertheless partial (cf. Section 2.10), and dually to the problem of characterizing those connected graphs which have a *rayless spanning tree* (RST for short) — a problem which, we will see, is closely related to Halin's.

For example the tree given in Figure 5 is a CST of the graph of Figure 1.

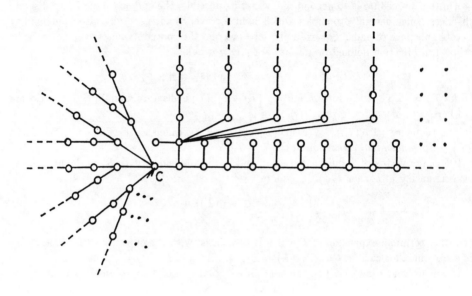

Figure 5

Definition 2.2 A subgraph H of G is *faithful* if we have the following equivalent properties:

(i) $(V(H), \mathsf{S}_H)$ is a uniform subspace of $(V(G), \mathsf{S}_G)$;

(ii) $(V(H), \sigma_H)$ is a proximity subspace of $(V(G), \sigma_G)$;

(iii) for any $A, B \in [V(H)]^\omega$, $A\sigma_G B$ implies $A\sigma_H B$.

A faithful subgraph — which is called a *reduced subgraph* by Jung [8], and an S-*subgraph* in [11] and [13] — is clearly terminally faithful.

2.2.2 We will give several characterizations of those connected graphs which have a *coterminal faithful tree* (CFT) or a *spanning faithful tree* (SFT). For example, the tree given in Figure 6 is an SFT of the graph in Figure 1; on the other hand, we can notice that the tree of Figure 5 is not faithful.

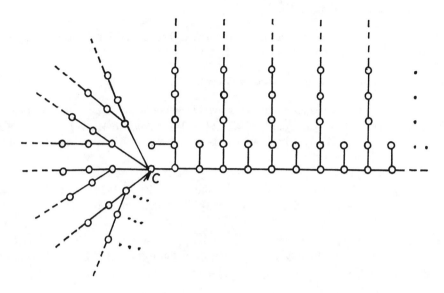

Figure 6

2.3   Multiplicity, neighbourhood, and valency of an end

For $\Omega \subseteq \mathbf{T}(G)$ we define $m(\Omega) := \sup\{|\mathbf{R}| : \mathbf{R}$ subset of pairwise disjoint elements of $\cup\Omega\}$. If $\Omega = \{\tau\}$, we write $m(\tau)$ for $m(\{\tau\})$, and we call it the *multiplicity* of τ.

For every subset Ω of $\mathbf{T}(G)$ the supremum $m(\Omega)$ is attained, i.e., there is a subset $\mathbf{R}$ of pairwise disjoint rays in $\cup\Omega$ such that $|\mathbf{R}| = m(\Omega)$. This was proved by Halin in [4, Satz 1] when $\Omega = \mathbf{T}(G)$, and in [5, Satz 1] when $|\Omega| = 1$, and by Polat in [14, Théorème 11.5] for any Ω.

Hence the multiplicity of an end τ is the maximum number of pairwise disjoint rays in τ.

Definition 2.3 Let $(x, \tau) \in V(G) \times \mathbf{T}(G)$. The vertex x is a *neighbour* of the end τ, and conversely τ is a *terminal neighbour* of x, if $x\hat{\delta}_G\tau$ (i.e., if x and τ cannot be separated by a finite subset of vertices different from x).

For example, a is the only neighbour of τ_ω in the graph of Figure 1.

We denote by V_τ and $\mathbf{T}_x$, respectively, the *neighbourhood* of τ and the *terminal neighbourhood* of x. Notice that $\mathbf{T}_x$ is a closed subset of $\mathbf{T}(G)$. The cardinal $v(\tau) := |V_\tau|$ is called the *valency* of τ.

More generally for $\Omega \subseteq \mathbf{T}(G)$ we let

$$V_\Omega := \bigcup_{\tau \in \Omega} V_\tau \text{ and } v(\Omega) := |V_\Omega|.$$

Proposition 2.1 $v(\tau) < \omega$ *implies* $m(\tau) \le \omega$.

This is a consequence of [14, Corollaire 12.5].

Lemma 2.1 $x\delta_G y$ for all $x, y \in V_\tau$.

Proof: If $x\hat{\delta}_G(\tau)$ and $y\hat{\delta}_G\tau$ then $x\delta_G y$, otherwise there is an $S \in [V(G) - \{x,y\}]^{<\omega}$ such that $C_{G-S}(\tau)$ contains x but not y, contrary to $y\hat{\delta}_G\tau$. □

Proposition 2.2 *If $m(\tau)$ is finite, then so is $v(\tau)$.*

Proof: Assume that τ has an infinite valency. We define by induction a family of paths $(W_n^p)_{n<\omega, p<\omega}$ such that the end-points a_n^p and b_n^p of W_n^p belong to V_τ, $W_n^p \cap W_{n'}^{p'} = \emptyset$ if $n \neq n'$, $W_n^p = \emptyset$ if $n > p$, W_n^n is reduced to a vertex if $n > 0$, and W_n^p is a subpath of W_n^{p+1} with $a_n^p = a_n^{p+1}$ whenever $n \leq p$.

Let $W_0^0 := \emptyset$. Suppose $(W_n^p)_{n<\omega}$ be defined for some $p \geq 0$. We then define W_n^{p+1} and the set A_n by induction on n. Let $W_0^{p+1} := \emptyset$ and $A_0 := V(\bigcup_{n<\omega} W_n^p)$. A_0 is finite since $W_n^p = \emptyset$ when $n > p$ by hypothesis. Assume that W_n^{p+1} and A_n are defined for some $n \geq 0$.
— If $n < p$, since A_n is finite, there exist $b_{n+1}^{p+1} \in V_\tau - A_n$ and a b_{n+1}^p, b_{n+1}^{p+1}-path P_{n+1} such that $V(P_{n+1}) \cap A_n = \{b_{n+1}^p\}$. We define $W_{n+1}^{p+1} := W_{n+1}^p \cup P_{n+1}$ and $A_{n+1} := A_n \cup V(P_{n+1})$.
— If $n = p$ then $W_{p+1}^p = \emptyset$ by hypothesis. Since A_n is finite there is $a_{p+1}^{p+1} \in V_\tau - A_n$. We define $W_{p+1}^{p+1} := \{a_{p+1}^{p+1}\}$ and $A_{p+1} := A_p \cup \{a_{p+1}^{p+1}\}$.
— If $n > p$ we let $W_{n+1}^{p+1} := \emptyset$.

Finally $(\bigcup_{p<\omega} W_n^p)_{n\geq 1}$ is an infinite family of pairwise disjoint rays which belong to τ since each of them contains infinitely many neighbours of τ. Therefore the multiplicity of τ is infinite. □

2.4   Multi-endings

The following definition of a multi-ending, somewhat stronger than the one given in [12, Definition 1.13], corresponds to that of a multi-ending called "faithful" (fidèle) in [14, Definition 8.1].

Definition 2.4 A *multi-ending* of a graph G is a subgraph M of G satisfying:

M1. M is a connected restriction of G;

M2. the boundary of M with every component of $G - M$ is finite;

M3. any infinite subset of $V(M)$ which is concentrated in G is also concentrated in M;

M4. M contains a ray;

M5. $V_{[R]_M} = V_{[R]_G}$ for any ray R of M;

M6. for any family $(R_i)_{i\in I}$ of pairwise disjoint rays of G such that $\{[R_i]_G : i \in I\} \subseteq \mathbf{T}_M(G)$, there is a family $(R_i')_{i\in I}$ of pairwise disjoint rays of M with $R_i \cap M \subseteq R_i'$, $i \in I$.

A multi-ending M is *discrete* if $\mathbf{T}_M(G)$ is a discrete subspace of $\mathbf{T}(G)$, and it is an *ending* if $|\mathbf{T}(M)| = 1$.

2.4.1 Consequences

By M3 a multi-ending is terminally faithful.

By M6, the intersection of M with every ray which is $\sim_G$-equivalent with a ray of M is infinite; on the other hand $m([R]_M) = m([R]_G)$ for any ray R of M.

Theorem 2.1 *If M is a multi-ending of a graph G, then $\mathbf{T}_M(G)$ is a closed set of the space $\mathbf{T}(G)$. Conversely, if G is connected, then for any non-empty closed subset Ω of $\mathbf{T}(G)$, there is a multi-ending M of G such that $\Omega = \mathbf{T}_M(G)$.*

This result, which establishes the links between the multi-endings of a graph and the closed subsets of its end space, is fundamental for the following theory.

2.4.2 Examples

2.4.2.1 Every ray (resp. union of rays) of a tree is an ending (resp. multi-ending) of this tree.

2.4.2.2 M_0 (Figure 7.a) is a discrete multi-ending of the graph G of Figure 1, while M_1 (Figure 7.b) is an ending of G. Notice that $M_0 \cup M_1$ is also a discrete multi-ending of G.

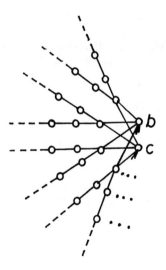

Figure 7.a

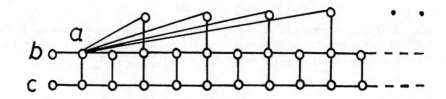

Figure 7.b

2.5   Dispersed sets. Uniformities $\hat{U}_G$ and U_G^*

Definition 2.5 A set A of vertices of G is *dispersed* if it has the following equivalent properties:
  (i) A is closed for $\mathrm{top}(\hat{S}_G)$;
  (ii) $A \, \hat{b}_G \{\tau\}$ for every $\tau \in \mathbf{T}(G)$;
  (iii) A has no concentrated subset.
We denote by $D(G)$ the set of all dispersed subsets of $V(G)$.

In particular every fragmented, and a fortiori every finite, subset of vertices is dispersed. In Figure 1, the set $\{y_n : n < \omega\}$ is dispersed, but not fragmented.
  For any $A \subseteq V(G)$ we define:
$\hat{A} := \cap\{\hat{S} : S \in [A]^{<\omega}\}$ and
$A^* := \cap\{S^* : S \in [A]^{<\omega}\} = \hat{A} \cap \mathbf{T}(G)^2$.

  The family $(\hat{A})_{A \in D(G)}$ *(resp.)* $(A^*)_{A \in D(G)}$ *is a base of a Hausdorff complete ultrauniformity on* $\hat{V}(G)$ *(resp.* $\mathbf{T}(G)$*), denoted by* $\hat{U}_G$ *(resp.* U_G^**), which is compatible with* $\mathrm{top}(\hat{S}_G)$ *resp.* $\mathrm{top}(S_G^*)$*).*

Lemma 2.2 *Let ρ be an equivalence relation on $\hat{V}(G)$ (resp. $\mathbf{T}(G)$) such that each class is closed and open for $\mathrm{top}(\hat{S}_G)$ (resp. $\mathrm{top}(S_G^*)$). Then there is a dispersed subset A such that $\hat{A} \subseteq \rho$ (resp. $A^* \subseteq \rho$).*

  Proof: By Proposition 1.1 it suffices to prove the result when ρ is an equivalence relation on $\mathbf{T}(G)$.
  (a) Let θ be a class of ρ. Since θ is closed there is a multi-ending M_θ of G such that $\mathbf{T}_{M_\theta}(G) = \theta$. We can suppose that each component of $G - M_\theta$ contains a ray. Let $B_\theta := \mathcal{B}(M_\theta)$. Since $\mathcal{B}(M_\theta, X)$ is finite for any component X of $G - M_\theta$, if B_θ has a

concentrated subset, this one must be concentrated in an end $\tau \in \theta$. On the other hand θ is open, thus for any $\tau \in \theta$ there exists $S \in [V(G)]$ such that $S^*(\tau) \subseteq \theta$, and this means that $B_\theta \cap C_{G-S}(\tau) = \emptyset$ since, by the assumption we made, each component of $G - M_\theta$ contains a ray. Hence B_θ is dispersed.

(b) Since all classes of ρ are closed and open, and pairwise disjoint, the union of any subset of classes of ρ is also closed and open. Therefore, by [12, Lemma 1.4.1], there is a dispersed subset A of $V(G)$ such that $A^* \subseteq \rho$. $\qquad \square$

As an immediate consequence we have:

Theorem 2.2 $\hat{\mathbf{U}}_G$ *(resp.* $\mathbf{U}_G^*$*) is the finest ultrauniformity compatible with* $\mathrm{top}(\hat{\mathbf{S}}_G)$ *(resp.* $\mathrm{top}(\mathbf{S}_G^*)$*).*

Theorem 2.3 *Let M be a multi-ending of G. The map $[R]_M \rightarrow [R]_G$, where R is any ray of M, is a uniform isomorphism between $(\mathbf{T}(M), \mathbf{U}_M^*)$ and the subspace $\mathbf{T}_M(G)$ of $(\mathbf{T}(G), \mathbf{U}_G^*)$.*

2.6 Expansion of a topological space

2.6.1 Notation:

If $(X_\xi)_{\xi < \alpha}$ is any sequence we will denote

$$X_{(\alpha)} := \bigcup_{\xi < \alpha} X_\xi.$$

Definition 2.6 An *expansion of length α* of a topological space T is a sequence $(T_\xi)_{\xi < \alpha}$ satisfying the following conditions: for any $\xi < \alpha$,

E1. $T_\xi \subseteq T_\eta$ for $\xi \leq \eta < \alpha$;

E2. T_ξ is a non-empty closed set;

E3. T_0 and $T_{\xi+1} - T_\xi$ (when $\xi + 1 < \alpha$) have only isolated points;

E4. if ξ is a limit ordinal, then $T_\xi = \overline{T_{(\xi)}}$;

E5. $T = T_{(\alpha)}$.

If all axioms but E5 are fulfilled, $(T_\xi)_{\xi < \alpha}$ is said to be a *partial expansion* of T.

The least ordinal α such that T has an expansion of length α is called the *degree* of T, and is denoted by $d(T)$. We set $d(\emptyset) = 0$.

2.6.2 Every T_1-space has an expansion. Therefore the end space of any graph has an expansion.

2.7 Terminal expansion of a graph

Definition 2.7 A *terminal expansion of length α* of a connected graph G is a sequence $(G_\xi)_{\xi < \alpha}$ of subgraphs of G satisfying the following conditions: for any $\xi < \alpha$,

TE1. $G_\xi \leq G_\eta$ for $\xi \leq \eta < \alpha$;

TE2. G_ξ is a multi-ending of G;

TE3. G_0 is discrete, and if $\xi + 1 < \alpha$, then for any $X \in \mathcal{C}_{G-G_\xi}$, $G_{\xi+1} \cap X \neq \emptyset$ implies that $G_{\xi+1} \cap X$ is a discrete multi-ending of X such that $V(x;G) \cap V(G_{\xi+1} \cap X) \neq \emptyset$ for all $x \in \mathcal{B}(G_\xi, X)$;

TE4. if ξ is a limit ordinal, then $\mathbf{T}_{G_{(\xi)}}(G) = \mathbf{T}_{G_\xi}(G)$, and for any $\eta < \xi$ and any $X \in \mathcal{C}_{G-G_\eta}$, $G_\xi \cap X \neq \emptyset$ implies $\mathbf{T}_{G_\xi}(G) \cap \mathbf{T}_X(G) \neq \emptyset$; moreover, for any $Y \in \mathcal{C}_{G-G_{(\xi)}}$, if $G_\xi \cap Y \neq \emptyset$, then $G_\xi \cap Y$ is connected with $V(x;G) \cap V(G_\xi \cap Y) \neq \emptyset$ for all $x \in \mathcal{B}(G_{(\xi)}, Y)$;

TE5. $G = G_{(\alpha)}$ and $\mathbf{T}(G) = \underset{\xi < \alpha}{\bigcup} \mathbf{T}_{G_\xi}(G)$.

If all axioms but TE5 are fulfilled, then $(G_\xi)_{\xi < \alpha}$ is said to be a *partial terminal expansion* of G.

Theorem 2.4 *Let G be a connected graph. If $(G_\xi)_{\xi < \alpha}$ is a (partial) terminal expansion of G, then $(\mathbf{T}_{G_\xi}(G))_{\xi < \alpha}$ is a (partial) expansion of $\mathbf{T}(G)$. Conversely, if $(\Omega_\xi)_{\xi < \alpha}$ is a (partial) expansion of $\mathbf{T}(G)$, then there is a (partial) terminal expansion $(G_\xi)_{\xi < \alpha}$ of G such that $\mathbf{T}_{G_\xi}(G) = \Omega_\xi$ for any $\xi < \alpha$.*

This result, which is a consequence of [12, 2.5.8 and 2.7] and of [14, 9.2], enables us to extend to any connected graph the relation we introduced in Section 6 between the terminal expansions of a tree and the expansions of its end space. As a trivial consequence of 2.6.2 and Theorem 2.4, we see that *any connected graph has a terminal expansion*.

2.7.1 Examples

2.7.1.1 The sequence $(T_n)_{n \leq \omega}$ defined in Section 6 is a terminal expansion of length $\omega + 1$ of the dyadic tree T.

2.7.1.2 If G is the graph of Figure 1 and M_0 and M_1 the multi-endings defined in Figure 7.a and Figure 7.b, respectively, then $(M_0, M_0 \cup M_1, G)$ and $(M_0 \cup M_1, G)$ are two terminal expansions of G of length 3 and 2, respectively.

2.8   Terminal degree of a graph

Definition 2.8 For a graph G, the ordinal $d(\mathbf{T}(G))$ will be called the *terminal degree* of G, and will also be denoted by $d_{\mathbf{T}}(G)$.

Obviously $d_{\mathbf{T}}(G) = \sup\{d_{\mathbf{T}}(X) : X \in \mathcal{C}_G\}$.

For example, if G is the graph of Figure 1, then $d_{\mathbf{T}}(G) = 2$ since on the one hand, G has a terminal expansion $(M_0 \cup M_1, G)$ of length 2, and on the other hand, $d_{\mathbf{T}}(G) \geq 2$ since $\mathbf{T}(G)$ is not discrete.

We will now give some of the principal properties of the terminal degree of a graph.

Theorem 2.5 $d_{\mathbf{T}}(G) \leq \omega_1$ *for any graph G. Conversely, for any ordinal $\alpha \leq \omega_1$ there is a graph G with $d_{\mathbf{T}}(G) = \alpha$.*

Proposition 2.3 *If G is a tree, then $d_{\mathbf{T}}(G) \le \omega + 1$. The terminal degree of the dyadic tree is $\omega + 1$.*

Theorem 2.6 *Let G be a graph. Then the following are equivalent:*
  (i) $d_{\mathbf{T}}(G) \le \omega$;
  (ii) $\mathbf{T}(G)$ is scattered (i.e., contains no non-empty subset which is dense in itself);
  (iii) G has no subdivision of the dyadic tree as a terminally faithful subgraph.

Theorem 2.7 *Let G be a graph. Then $d_{\mathbf{T}}(G) \le \omega + 1$ if and only if there is a countable family of closed scattered subsets of $\mathbf{T}(G)$ whose union is dense.*

Corollary 2.1 *If a graph G is such that $m(\mathbf{T}(G)) \le \omega$ (in particular, if G is countable), then $d_{\mathbf{T}}(G) \le \omega + 1$.*

Proof: Let Δ be a maximal set of pairwise disjoint rays of G. The set $\mathbf{T}_{\cup\Delta}(G)$ is countable since $|\Delta| \le m(\mathbf{T}(G)) \le \omega$, and thus it is scattered; furthermore, it is dense in $\mathbf{T}(G)$ since Δ is maximal. Hence $d_{\mathbf{T}}(G) \le \omega + 1$ by Theorem 2.7.   □

The following results will be useful to give certain characterizations of those graphs which have a spanning faithful tree. First, notice that, as a consequence of Proposition 1.1, we have $d_{\mathbf{T}}(G^+) = d(\hat{V}(G))$.

Lemma 2.3 *Let α be a limit ordinal, and $(G_\xi)_{\xi<\alpha}$ a partial terminal expansion of a connected graph G. For any $X \in \mathcal{C}_{G-G_{(\alpha)}}$ with $B := \mathcal{B}(X, G)$ infinite, there is an end*
$$\tau \in \mathbf{T}_{G_{(\alpha)}}(G) - \bigcup_{\xi<\alpha} \mathbf{T}_{G_{(\xi)}}(G) \text{ such that } B \subseteq V_\tau.$$

Proof: By [12, Lemma 2.5.6] there exists a $\tau \in \mathbf{T}_{G_{(\alpha)}}(G) - \bigcup_{\xi<\alpha} \mathbf{T}_G(G)$ such that B is concentrated in τ. We will show that $B \subseteq V_\tau$. Let $x \in B$; there is a $\xi < \alpha$ such that $x \in V(G_\xi)$. Then for $\eta > \xi$, $x \in \mathcal{B}(G_\eta, \mathcal{C}_{G-G_\eta}(\tau))$, thus by TE3 $V(x; G) \cap V(K_\eta) \ne \emptyset$ where $K_\eta := \mathcal{C}_{G-G_\eta}(\tau) \cap G_{\eta+1}$. And, since τ is in the closure of $\bigcup_{\xi<\eta<\alpha} \mathbf{T}_{K_\eta}(G)$, this implies that x is a neighbour of τ.   □

Theorem 2.8 *If the set $\Sigma := \{\tau \in \mathbf{T}(G) : \{x \in V_\tau : \mathbf{T}_x \text{ is not scattered}\} \text{ is infinite}\}$ has a countable cover by closed scattered sets, then $d(\hat{V}(G)) \le \omega + 1$.*

Proof: (a) Let $(\Sigma^n)_{n<\omega}$ be a cover of Σ by closed scattered subsets of $\mathbf{T}(G)$. Let $n < \omega$. Since Σ^n is scattered, $d(\Sigma^n) \le \omega$ by Theorem 2.6. Let $(\Sigma^n_i)_{i<\omega}$ be an expansion of Σ^n, and put $\Sigma_n := \bigcup_{i+j=n} \Sigma^j_i$, $n < \omega$. Then $(\Sigma_n)_{n<\omega}$ is a partial expansion of the closure of $\bigcup_{n<\omega} \Sigma^n$, such that $\Sigma \subseteq \bigcup_{n<\omega} \Sigma_n$.

(b) For any $x \in V(G)$ with $d(\mathbf{T}_x) \le \omega$, let $(\mathbf{T}^n_x)_{n<\omega}$ be an expansion of $\mathbf{T}_x$.

(c) We define by induction a partial terminal expansion $(G_n)_{n<\omega}$ of G so that, for any $n < \omega$ with $G_n \ne G$, $\mathbf{T}_{G_{n+1}}(G) \cap \mathbf{T}_X(G) \ne \emptyset$ for all $X \in \mathcal{C}_{G-G_n}$.

Let G_0 be any multi-ending of G such that every component of $G - G_0$ has a ray. Let $n < \omega$. Assume that the partial terminal expansion $(G_i)_{i\le n}$ is defined, so that every component of $G - G_n$ has a ray.

For $X \in \mathcal{C}_{G-G_n}$ let τ_X be any element of $\mathbf{T}_X(G)$, and for $x \in \mathcal{B}(G_n, X)$ let n_x be the least integer such that $x \in V(G_{n_x})$. Then the set $\Omega^X_n := \cup\{\mathbf{T}^{n-n_x}_x \cap \mathbf{T}_X(G) : x \in \mathcal{B}(G_n, X)\}$

is a closed discrete subspace of $\mathbf{T}(G)$, since it is the union of a finite family of discrete closed subsets of $\mathbf{T}(G)$.

Define $\Theta_n := \cup\{\Omega_n^X \cup \{\tau_X\} : X \in \mathcal{C}_{G-G_n}\} \cup (\Sigma_n \cap \mathbf{T}_{G-G_n}(G))$. All points of Θ_n are clearly isolated in Θ_n. By using [12, Lemma 2.6] and [14, Lemma 8.3] we can then define the multi-ending G_{n+1} with $\mathbf{T}_{G_{n+1}}(G) = \Theta_n$ and so that every component of $G - G_{n+1}$ has a ray.

Suppose $G_{(\omega)} \neq G$ and let $X \in \mathcal{C}_{G-G_{(\omega)}}$, and $B := \mathcal{B}(G_{(\omega)}, X)$. Assume B finite. Then $B \subseteq V(G_n)$ for some $n < \omega$. By the properties of $(G_n)_{n<\omega}$, $\mathbf{T}_{G_{n+1}}(G) \cap \mathbf{T}_X(G) \neq \emptyset$, thus $X \notin \mathcal{C}_{G-G_{n+1}}$, a contradiction. Hence B is infinite. Them, by Lemma 2.3, there is $\tau \in \mathbf{T}_{G_{(\omega)}}(G) - \bigcup_{n<\omega} \mathbf{T}_{G_n}(G)$ with $B \subseteq V_\tau$. This end $\tau \notin \Sigma$ since by the construction, $\Sigma \subseteq \bigcup_{n<\omega} \mathbf{T}_{G_n}(G)$.

Thus there is $x \in B$ with $d(\mathbf{T}_x) \leq \omega$. Let n and p be the least integers such that $x \in V(G_n)$ and $\tau \in \mathbf{T}_x^p$. Then $\tau \in \mathbf{T}_{G_{n+p}}(G)$, contrary to $\tau \notin \bigcup_{n<\omega} \mathbf{T}_{G_n}(G)$. Therefore $G = G_{(\omega)}$. $\square$

Remark 2.1 By the preceding proof, if $\mathbf{T}_x$ is scattered for all $x \in V(G)$, then there exists a terminal expansion $(G_n)_{n\leq\omega}$ of G such that $V_\tau = \emptyset$ for all $\tau \notin \bigcup_{n<\omega} \mathbf{T}_{G_n}(G)$.

In particular we get:

Corollary 2.2 $d(\hat{V}(G))) \leq \omega + 1$ whenever G satisfies one of the following conditions:
  (i) $|\mathbf{T}_x| \leq \omega$ for every $x \in V(G)$;
  (ii) $\{\tau \in \mathbf{T}(G) : v(\tau) \geq \omega\}$ is countable;
  (iii) $\{\tau \in \mathbf{T}(G) : m(\tau) \geq \omega\}$ is countable.

Proof: (i) If every vertex has a countable terminal neighbourhood, then the subset Σ of $\mathbf{T}(G)$, defined in Theorem 2.8, is empty.
  (ii) In this case, Σ has an obvious countable cover by singletons.
  (iii) This is a consequence of (ii) since $m(\tau) < \omega$ implies $v(\tau) < \omega$ by Proposition 2.2.$\square$

2.9   Terminal expansions of higher order

We will now generalize the concept of terminal expansion in order to extend certain inductive proofs which apply to graphs of terminal degree less than ω_1 to the case of graphs of terminal degree ω_1.

Definition 2.9 Let α and ϵ be ordinals, G a connected graph. We define by induction the concepts of *terminal expansion of order ϵ and length α* and of *terminal degree of order ϵ* of G, denoted by $d_\mathbf{T}^{(\epsilon)}(G)$.

  (i) $\epsilon = 0$: A terminal expansion of order 0 of G is a terminal expansion of G, and $d_\mathbf{T}^{(0)}(G) = d_\mathbf{T}(G)$;

  (ii) $\epsilon > 0$: A terminal expansion of order ϵ and length α of G is a sequence $(G_\xi)_{\xi<\alpha}$ satisfying all the axioms (TE1 — TE5) of Definition 2.7, where "discrete multi-ending" is replaced by "multi-ending whose terminal degree of order δ, for some $\delta < \epsilon$, is strictly less than ω_1". $d_\mathbf{T}^{(\epsilon)}(G)$ is the least ordinal α such that G has a terminal expansion of order ϵ and length α.

The basic properties of terminal expansions obviously hold for terminal expansions of higher order. Moreover we have:

Proposition 2.4 *Let $\delta < \epsilon$. Then*

$$d_{\mathbf{T}}^{(\epsilon)}(G) \leq d_{\mathbf{T}}^{(\delta)}(G) \leq d_{\mathbf{T}}(G) \leq \omega_1.$$

In particular, $d_{\mathbf{T}}^{(\epsilon)}(G) = 1$ if $d_{\mathbf{T}}^{(\delta)}(G) < \omega_1$.

Definition 2.10 A graph G is *terminally inaccessible* if $d_{\mathbf{T}}^{(\epsilon)}(G) = \omega_1$ for every ordinal ϵ. Otherwise G is called *terminally accessible*; and the least ordinal ϵ such that $d_{\mathbf{T}}^{(\epsilon)}(G) < \omega_1$ is the *order of terminal accessibility* of G.

The problem of the existence of terminally inaccessible graphs is still unsolved.

Theorem 2.9 *Let G and H be two connected graphs such that there is a uniform monomorphism from $(\mathbf{T}(G), \mathbf{U}_G^*)$ into $(\mathbf{T}(H), \mathbf{U}_H^*)$. Then $d_{\mathbf{T}}^{(\epsilon)}(G) \leq d_{\mathbf{T}}^{(\epsilon)}(H)$ for every ordinal ϵ.*

We will now apply the preceding concepts to the problems of the existence of the special trees (CST, RST, CFT, SFT) we introduced in Section 2.2.

2.10 Coterminal spanning trees (CST). Rayless spanning trees (RST)

Theorem 2.10 *Let G be a one-ended connected graph. Then G has an RST if and only if G has a CST with $V_{\mathbf{T}(G)} \neq \emptyset$.*

Halin proved in [2, Satz 3] that every countable one-ended connected graph has a CST. We obtain here a stronger result than that of Halin (Theorem 2.11). Nevertheless, this result is still only partial and the problem of the existence of a CST for an arbitrary graph is still open.

Lemma 2.4 *Let G be a one-ended connected graph such that $V_{\mathbf{T}(G)} \neq \emptyset$. Then G has a CST (resp. an RST) if and only if G has a coterminal (resp. rayless) tree containing $V_{\mathbf{T}(G)}$.*

Proof: The sufficiency is the only part to be proved. Let T be a coterminal (resp. rayless) tree containing $V_{\mathbf{T}(G)}$. Contract T onto a single vertex $t \notin V(G)$. Let G_T be the resulting graph, i.e., $V(G_T) = V(G-T) \cup \{t\}$ and $E(G_T) = (E(G) \cap [V(G-T)]^2) \cup \{\{x, t\} : x \in V(G-T) \text{ and } \{x, y\} \in E(G) \text{ for some } y \in V(T)\}$.

For every $\tau \in \mathbf{T}(G_T)$ we have $V_\tau = \{t\}$; thus by Corollary 2.2(ii) $d_{\mathbf{T}}(G_T) \leq \omega + 1$ and by Proposition 2.1 $m(\tau) \leq \omega$. Therefore by [14, Théorème 10.9], G_T has an RST T'.

For any $x \in V(t; T')$ let e_x be an edge of G incident with x and with a vertex of T. Then

$$(T' - \{t\}) \cup T \cup \bigcup \{\langle e_x \rangle : x \in V(t; T')\}$$

is clearly a CST (resp. an RST) of G. $\square$

Lemma 2.5 *Let G be a one-ended graph such that $0 < v(\mathbf{T}(G)) \leq \omega$. Then $V_{\mathbf{T}(G)}$ is contained in a finite path or a ray of G.*

Proof: Let $V_{\mathbf{T}(G)} = \{x_n : n < \omega\}$. We define by induction a sequence $(W_n)_{n<\omega}$ of finite paths of G such that $x_n \in V(W_n)$ and $W_n \subseteq W_{n+1}$.

Let $W_0 = \langle x_0 \rangle$, and $n \geq 0$. Suppose that $W_n = \langle y_0, \ldots, y_k \rangle$ is defined such that $y_0 = x_0$, $y_k = x_i$ for some $i \leq n$, and $x_n = y_j$ for some $j \leq k$. If $x_{n+1} \in V(W_n)$, then $W_{n+1} := W_n$. Assume $x_{n+1} \notin V(W_n)$, $y_k \delta_G x_{n+1}$ since $y_k, x_{n+1} \in V_{\mathbf{T}(G)}$ thus, since W_n is finite, there is a $y_k x_{n+1}$-path P internally disjoint from W_n. Then define $W_{n+1} := W_n \cup P$.

Finally, $\bigcup_{n<\omega} W_n$ is a path or a ray of G which contains $V_{\mathbf{T}(G)}$. $\square$

Theorem 2.11 *Every one-ended connected graph G with $\min\{m(\mathbf{T}(G)), v(\mathbf{T}(G))\} \leq \omega$ has a CST.*

This is a consequence of Lemma 3.4 of [12] and of the two preceding lemmas.

The next result reduces the problems of the existence of a CST or RST for terminally accessible graphs to the corresponding problems for one-ended connected graphs. The proof proceeds by induction on the terminal degree of higher order.

Theorem 2.12 *Let G be a terminally accessible connected graph. Then G has a CST (resp. RST) if and only if every one-ended connected restriction of G has a CST (resp. and every end of G has a non-empty neighbourhood).*

Combining this result with Theorem 2.11 we obtain:

Corollary 2.3 *Let G be a terminally accessible connected graph. If every end of G has countable multiplicity or countable neighbourhood, then G has a CST; moreover, if the neighbourhood of every end is non-empty, then G has an RST.*

2.11   Coterminal faithful trees (CFT). Spanning faithful trees (SFT)

We will give several characterizations of the graphs having a CFT or an SFT.

Remark 2.2 We can easily observe that a *connected graph G has an SFT if and only if G^+ has a CFT* (1.3.5). Thus, since the topological spaces induced by the uniformities $\mathbf{S}^*_{G^+}$ and $\hat{\mathbf{S}}_G$ are homeomorphic (Proposition 1.1), any topological characterization of the graphs having a CFT corresponds to a similar characterization of those having an SFT.

Theorem 2.13 *Let G be a connected graph. The following properties are equivalent:*
  (i) G has a CFT;
  (ii) there is a tree T such that $(\mathbf{T}(G), \mathbf{S}^*_G)$ is uniformly isomorphic to $(\mathbf{T}(T), \mathbf{S}^*_T)$;
  (iii) there is a tree T such that $(\mathbf{T}(G), \sigma^*_G)$ is proximally isomorphic to $(\mathbf{T}(T), \sigma^*_T)$;
  (iv) there is a tree T such that $(\mathbf{T}(G), \mathbf{U}^*_G)$ is uniformly isomorphic to $(\mathbf{T}(T), \mathbf{U}^*_T)$;
  (v) there is a countable family $(A_n)_{n<\omega}$ of dispersed subsets such that $\bigcap_{n<\omega} A^*_n = id_{\mathbf{T}(G)}$;
  (vi) $(\mathbf{T}(G), top(\mathbf{S}^*_G))$ is first countable with $d(\mathbf{T}(G)) \leq \omega + 1$;
  (vii) $top(\mathbf{S}^*_G)$ is ultrametrizable.

Proof: The equivalence of the first six properties was proved in [11, Corollaire 6.17] and [13, Théorème 5.2]. We will prove that (i) $\Rightarrow$ (vii) $\Rightarrow$ (v).

(i) $\Rightarrow$ (vii): Let T be a CFT of G. Take any vertex a of T, and for every $n < \omega$, let $A_n = \{x \in V(T) : d_T(a,x) \leq n\}$, where d_T is the natural distance on $V(T)$. Then A_n is a dispersed subset of $V(T)$, and the family $(A_n^*)_{n<\omega}$ generates an ultrametrizable uniformity (i.e., a uniformity having a countable base of equivalent relations) which is clearly compatible with top$(\mathbf{S}_T^*)$ since T is a tree. Hence top$(\mathbf{S}_G^*)$ is ultrametrizable since the topological spaces $\mathbf{T}(G)$ and $\mathbf{T}(T)$ are homeomorphic by [11, 5.12].

(vii) $\Rightarrow$ (v): Suppose now that top$(\mathbf{S}_G^*)$ is ultrametrizable. Then the ultrametrizable uniformity compatible with top$(\mathbf{S}_G^*)$ has a countable base $(\rho_n)_{n<\omega}$ of equivalence relations. By Lemma 2.2, for any $n < \omega$, there is a dispersed subset A_n of $V(G)$ such that $A_n^* \subseteq \rho_n$. Hence $\bigcap_{n<\omega} A_n^* \subseteq \bigcap_{n<\omega} \rho_n = id_{\mathbf{T}(G)}$ since the space $\mathbf{T}(G)$ is Hausdorff. $\qquad\square$

We do not know if the condition "top$(\mathbf{S}_G^*)$ is metrizable" implies the existence of a CFT. This might be an interesting question to study, since it is not even known if any metrizable zero-dimensional space is ultrametrizable.

Theorem 2.14 *Let G be a connected graph. The following properties are equivalent:*
(i) G has an SFT;
(ii) there is a tree T such that $(V(G), \mathbf{S}_G)$ is uniformly isomorphic to $(V(T), \mathbf{S}_T)$;
(iii) there is a tree T such that $V(G), \sigma_G)$ is proximally isomorphic to $(V(T), \sigma_T)$;
(iv) $V(G)$ has a countable cover by dispersed subsets:
(v) $(\hat{V}(G), top(\hat{\mathbf{S}}_G))$ is first countable with $d(\hat{V}(G)) \leq \omega + 1$;
(vi) $d(\hat{V}(G)) \leq \omega + 1$ with $m(\tau) + v(\tau) \leq \omega$ for all $\tau \in \mathbf{T}(G)$;
(vii) top$(\hat{\mathbf{S}}_G)$ is ultrametrizable.

The equivalence of the first six properties was proved in [11, Corollaire 6.19] and [13, 5.2 and 5.3].

(vii) is a consequence of Theorem 2.13(vii) and of Remark 2.2. Notice that the equivalence (i) $\Leftrightarrow$ (iv) of Theorem 2.14 was obtained by Jung [8, Theorem 5].

As a trivial consequence of Proposition 2.1, Corollary 2.2(ii) and Theorem 2.14(vi) we have:

Corollary 2.4 *If the multiplicity of each end of a connected graph G is finite, then G has an SFT.*

Theorem 2.15 *Let G be a connected graph such that $m(\mathbf{T}(G)) + v(\mathbf{T}(G)) \leq \omega$. Then G as an SFT.*

This condition, which is one of the simplest to check, is unfortunately not sufficient in general. However, we have:

Theorem 2.16 *Let G be a connected graph such that $\mathbf{T}(G)$ is compact. Then G has an SFT if and only if $m(\mathbf{T}(G)) + v(\mathbf{T}(G)) \leq \omega$.*

References

[1] H. Freudenthal, Über die Enden diskreter Räume und Gruppen. *Comment. Math. Helv.* **17**(1944), 1–38.

[2] R. Halin, Über unendliche Wege in Graphen. *Math. Ann.* **157**(1964), 125–137.

[3] R. Halin, Charakterisierung der Graphen ohne unendliche Wege. *Arch. Math. (Basel)* **16**(1965), 227–231.

[4] R. Halin, Über die Maximalzahl fremder unendlicher Wege in Graphen. *Math. Nachr.* **30**(1965), 63–85.

[5] R. Halin, Die Maximalzahl fremder zweiseitig unendlicher Wege in Graphen. *Math. Nachr.* **44**(1970), 119–127.

[6] R. Halin, A note on Menger's Theorem for infinite locally finite graphs. *Abh. Math. Sem. Univ. Hamburg* **40**(1974), 111–114.

[7] H. A. Jung, Wurzelbäume und unendliche Wege in Graphen. *Math. Nachr.* **41**(1969), 1–22.

[8] H. A. Jung, Connectivity in infinite graphs. In: *Studies in Pure Mathematics*, pp. 137–143. Editor, L. Mirsky. New York–London: Academic Press 1971.

[9] S. A. Naimpally and B. D. Warrack, *Proximity Spaces*. Cambridge: Cambridge University Press 1970.

[10] N. Polat, Aspects topologiques de la séparation dans les graphes infinis. I. *Math. Z.* **165**(1979), 73–100.

[11] N. Polat, Aspects topologiques de la séparation dans les graphes infinis. II. *Math. Z.* **165**(1979), 171–191.

[12] N. Polat, Développements terminaux des graphes infinis. I. Arbres maximaux coterminaux. *Math. Nachr.* **107**(1982), 283–314.

[13] N. Polat, Développements terminaux des graphes infinis. II. Degré de développement terminal. *Math. Nachr.* **110**(1983), 97–125.

[14] N. Polat, Développements terminaux des graphes infinis. III. Arbres maximaux sans rayon, cardinalité maximum des ensembles disjoints de rayons. *Math. Nachr.* **115**(1984), 337–352.

[15] B. Zelinka, Uneigentliche Knotenpunkte eines Graphen. *Časopis Pěst. Mat.* **95**(1970), 155–169.

DENDROIDS, END-SEPARATORS, AND ALMOST CIRCUIT-CONNECTED TREES

G. SABIDUSSI

Département de mathématiques et de statistique

Université de Montréal

C.P. 6128

Succursale A

Montréal

Québec, H3C 3J7

Canada

Abstract

An and-separator of an infinite connected graph G is a subgraph S of G such that no component of the edge-complement G\S contains two rays belonging to different ends of G. We show that the minimal end-separators of G are determined by the dendroids of G in the sense that every minimal end-separator is contained in a dendroid, and that conversely, the edges of a dendroid whose canonical circuit is terminally faithful in G (if such edges exist) form a minimal end-separator. Since any dendroid of an infinite graph is determined by a spanning tree and a dendroid of the latter, one is naturally led to a class of trees (called almost circuit-connected, a.c.c) defined by the property that the removal of all circuits (2-way infinite paths) results in a locally finite graph. We show that every locally cofinite dendroid of such a tree T induces a minimal end-separator in any graph spanned by T, and that this universal property characterises a.c.c. trees. Several other characterisations of these trees are also given.

1   Introduction, definitions, notation

In this paper we explore the relationship between the dendroids and end-separators of infinite graphs. A dendroid of a graph G is a minimal subgraph containing at least one edge of every (finite or infinite) circuit in G; an end-separator is a subgraph whose edge-complement has no component containing two rays (= one-way infinite paths) which belong to different ends of G. As in all situations pertaining to separation, the question of the existence of minimal end-separators is of fundamental concern.

The point of departure of this investigation is the observation that any dendroid of a graph is an end-separator, and that in the case of trees, minimal end-separators and dendroids are the same thing. For arbitrary graphs we establish the precise link between the two types of subgraphs by showing that every minimal end-separator is contained in a dendroid, and that every dendroid contains at most one minimal end-separator (Section 3). Our basic tool is a result of Bean [2] which describes the dendroids of a connected graph G in terms of its spanning trees and the dendroids of the latter. Attention is therefore directed

G. Hahn et al. (eds.), Cycles and Rays, 221–236.

to the question as to what spanning trees give rise to dendroids which induce minimal end-separators. There are various ways in which this question can be made precise. We shall concentrate on trees which force the existence of a minimal end-separator in any graph G which they span. It turns out that this universal property is enjoyed by a surprisingly large class of trees (which we call almost circuit-connected). Intrinsically they are characterised by the property that the removal of the edges of all double rays (= two-way infinite paths) results in a graph all of whose components are finite. Various properties of these trees are considered in Section 4. In Section 5 it is shown that almost circuit connected trees induce minimal end-separators via their locally cofinite dendroids. Section 2 contains technical results about dendroids of trees.

We consider only graphs without loops, multiple edges or *isolated vertices*. Whenever convenient a graph may therefore be identified with its set of edges. Thus we shall write $e \in G$ for $e \in E(G)$. The union and intersection of graphs is defined as the union, resp. intersection, of their edge-sets. Consequently, disjointness of graphs means that they are *edge-disjoint*. Given two graphs G, H we denote by $G \backslash H$ the graph whose edge-set is $E(G) \backslash E(H)$. For a vertex $x \in V(G)$ the degree of x is denoted by $d(x; G)$. A subgraph $H \subset G$ is *locally cofinite* in G if $G \backslash H$ is locally finite. Note that if G is locally finite, then every subgraph is locally cofinite in G.

A *ray* is a one-way infinite path, usually denoted by $R = (x_0, x_1, \ldots)$, where $x_0, x_1, \ldots$ are the vertices of R. Following Halin [3], we say that two rays $R_1, R_2 \subset G$ are equivalent in G, $R_1 \sim_G R_2$, if there is a ray $R \subset G$ such that $V(R) \cap V(R_i)$ is infinite, $i = 1, 2,$. When no confusion is likely we write $\sim$ for $\sim_G$. The equivalence classes modulo $\sim_G$ are the *ends* of G. Denote the set of ends by $\mathcal{E}(G)$. If H is a subgraph of G there is a natural map $\mathcal{E}(H) \to \mathcal{E}(G)$ which associates with every end of H the unique end of G containing it. In general, this map is neither one-one nor onto (in the example of Figure 1 it fails to be one-one). If the map $\mathcal{E}(H) \to \mathcal{E}(G)$ is one-one [onto], then H is called *terminally faithful* [*terminally complete*] in G (for details see [3]). If the rays of H belong to only finitely many ends of G, then H is *terminally finite* in G. If H has a ray and all rays of H are contained in a single end of G, then H is *1-ended* in G. Reference to G will often be omitted. G is 1-ended in itself means that G has exactly one end.

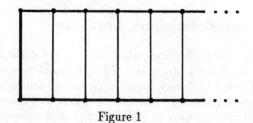

Figure 1

Definition 1.1 An *end-separator* of G is a subgraph $A \subset G$ such that any component of $G \backslash A$ is either rayless or 1-ended in G, i.e., no component of $G \backslash A$ contains two rays which are inequivalent in G.

Observe that it can happen that an end-separator (even a minimal one) contains infinitely many edges of every ray in a given end ϵ, so that in $G \backslash A$ there is no ray representing ϵ (see Figure 2).

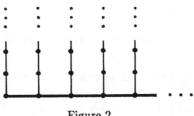

Figure 2

A *circuit* is a non-empty, connected, 2-regular graph, i.e., either a *cycle* or a *double ray* (two-way infinite path). Any double ray Z in a graph G is either 1-ended or terminally faithful. In the latter case it is more natural to say that Z is *2-ended*.

As is well known, the circuits of a graph form a binary exchange system. The matroid-theoretic notion of a dendroid therefore applies to this system.

Definition 1.2 (i) A graph is *circuit-free* if it contains no (finite or infinite) circuit. (ii) A *dendroid* of a graph G is a minimal subgraph D of G such that $G \backslash D$ is circuit-free.

Dendroids were originally introduced by Tutte [4]. The word if not the concept has disappeared from the literature, the reason being that for (finite) matroids the dendroids are just the dual bases. However, since for infinite exchange systems, especially non-finitary ones such as the circuit systems of graphs, there is no satisfactory notion of duality, it seemed appropriate to revert to Tutte's original highly graphic term, following in this respect Bean [1, 2].

The basic result concerning the structure of dendroids in a graph (as well as their existence) was established by Bean [2], Theorems 3.3 and 3.8:

Lemma 1.3 *Let G be connected. Then $D \subset G$ is a dendroid of G if and only if there is a spanning tree T of G and an dendroid D_T of T such that*

$$D = G \backslash T \cup D_T.$$

We shall say that D is *based* on T. In general, the spanning tree on which a dendroid is based is not unique. Note that $G \backslash T$ is a dendroid for the exchange system formed by the (finite) cycles of G.

Given a dendroid D of G and $e \in D$ there is a unique circuit C in G such that $C \cap D = e$ (the *canonical circuit* determined by D_T and e). We shall denote this circuit by $C(e, D)$. It is a consequence of (1.3) that if $D = G \backslash T \cup D_T$ and $e \in D$, then $C(e, D)$ coincides with the canonical circuit $C(e, D_T)$ in T.

Notation 1.4 Given a dendroid D of G put

$$D^{(0)} := \{e \in D : C(e, D) \text{ is finite}\},$$

$$D^{(i)} := \{e \in D \backslash D^{(0)} : C(e, D) \text{ is } i\text{-ended in } G\}, \quad i = 1, 2,$$

$$D_\infty := D \backslash D^{(0)} = D^{(1)} \cup D^{(2)}.$$

By the preceding remark, if $D = G\backslash T \cup D_T$, then $D_T \subset D_\infty$, $D^{(0)} \subset G\backslash T$, and it is easily seen [1] that

$$D_\infty = D_T \cup \{e \in G\backslash T : C_e \cap D_T \neq \emptyset\},$$

where $C_e := C(e, G\backslash T)$ is the unique cycle in G with $C_e \cap G\backslash T = e$.

The following property of trees will be important in the sequel.

Definition 1.5 Let T be a tree. By T_*, the *core* of T, we denote the union of all double rays contained in T. T is *circuit-connected* ([2] 3.5) if it is non-empty and $T = T_*$.

Note that the circuits of T are precisely those of T_*. Hence any dendroid of T is contained in T_*.

If $x, y \in V(T)$ the unique path in T joining x and y will be denoted by T_{xy}.

Given $A, B \subset V(G)$, an AB-path $P \subset G$ is an *AB-chord* if the only vertices of P belonging to $A \cup B$ are its endpoints. If X, Y are subgraphs of G, then an *XY-chord* is a $V(X), V(Y)$-chord.

Throughout the paper, G stands for a *connected infinite graph*.

2 Dendroids of trees

This section contains an assortment of elementary facts about dendroids in arbitrary trees which will be needed in later sections. The first result, establishing the existence of dendroids, is due to Bean [2], Theorem 3.8. We present here a slightly stronger version. The proof is essentially Bean's with appropriate modifications.

Lemma 2.1 *Any tree T has a dendroid D which is locally cofinite in T_*. Moreover, if Z is a double ray in T and $a \in Z$, then D can be so chosen that $a \in D$ and $C(a, D) = Z$.*

Proof: If T contains no double ray, i.e., if $T_* = \emptyset$, then $D = \emptyset$ and there is nothing to prove. Assume therefore that $T_* \neq \emptyset$. Fix $u \in V := V(T_*)$ and consider the natural partial order $\leq$ on V with u as least element. For $x \in V$ let Σ_x be the set of immediate successors of x with respect to $\leq$. Then $\Sigma_x \neq \emptyset$ by circuit-connectedness of T_*. Let σ be a choice function for the family $(\Sigma_x)_{x \in V}$ and put $E_x := \{[x, y] : y \in \Sigma_x, y \neq \sigma(x)\}$. Then $D := \bigcup \{E_x : x \in V\}$ is a dendroid of T, and for any $x \in V$, D contains all but at most two edges of T_* incident with x. Thus $T_* \backslash D$ is locally finite (in fact, the degree of any vertex in $T_* \backslash D$ is either 1 or 2).

The canonical circuits with respect to D are obtained as follows. For $x \in V$, $R_x := (x, \sigma(x), \sigma^2(x), \ldots)$ is a ray in $T_* \backslash D$. If $e = [x, y] \in D$, then $C(e, D) = R_x \cup e \cup R_y$.

Finally, if a double ray Z and $a = [u, v] \in Z$ are given, take u as the root of T_* and let σ be any choice function such that $v \neq \sigma(u) \in V(Z) \cap \Sigma_u$ and $\sigma(x) \in V(Z) \cap \Sigma_x$ for $x \in V \backslash \{u\}$. Then $a \in D$ and $C(a, D) = Z$. $\square$

Lemma 2.2 *Let T be a tree containing a ray, D a dendroid of T. Then given any $x \in V(T)$ there is a unique ray $R_x \subset T \backslash D$ which starts at x.*

Proof: The uniqueness of R_x follows from the circuit-freeness of $T \backslash D$. To show that R_x exists note that since $T \backslash D$ spans T it is sufficient to show that every component of

$T\backslash D$ contains a ray. If T is circuit-free, then $D = \emptyset$ and by hypothesis T contains a ray. We may therefore assume that $D \neq \emptyset$.

Take any component K of $T\backslash D$. Since D is non-empty there is an edge $e = [u, v] \in D$ such that $u \in V(K)$. $C(e, D)\backslash e$ consists of two rays one of which starts at u and hence is contained in K. □

Note that if $D = G\backslash T \cup D_T$ is a dendroid of an arbitrary graph G and if the tree T contains a ray, then the above result implies that given any $x \in V(G)$ there is a ray $R_x \subset G\backslash D = T\backslash D_T$ which starts at x.

Corollary 2.3 *Let T be a tree containing a double ray, D a dendroid of T. Then any ray $R \subset T\backslash D$ has a subray which is contained in the canonical circuit $C(e, D)$ of some edge $e \in D$.*

Proof: Let K be the component of $T\backslash D$ containing R. Since $D \neq \emptyset$ it follows from the proof of (2.2) that if e is any edge of D having an endpoint in K, then K contains a subray of $C(e, D)$. Moreover, any two rays in K have a common subray. □

Given a dendroid of a tree T, call two edges $a, b \in D$ *neighbours* if $C(a, D) \cap C(b, D) \neq \emptyset$. We then have:

Lemma 2.4 *Let D be a dendroid of a tree T, $a, b \in D$. Then the following are equivalent:*
(i) a and b are neighbours;
(ii) $C(a, D) \cap C(b, D)$ is a ray;
(iii) there are no edges of D between a and b, i.e., a, b are the only edges of D on the shortest path $T(a, b) \subset T$ which contains a and b.

Proof: (iii) $\Rightarrow$ (ii). Suppose $a = [u, v]$, $b = [x, y]$ are the only edges of D on the path $T(a, b)$. Let the notation be so chosen that u, v, x, y follow each other on $T(a, b)$ in this order, i.e., the path T_{vx} is (edge-)disjoint from D. Counting from u in the direction of v let w be the last vertex of $C(a, D)$ belonging to $T(a, b)$, and R the ray in $C(a, D)$ starting at w and not containing a. Define S and z similarly, using b. Then $w, z \in V(T_{vx})$, because otherwise at least one of $C(a, D)$, $C(b, D)$ would contain both a and b which is impossible. It follows that $w = z$ and $R = S$, otherwise $R \cup T_{wz} \cup S$ is a double ray missing D. Thus $C(a, D) \cap C(b, D) = R$.

(ii) $\Rightarrow$ (i) is obvious. (i) $\Rightarrow$ (iii) also is immediate, for if there is an edge $e \in D$ between a and b, then $C(a, D)$ and $C(b, D)$ are in different components of $T\backslash e$. □

Corollary 2.5 *Let $a, b \in D$. Then there is a sequence of edges $a_0, \ldots, a_n \in D$ such that $a_0 = a$, $a_n = b$, and a_{i-1}, a_i are neighbours, $i = 1, \ldots, n$.*

Proof: For $a_0, \ldots, a_n$ take the edges of $T(a, b) \cap D$ in their order on $T(a, b)$. □

3 Extending minimal end-separators

The obvious approach to finding a minimal end-separator would be to show that the set of end-separators in a graph G is inductive with respect to inclusion. However, this is easily seen to be false, even under very restrictive additional conditions, such as local finiteness. For example, if G is a double ray, then any ray in G is an end-separator; hence G contains decreasing sequences of end-separators having empty intersection. In this section we show that minimal end-separators can be obtained in a natural way via dendroids: every minimal end-separator of G can be extended to a dendroid (Theorem (3.2)). In the opposite direction, it is not true in general that every dendroid of G contains a minimal end-separator. If it does, the latter is unique and consists of those edges of the dendroid whose canonical circuit is 2-ended in G (Lemma (3.3)).

Given a minimal end-separator M of G, denote the components of $G\backslash M$ by H_i, $i \in I$. Also, for $e \in M$ we shall use the notation $M_e := M\backslash e$.

Lemma 3.1 *Suppose G contains a ray, and let M be a minimal end-separator of G, Then:*
(i) $G\backslash M$ spans G;
(ii) no edge $e \in M$ has both endpoints in the same component H_i;
(iii) every H_i is 1-ended in G

Proof: If $M = \emptyset$ there is nothing to prove. We may therefore assume that $M \neq \emptyset$.

(i): For $x \in V(G)$ let E_x be the set of edges of G incident with x. Suppose $E_x \subset M$ for some x. Take any $e = [x,y] \in E_x$. Then $y \in V(H_i)$ for some i otherwise M_e would be an end-separator. Clearly $H_i \cup e$ is the component of $G\backslash M_e$ containing y; all other components of $G\backslash M_e$ are components of $G\backslash M$. By the minimality of M, $H_i \cup e$ contains two rays which are inequivalent in G, whereas H_i is either rayless or 1-ended, clearly an impossibility.

(ii): If the endpoints of $e \in M$ are in the same H_i, then $H_i \cup e$ is a component of $G\backslash M_e$, and this gives rise to the same contradiction as in (i).

(iii): Consider any component H_i. Since G is connected and $M \neq \emptyset$ there is an edge $e = [x,y] \in M$ with $x \in V(H_i)$. By (i) and (ii), $y \in V(H_j)$ for some $j \neq i$. One of the components of $G\backslash M_e$ is $H := H_i \cup e \cup H_j$, the others are the components H_k, $k \neq i,j$. By minimality of M it follows that H contains two rays R, S which are inequivalent in G. Clearly there are vertex-disjoint subrays R', S' not containing e, and one of these, say R', is contained in H_i.   □

Theorem 3.2 *Any minimal end-separator of G can be extended to a dendroid of G.*

Proof: We may assume that G contains a ray, otherwise there is nothing to prove. By Lemma (3.1)(iii), if M is a minimal end-separator of G, then every component H_i contains a ray R_i. Let T_i be a spanning tree of H_i extending R_i, and D_i a dendroid of H_i based on T_i. We claim that

$$D := M \cup \bigcup_{i \in I} D_i$$

is a dendroid of G. If C is a circuit in G, then it either meets M or is contained in some H_i, in which case it meets D_i. Thus D meets every circuit. To show that it is minimal with this property, take any $e \in D$. If $e \in H_i$ for some $i \in I$, then taking the canonical circuit $C(e, D_i)$ in H_i we have

$$C(e, D_i) \cap D\backslash e = C(e, D_i) \cap D_i\backslash e = \emptyset.$$

If $e = [x,y] \in M$, then by Lemma (3.1)(i),(ii), $x \in V(H_i)$, $y \in V(H_j)$ for some $i \neq j$. Since $T_i \supset R_i$, application of Lemma (2.2) to H_i yields a ray $S_x \subset H_i\backslash D_i$ starting at x; similarly one obtains a ray $S_y \subset H_j\backslash D_j$ starting at y. Then $S_x \cup e \cup S_y$ is a double ray in G which misses $D\backslash e$.   □

Lemma 3.3 *Let A be an end-separator of G contained in a dendroid D, and let $e \in D$. Then $A \backslash e$ is an end-separator of G if and only if $e \notin D^{(2)}$. That is to say, $D^{(2)}$ is the intersection of all end-separators of G contained in D.*

Proof: Suppose $e \in D^{(2)}$. Then $C(e, D) \backslash e = R \dot\cup S$, where R, S are two rays which are inequivalent in G. Since $C(e, D) \subset G \backslash (D \backslash e)$, R and S are in the same component of $G \backslash (D \backslash e)$, so that $D \backslash e$ is not an end-separator.

Conversely, suppose $e \in A \backslash D^{(2)}$. Let K be the component of $G \backslash (A \backslash e)$ which contains e and hence $C(e, D)$. Except for K, any component of $G \backslash (A \backslash e)$ is a component of $G \backslash A$. Thus, if $A \backslash e$ is not an end-separator, then K has two rays S_0, S_1 which are inequivalent in G and may be assumed not to contain e. If $C(e, D)$ is finite, then $K \backslash e$ is connected, hence S_0, S_1 are in the same component of $G \backslash A$, but this is impossible. Therefore $C(e, D) \backslash e = R_0 \dot\cup R_1$, and $R_0 \sim R_1$ because $e \notin D^{(2)}$. The notation can be so chosen that the rays R_i and S_i are in the same component K_i of $G \backslash A$, $i = 0, 1$. Since K_0, K_1 are 1-ended, we have $S_0 \sim R_0 \sim R_1 \sim S_1$, a contradiction. □

Summing up the preceding results, we have:

Theorem 3.4 *Any minimal end-separator of a connected graph G is contained in a dendroid of G. Conversely, any dendroid D of G either contains no minimal end-separator of G or exactly one, viz. $D^{(2)}$.*

The study of minimal end-separators is thus completely reduced to that of dendroids. Since any dendroid of G is based on a spanning tree T of G, both the intrinsic properties of the tree and its relation to G will influence the existence and properties of the corresponding minimal end-separators. In this connection it is of interest to consider the case where the end-structure of T is most closely tied to that of G.

Proposition 3.5 *A spanning tree T of G is terminally faithful if and only if given any dendroid D of G based on T and any $e \in D_\infty$, $C(e, D)$ is 2-ended in G, i.e., $D^{(2)} = D_\infty$.*

Proof: Let $e \in D_\infty$. Then $C(e, D) \backslash e = R \dot\cup S \subset T$, and the rays R, S are inequivalent in T. Hence if T is terminally faithful, they are also inequivalent in G, i.e., $C(e, D)$ is 2-ended.

Conversely, suppose T is not terminally faithful. Then T contains a 1-ended double ray Z. Choose any $e \in Z$. By Lemma (2.1), T has a dendroid D_T such that Z is the canonical circuit in T through e w.r. to D_T. Taking $D = G \backslash T \cup D_T$ we then have $C(e, D) = C(e, D_T) = Z$. Hence $e \in D_\infty$ and $C(e, D)$ is 1-ended. □

It should be noted that even if T is not terminally faithful there may exist dendroids based on T for which $D^{(2)} = D_\infty$.

4 Almost circuit-connected trees

We now introduce a class of trees T which have the following universal property with respect to terminal completeness:

> If G is any graph containing T as a terminally finite spanning tree, then T is terminally complete in G.

Recall that *terminally finite* means that the rays of T belong to only finitely many ends of G. It was shown by Halin [3], Satz 5, that infinite locally finite trees have this universal property (even without the restriction of terminal finiteness). We extend it here as far as it will go. The motivation for considering these more general trees is that they induce a minimal end-separator in every graph they span (see Section 5).

Definition 4.1 A tree T is *almost circuit-connected (a.c.c.)* if it contains a ray, and $T \backslash T_*$ is locally finite.

If T is any tree with $T_* \neq \emptyset$, then $T \backslash T_*$ is rayless. Hence by König's theorem, any infinite component K of $T \backslash T_*$ must contain a vertex x with $d(x; K)$ infinite. This means that T is a.c.c. if and only if $T_* \neq \emptyset$ and every component of $T \backslash T_*$ is finite, or $T_* = \emptyset$ and T contains a ray R such that every component of $T \backslash R$ is finite. In the latter case T is locally finite and 1-ended in itself. Clearly any circuit-connected tree is a.c.c.

Given an arbitrary tree T it will be convenient to modify the notion of the core of T by setting

$$T_{(*)} = \begin{cases} T_*, & \text{if } T \text{ contains a double ray,} \\ R & \text{(an arbitrary, but fixed ray),} \\ & \text{if } T \text{ is 1-ended in itself} \\ \emptyset, & \text{if } T \text{ is rayless.} \end{cases}$$

In terms of this notation we have:
(i) If T is any tree, then any component of $T \backslash T_{(*)}$ is rayless;
(ii) T is a.c.c. if and only if it contains a ray and every component of $T \backslash T_{(*)}$ is finite.

König's theorem immediately gives:

Remark 4.2 *Any infinite, locally finite tree is a.c.c.*

Before dealing with the terminal completeness property of a.c.c. trees we characterise them by two simple intrinsic properties which are of relevance in the sequel.

Lemma 4.3 *A tree T containing a ray has a locally cofinite dendroid if and only if T is a.c.c.*

Proof: Suppose D is a dendroid of T such that $T \backslash D$ is locally finite. Since $T \backslash D \supset T \backslash T_*$ the latter is also locally finite. The converse follows immediately from Lemma (2.1). □

Lemma 4.4 *(i) Let T be a tree containing a ray. Then T is a.c.c. if and only if for any finite $F \subset V(T)$, $T - F$ has only finitely many finite components.*
(ii) If T is a.c.c. and $F \subset V(T)$ is finite, then any infinite component of $T - F$ contains a ray.

Here we make an exception to our basic convention about graphs by allowing *isolated vertices*. Thus, $T - F$ is the induced subgraph of T with $V(T - F) = V(T) \backslash F$. As such it may have isolated vertices, and these are counted among the finite components.

Proof: (i) Necessity. Suppose T is a.c.c. If $T_* = \emptyset$, then T is locally finite, hence $T - F$ has only finitely many components. We may therefore suppose that $T_* \neq \emptyset$. In this case there are only finitely many components of $T - F$ which are vertex-disjoint from T_*. Hence if $T - F$ has infinitely many finite components, infinitely many of them contain a vertex of T_*. Denote these components by T_i, $i \in I$, and in each choose a vertex $v_i \in V(T_*)$. Then there is a double ray $Z_i \subset T$ such that $v_i \in V(Z_i)$. T_i being finite there is a shortest segment P_i of Z_i whose endpoints x_i, y_i belong to F and which contains v_i. By finiteness of F there exist two distinct vertices $x, y \in F$ and an infinite $J \subset I$ such that $x = x_i$, $y = y_i$ for every $i \in J$. Thus $P_i = P_j$ for any $i, j \in J$, a contradiction.

Sufficiency. If T is not a.c.c., then there is a component K of $T \backslash T_{(*)}$ and a vertex $x \in V(K)$ such that $d(x; K)$ is infinite. Then $T - \{x\}$ has $d(x; K)$ finite components (indeed all but one component of $T - \{x\}$ is finite).

(ii) Assume first that T is circuit-connected. If $T - F$ has an infinite rayless component T_0, then T_0 has infinitely many pendant vertices. By circuit-connectedness of T, they each have a neighbour in F. Since F is finite there are two distinct pendant vertices of T_0 which have a common neighbour in F, an absurdity.

If T is arbitrary a.c.c., then given any infinite component K of $T - F$, we have that $K \cap T_{(*)}$ is an infinite component of $T_{(*)} - F$, and the previous case applies. $\quad\square$

Consider the following condition on a graph G:

(∗)   Given $U, F \subset V(G)$, where U is infinite and F is finite, there is a ray $R \subset G$ which starts at a vertex in U and avoids F.

It will be seen that for trees this amounts to being a.c.c.

Lemma 4.5 *Let G be a connected graph having a terminally finite subgraph X whose components satisfy (∗). Then given any ray R in G with $V(R) \subset V(X)$ there is a ray S in X such that $R \sim_G S$.*

Proof: We distinguish two cases according as $V(R)$ meets infinitely many or only finitely many components of X.

Case 1: W.l.o.g. we may suppose that there is an infinite sequence $u_0, u_1, \ldots \in V(R)$ and distinct components $X_0, X_1, \ldots$ of X such that $u_i \in V(X_i)$. By hypothesis, X_i contains a ray R_i and we may assume that R_i starts at u_i. Since the rays of X belong to only finitely many ends of G, we may further assume that all rays R_i are equivalent in G.

Take any finite $F \subset V(G)$. Choose n sufficiently large such that R_n avoids F. Since $R_0 \sim_G R_n$ there is an $R_0 R_n$-chord $P \subset G$ which avoids F; let v be the endpoint of P in R_n. Then the segment of R_n from u_n to v followed by P is an RR_0-path in G which avoids F. Thus $R \sim_G R_0 \subset X$.

Case 2: $V(R)$ meets only finitely many components of X. It suffices to consider the case where some component Y of X contains an infinite subset U of $V(R)$. X being terminally finite in G, so also is Y, hence there are rays $S_1, \ldots, S_r \subset Y$ such that every ray in Y is equivalent in G to one of the S_i's.

Put $W := V(S_1) \cup \ldots \cup V(S_r)$.

Construct an infinite sequence Q_n, $n \in \mathbf{N}$, of vertex-disjoint UW-chords in G as follows. For Q_0 take any UW-chord in G, and suppose $Q_0, \ldots, Q_{n-1}$ have already been constructed. Put $F := V(Q_0) \cup \ldots \cup V(Q_{n-1})$. By (∗) there is a ray S in Y which avoids F and starts at some $u_k \in U \backslash F$. Since $S \sim_G S_i$ for some i, there is an SS_i-chord $P_n \subset G$, avoiding F, with endpoints $v \in V(S)$, $w \in V(S_i)$. Then the segment of S from u_k to v together with P_n forms the desired UW-chord Q_n.

Since there are only finitely many S_i's infinitely many of the chords Q_n lead from U to some S_{i_0}, $1 \leq i_0 \leq r$. Thus $R \sim_G S_{i_0} \subset X$.   □

For reference in Section 5 we note that the preceding immediately implies:

Corollary 4.6 *Let X be a 1-ended subgraph of a connected graph G, and suppose that every component of X satisfies* (*)*. Then the subgraph of G induced by X is likewise 1-ended in G.*

Theorem 4.7 *For a tree T containing a ray the following are equivalent:*

(i) T is a.c.c.;

(ii) T satisfies (*)*;*

(iii) T is terminally complete in any graph G in which it is contained as a terminally finite spanning tree.

Proof: (i) $\Rightarrow$ (ii): We distinguish three cases.

Case 1: T is circuit-connected. Let $U, F \subset V(T)$ be infinite and finite, respectively. By Lemma (4.4) there is an infinite component T_0 of $T - F$ such that $U \cap V(T_0) \neq \emptyset$, and T_o contains a ray. Hence T_0 contains a ray starting in U.

Case 2: $T \neq T_* \neq \emptyset$. Given an infinite $U \subset V(T)$ let T_i, $i \in I$, be the components of $T \backslash T_*$ containing a vertex in U, and put $U_i := U \cap V(T_i)$, $i \in I$. By hypothesis, U_i is a non-empty finite set, hence I is infinite. Let w_i be the common vertex of T_i and T_*, $i \in I$.

Now consider $W := \{w_i : i \in I\} \cup V(T_*)$. Given any finite $F \subset V(T)$ let

$$F' := (F \cap V(T_*)) \cup \{w_i : F \cap V(T_i) \neq \emptyset\}.$$

This is a finite subset of $V(T_*)$, hence by Case 1, T_* contains a ray R avoiding F' and starting at some w_i for which T_i is vertex-disjoint from F.

Case 3: T is 1-ended in itself. Then T is locally finite, and any connected, infinite, locally finite graph trivially satisfies (*).

(ii) $\Rightarrow$ (iii): This is an immediate consequence of Lemma (4.5).

(iii) $\Rightarrow$ (i): Suppose T is not a.c.c. Then there is a component K of $T \backslash T_{(*)}$ and a vertex $w \in V(K)$ with $d(w; K)$ infinite. Let $(u_n)_{n \in \mathbf{N}}$ be a sequence of distinct neighbours of w in K.

Let $R_i = (x_{i0}, x_{i1}, \ldots)$, $i \in I$, be a family of rays in $T_{(*)}$, one from each end of T. Fix some $k \in I$ and consider the graph

$$G := T \cup \{[u_n, u_{n+1}] : n \in \mathbf{N}\} \cup A,$$

where

$$A = \{[x_{in}, x_{kn}] : i \in I \backslash \{k\}, \ n \geq r_i\}.$$

Here r_i is some number sufficiently large such that $x_{in} \neq x_{kn}$ for $n \geq r_i$ (such an r_i exists since R_i and R_k are in different ends). T is a 1-ended spanning tree of G and $R := (u_0, u_1, \ldots)$ is a ray in G.

Note that every edge in A has both endpoints in $T_{(*)}$. This means that any path in G which joins $u_n \in V(R)$ with a vertex belonging to some R_i must pass through w. Therefore R cannot be equivalent to any R_i (indeed R represents a free end of G which does not exist in T). Thus T is not terminally complete in G.   □

5 Existence of minimal end-separators

In this section we show that an a.c.c. tree T induces a minimal end-separator in any graph which is spanned by T. By the results of Section 3 we have to consider the 2-ended part $D^{(2)}$ of the dendroids associated with T. If $D^{(2)}$ is to be an end-separator, then it must meet every 2-ended double ray in G. Without making any assumptions on T we first consider what happens when $D^{(2)}$ fails to do this.

Lemma 5.1 *Let G be any graph, D a dendroid of G. If Z is a 2-ended double ray in G which misses $D^{(2)}$, then $Z \cap D$ is infinite.*

Proof: Suppose $Z \cap D$ is finite. Then we may assume that among all 2-ended double rays missing $D^{(2)}$, Z is so chosen as to minimize $|Z \cap D|$.

Take any $e \in Z \cap D$. Since Z misses $D^{(2)}$ we have that $Z_e := C(e, D)$ likewise misses $D^{(2)}$. If Z_e is finite, then $Z + Z_e$ contains a double ray Z', and

$$Z' \cap D \subset (Z \cap D) + (Z_e \cap D) = (Z \cap D) \backslash e.$$

Also, Z' is 2-ended since it differs from Z only by finitely many edges. This contradicts the minimality of $|Z \cap D|$.

It follows that Z_e is infinite, $Z_e \backslash e = R_0 \dot\cup R_1$, say, and since $e \notin D^{(2)}$, the two rays R_0, R_1 are equivalent in G. Also, $R_i \cap D = \emptyset$. Let $Z \backslash e = S_0 \dot\cup S_1$. Then $S_0 \not\sim_G S_1$.

We now distinguish a number of cases. In each we shall construct a 2-ended double ray contained in $Z \cup Z_e$ (and hence disjoint from $D^{(2)}$) having fewer edges in D than Z, thus giving rise to a contradiction.

Case 1: $V(Z_e) \cap V(S_i)$ is finite, $i = 0, 1$.

Counting from e let x_i be the last vertex of S_i belonging to Z_e. Clearly $x_0 \neq x_1$, and $x_i \in V(R_k)$ for a unique $k = k(i)$. Let R'_k, S'_i be the subrays of R_k, S_i, respectively, starting at x_i. Then $R'_k \cup S'_i$ is a double ray not containing e, hence $S'_i \cap D \neq \emptyset$, $i = 0, 1$. This means that $|S'_i \cap D| < |Z \cap D| - 1$, $i = 0, 1$.

Suppose that x_0 and x_1 are not on the same R_k, w.l.o.g. $x_0 \in V(R_0)$, $x_1 \in V(R_1)$. By W denote the segment of Z_e from x_0 to x_1. Then $e \in W$ and

$$Z'_0 := R'_0 \cup S'_0, \quad Z'_1 := R'_0 \cup W \cup S'_1$$

are two double rays containing fewer edges of D than Z. At least one of Z'_0, Z'_1 is 2-ended, for if both are 1-ended, then $R'_0 \sim_G S'_0$ and $R'_0 \sim_G S'_1$, and therefore $S_0 \sim_G S_1$, a contradiction.

Now suppose that x_0 and x_1 are on the same R_k, say on R_0. Let S'_i, $i = 0, 1$, and W be as before. Note that this time, $e \notin W \subset R_0$. Then $Z' := S'_0 \cup W \cup S'_1$ is a 2-ended ray such that

$$Z' \cap D = (S'_0 \cup S'_1) \cap D \subset (Z \cap D) \backslash e.$$

Case 2: $V(Z_e) \cap V(S_0)$ is infinite, $V(Z_e) \cap V(S_1)$ is finite.

Since all but finitely many edges of $Z_e \cup S_0$ belong to $G \backslash D \subset T$ it follows that $Z_e \cap S_0$ contains a ray. Let S''_0 be the unique maximal ray in $Z_e \cap S_0$, x_0 its origin, and suppose the notation is so chosen that $S''_0 \subset R_0$. As in Case 1, let x_1 be the last vertex of S_1 on Z_e, and S'_1 the subray of S_1 starting at x_1. Suppose x_1 belongs to R_k; denote by R'_k the subray of R_k starting at x_1. By the same argument as in Case 1, $S'_1 \cap D \neq \emptyset$.

If $x_1 \in V(R_1)$ consider the double ray $Z' := R_1' \cup S_1'$. Then

$$Z' \cap D = S_1' \cap D \subset (Z \cap D)\backslash e$$

and $R_1' \sim R_0 \sim S_0'' \nsim S_1'$, so that Z' is 2-ended.

If $x_1 \in V(R_0)$, then it must lie between x_0 and e on R_0 because x_0 is the origin of the ray $S_0'' \subset Z \cap Z_e$. Let W be the segment of R_0 between x_0 and x_1 and take $Z' := S_0'' \cup W \cup S_1'$. Clearly Z' is 2-ended and $Z' \cap D \subset (Z \cap D)\backslash e$.

Case 3: $V(Z_e) \cap V(S_i)$ is infinite, $i = 0, 1$. In this case there are subrays $S_i'' \subset Z_e \cap S_i$, $i = 0, 1$, but this is impossible since Z is 2-ended whereas Z_e is 1-ended. Hence this case cannot arise.   □

We can now state the main result of this paper.

Theorem 5.2 *Let T be an a.c.c. tree, D_T a locally cofinite dendroid of T. If G is any graph spanned by T, then $D := G\backslash T \cup D_T$ has the property that $D^{(2)}$ is a minimal end-separator of G.*

Proof: We have to show that $D^{(2)}$ is an end-separator. Suppose by way of contradiction that $D^{(2)}$ misses some 2-ended double ray Z. By Lemma (5.1), $Z \cap D$ is infinite. Denote its edges consecutively by $e_i = [u_i, v_i]$, $i \in I$, where $I = \mathbf{N}$ or $\mathbf{Z}$ according as $Z\backslash D$ does or does not contain a ray, and let the notation be so chosen that u_i precedes v_i in the natural order on Z. Put

$$X := Z\backslash D \cup \bigcup_{i \in I} C(e_i, D)\backslash e_i,$$

and let X_i be the component of X containing u_i, $i \in I$. If $v_i \neq u_{i+1}$ then X contains the path in Z from v_i to u_{i+1}, hence we have that the component of X containing v_i is X_{i+1}, and trivailly this also holds if $v_i = u_{i+1}$.

It is immediate from the definition of X that $V(X) \supset V(Z)$ and

$$X \subset T\backslash D.$$

Therefore every X_i is either rayless or 1-ended in G.

If $e_i \in D^{(0)}$, then $C(e_i, D)\backslash e_i$ is a path in X joining u_i and v_i, so that $X_{i+1} = X_i$. If $e_i \in D^{(1)}$, write $C(e_i, D)\backslash e_i = R_i \dot\cup S_i$ with the rays R_i, S_i originating at u_i, v_i, respectively. Then $R_i \subset X_i$, $S_i \subset X_{i+1}$, and this implies that $X_{i+1} \neq X_i$ (otherwise there would be a contradiction against the circuit-freeness of X_i). Moreover, $R_i \sim_G S_i$.

It follows from this that if $Z \cap D_\infty \neq \emptyset$, then X is disconnected, and every component X_i, as well as X itself, is 1-ended in G. On the other hand, if $Z \cap D_\infty = \emptyset$, then X is connected. In either case every component of X is infinite.

Since $X \subset T\backslash D = T\backslash D_T$ and D_T is locally cofinite in T, we have that X is locally finite. Therefore every component X_i satisfies $(*)$, hence by Corollary (4.6) the induced subgraph $G|X$ is 1-ended. But $Z \subset G|X$, so that Z is 1-ended, a contradiction.   □

Corollary 5.3 *If G admits a locally finite spanning tree T, then for any dendroid D based on T, $D^{(2)}$ is a minimal end-separator of G.*

Proof: Every subgraph of a locally finite graph is locally cofinite.   □

Thus, for graphs with locally finite spanning trees the property of $D^{(2)}$ being an end-separator is independent of the particular dendroid but resides in the tree on which it is based. For graphs with a.c.c. spanning trees this is no longer the case. The same spanning tree will in general admit dendroids for which $D^{(2)}$ may or may not be an end-separator. This situation is illustrated in Figures 3 and 4 showing, respectively, a locally cofinite dendroid and one which is not, for an a.c.c. spanning tree (which by Theorem (4.7) is terminally complete). The dendroid in Figure 3 gives rise to an end-separator, whereas that in Figure 4 does not. Edges are labelled 0,1,2 according as they belong to $D^{(0)}$, $D^{(1)}$, or $D^{(2)}$.

It should also be noted that the converse of Theorem (5.2) is not true. The simplest counterexamples are trees. It is easy to see that any a.c.c. tree T which is not locally finite has a dendroid which is not locally cofinite (in the proof of (2.1) take u to be a vertex of infinite degree and use the family $(\Sigma_x)_{x \in V \setminus u}$). In a tree, $D^{(2)} = D$ for any dendroid, and dendroids always are end-separators.

The most interesting consequence of Theorem (5.2) is:

Theorem 5.4 *Any graph admitting an a.c.c. spanning tree has a minimal end-separator.*

Proof: (4.3), (5.2). □

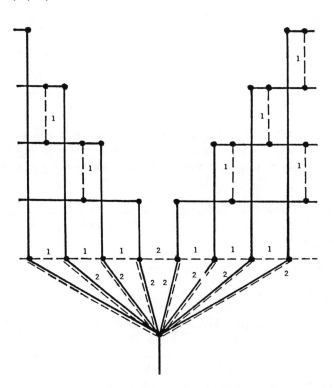

Figure 3 (solid lines: T; dashed lines: D)

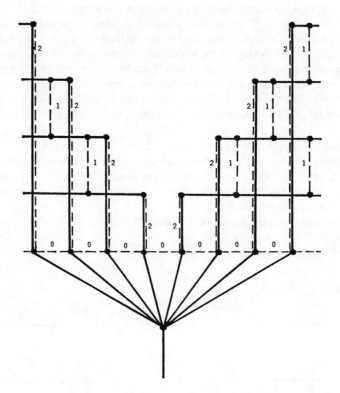

Figure 4 (solid lines: T; dashed lines: D)

One consequence of Theorem (5.4) is that if G has more than one end and admits an a.c.c. spanning tree, then G has at least one dendroid for which $D^{(2)}$ is non-empty. We now show that this has nothing to do with D, but is in reality a characteristic property of a.c.c. trees.

Theorem 5.5 *Let T be a tree containing a double ray. Then a n.a.s.c. for T to be a.c.c. is that whenever T spans a graph G having more than one end, then $D^{(2)} \neq \emptyset$ for any dendroid D of G based on T.*

To begin with, note that (without any assumptions on T):

Lemma 5.6 *For any dendroid of G, $D^{(2)} = \emptyset$ if and only if $D^{(2)} \cap D_T = \emptyset$.*

Proof: Take any $e \in D_\infty \backslash D_T$ and denote by $a_1, \ldots, a_n$ the edges of D_T on the cycle $C_e := C(e, G\backslash T)$ (in cyclic order so that e lies between a_n and a_1). By Lemma (2.4), $C(a_i, D) \cap C(a_{i+1}, D) =: R_i$ is a ray, $i = 1, \ldots, n-1$. Let R_0 be the ray in $C(a_1, D)$ which starts on C_e and is disjoint from R_1; similarly, let R_n be the ray in $C(a_n, D)$ starting on C_e and disjoint from R_{n-1}. Then R_0 and R_n are contained in $C(e, D)$. If $e \in D^{(2)} \backslash D_T$, then $R_0 \not\sim_G R_n$. This implies that there is an i, $1 \leq i \leq n$, such that $R_{i-1} \not\sim_G R_i$, i.e., $a_i \in D^{(2)} \cap D_T$.   □

Lemma 5.7 *Let D be a dendroid of G and suppose that D is based on a spanning tree T of G such that $D^{(2)} \cap D_T = \emptyset$. Then T is rayless or 1-ended in G.*

Proof: We may assume that $D_T \neq \emptyset$, i.e., that T has a double ray, otherwise there is nothing to prove. By hypothesis, $C(e, D)$ is 1-ended for every $e \in D_T$.

We show first that any two rays in $T \backslash D_T$ are equivalent in G. If R, S are two such rays, then, taking subrays if necessary, we may assume by Corollary (2.3) that $R \subset C(a, D)$, $S \subset C(b, D)$ for some $a, b \in D_T$. By Corollary (2.5) there exist $a_0, \ldots, a_n \in D_T$ such that $a_0 = a$, $a_n = b$, and $C(a_{i-1}, D) \cap C(a_i, D) =: R_i$ is a ray, $i = 1, \ldots, n$. Then $R \sim R_1 \sim \ldots \sim R_n \sim S$.

Now let R be any ray in T no subray of which is contained in $T \backslash D_T$. Then $R \cap D_T$ is infinite; starting from the origin of R denote the edges of $R \cap D_T$ consecutively by $e_0, e_1, \ldots$. By Lemma (2.4)(ii), $S_i := C(e_i, D) \cap C(e_{i+1}, D)$ is a ray in T, $i \in \mathbf{N}$. Note that S_i starts at some vertex $x_i \in V(R)$, is edge-disjoint from R, and S_i, S_j are vertex-disjoint for $i \neq j$. Since $S_i \subset T \backslash D_T$, we have that $S_0 \sim S_i$ for any i.

Let F be a finite subset of $V(G)$. Choose i sufficiently large so that $V(S_i) \cap F = \emptyset$. Since $S_0 \sim S_i$, there is an $S_0 S_i$-chord P in G which avoids F. Let y_i be the endpoint of P on S_i. Then P together with the segment of S_i between y_i and x_i is an $S_0 R$-path avoiding F. Therefore $S_0 \sim R$, and hence any two rays in T are equivalent in G. $\square$

Proof of (5.5): It T is a.c.c., then by Lemma (5.7) it must be 1-ended in G. Hence by Corollary (4.6), G is 1-ended in itself, contrary to hypothesis.

Conversely, if T is not a.c.c. consider the graph G constructed in the proof of Theorem (4.7) (iii) $\Rightarrow$ (i). Take any dendroid D_T of T and consider

$$D := G \backslash T \cup D_T = R \cup A \cup D_T.$$

Then it is easy to check that

$$D^{(0)} = R, \quad D^{(1)} = A \cup D_T,$$

whence $D^{(2)} = \emptyset$. $\square$

Acknowledgements

The author wishes to thank F. Laviolette and N. Polat for a number of useful comments. Support of the Natural Sciences and Engineering Research Council of Canada, Grant A-7315, is gratefully acknowledged.

References

[1] Bean, D. W. T., Infinite Exchange Systems, Ph.D. Thesis, McMaster University, 1967.

[2] Bean, D. W. T., Minimal edge-sets meeting all circuits of a graph, in: *Proc. Louisiana Conference on Combinatorics, Graph Theory and Computing* (R.C. Mullin, K.B. Reid, D.P. Roselle, eds.), LSU, Baton Rouge, 1970, 1–45.

[3] Halin, R., Über unendliche Wege in Graphen, *Math. Ann.* **157**(1964), 125–137.

[4] Tutte, W. T., Lectures on matroids, *J. Res. Nat. Bur. Standards* **69B**(1965), 1–47.

PARTITION THEOREMS FOR GRAPHS RESPECTING THE CHROMATIC NUMBER

N. SAUER, M. EL-ZAHAR

Department of Mathematics and Statistics
University of Calgary
Calgary, Alberta, T2N 1N4
Canada

Abstract

We exhibit general results which allow us to deduce the following: For every family $G_1, G_2, \ldots, G_m$ of K_r-free graphs there exists a K_r-free graph G with $\chi(G) = \chi(G_1) + \chi(G_2) + \ldots + \chi(G_m) - m + 1$ such that for every partition of the vertices of G, $V(G) = V_1 \dot\cup V_2 \dot\cup \ldots \dot\cup V_m$ there exists an i and an embedding $\alpha : G_i \to H$ such that $\alpha(G_i) \subseteq V_i$. And for each n-chromatic K_r-free graph G, there exists an n-chromatic K_r-free graph H such that for every partition of the vertices of H, $V(H) = V_1 \dot\cup \ldots \dot\cup V_{n+1}$, there exists an i with $1 \leq i \leq n+1$ and an embedding $\sigma : G \to H$ with $\sigma(G) \cap V_i = \emptyset$.

We will then apply those results to get similar ones for partially ordered sets.

For the language $L = \{=, \sim, \approx\}$ of three binary relation symbols with equality $=$, the theory T_n says that $\sim$ is symmetric and anti-reflexive and $\approx$ is an equivalence relation with n classes, each of which is an independent set under $\sim$. An n-colored (or n-partite) graph $G = (V, E, P)$ is a model of the theory T_n. The edge set E realizes $\sim$ and the partition P realizes $\approx$. If $G = (V, E, P)$ is an n-colored graph we will assume that the classes of P are numbered from 1 to n and denote them by $L_1(G), L_2(G), \ldots, L_n(G)$.

Let A_n^r be the set (up to isomorphism) of n-colored graphs $G = (V, E, P)$ such that (V, E) is K_r-free, i.e., contains no complete subgraph on r vertices. Clearly A_n^r is closed under taking substructures and satisfies the following amalgamation property:

(AP) for every $G_0, G_1, G_2 \in A_n^r$ and every embeddings $\sigma_i : G_0 \to G_i$ $(i = 1, 2)$, there exist $G \in A_n^r$ and embeddings $\tau_i : G_i \to G$ $(i = 1, 2)$ such that $\tau_1 \circ \sigma_1 = \tau_2 \circ \sigma_2$.

According to Fraïssé [1], there exists a countable homogeneous structure B_n^r whose age is A_n^r (the age of a relational structure R is the class of all finite substructures of R up to isomorphism). We call B_n^r the universal n-partite K_r-free graph. The universal n-partite graph B_n is B_n^r for $r > n$. It follows that B_n^r satisfies the following embedding property:

(EP) for every finite n-colored K_r-free graph G and every $x \in V(G)$, each embedding $\sigma : G - x \to B_n^r$ extends to an embedding $\sigma' : G \to B_n^r$.

Further properties of the graphs B_n^r are:

(i) If $r_1 \leq r$ and $n_1 \leq n$ then B_n^r embeds $B_{n_1}^{r_1}$.

G. Hahn et al. (eds.), Cycles and Rays, 237–242.

(ii) For a subset $X \subseteq V(B_n^r)$, let $I(X) = \{y \in V(B_n^r) : (x, y) \notin E(B_n^r) \text{ for each } x \in X\}$. Then $I(X) \cong B_n^r$ whenever X is finite.

A relational structure R is called indivisible if for every partition $R = C \dot\cup D$, one of the classes C and D contains an isomorphic copy of R. The age of R is called indivisible if for every partition $R = C \dot\cup D$ one of C and D has the same age as R.

It is clear that neither B_n^r nor its age is indivisible. However a weaker form of indivisibility is true.

Theorem 1 *Let $n = n_1 + n_2 - 1$. Then for every partition of the vertices of B_n^r into two classes $B \dot\cup R$ either $Age(B_{n_1}^r) \subseteq Age(B)$ of R embeds $B_{n_2}^r$.*

Proof: We use induction on n. The statement is obviously true for $n = 1$. Assume that there exists an n_1-colored graph G such that G cannot be embedded into B. Let $x_1, x_2, \ldots$ be an ordering of the vertices of $B_{n_2}^r$ and denote by H_m the n_2-colored subgraph of $B_{n_2}^r$ spanned by $x_1, \ldots, x_m$. Assume by the way of contradiction that there is an m and an embedding $\sigma : H_m \to B_n^r$ such that $\sigma(H_m) \subseteq R$ but σ has no extension $\sigma' : H_{m+1} \to B_n^r$ with $\sigma'(x_{m+1}) \in R$. Write $\sigma = (\alpha, \beta)$ where α is the graph embedding part of σ and $\beta : \{1, \ldots, n_2\} \to \{1, \ldots, n\}$.

With no loss of generality we can assume that β is the inclusion map and $x_{m+1} \in L_1(B_{n_2}^r)$. As we noted before, the set $I(\sigma(H_m))$ spans a copy of B_n^r. Consequently, there is an embedding $\tau : B_{n-1}^r \to I(\sigma(H_m)) - L_1(B_n^r)$. Denote by H the $(n_1 - 1)$-colored graph $G - L_1(G)$ and write $L_1(G) = \{y_1, \ldots, y_t\}$. By the induction hypothesis either H can be embedded into $\tau(B_{n-1}^r) \cap B$ or $\tau(B_{n-1}^r) \cap R$ embeds $B_{n_2}^r$. We can assume that there is an embedding $\theta : H \to \tau(B_{n-1}^r) \cap B$. Let F be the n-colored graph obtained from a copy of H_m and a copy of G by connecting each y_i $(i = 1, \ldots, t)$ to H_m in the same way x_{m+1} is connected and adding no further edges. The unary relations are defined on F such that each vertex $v \in H$ (respectively $x_i \in H_m$) has the same unary relation it gets by θ (respectively σ) and $y_1, \ldots, y_t$ have the unary relation U_1. Note that F does not contain at $(r + 1)$-clique and that each color-class is independent, i.e., $F \in Age(B_n^r)$. The map $\theta \cup \sigma : F - \{y_1, \ldots, y_t\} \to B_n^r$ has an extension $\epsilon : F \to B_n^r$. If for some $i \in \{1, \ldots, t\}$, $\epsilon(y_i) \in R$ then $\sigma(H_m) \cup \{\epsilon(y_i)\}$ is an isomorphic copy of H_{m+1} in R. Thus $\epsilon(y_i) \in B$ for each i. But then $\theta(H) \cup \{\epsilon(y_1), \ldots, \epsilon(y_t)\}$ is an isomorphic copy of G in B which is a contradiction. This completes the proof of the theorem. $\square$

An immediate consequence of Theorem 1 is the following theorem which follows by induction on m.

Theorem 2 *For every family of finite K_r-free graphs $G_1, \ldots, G_m$ there exists a finite K_r-free graph G with $\chi(G) = \chi(G_1) + \ldots + \chi(G_m) - m + 1$ such that for every partition of the vertices of G, $V(G) = V_1 \dot\cup \ldots \dot\cup V_m$ there exists an $i \in \{1, \ldots, m\}$ and an embedding $\alpha : G_i \to H$ such that $\alpha(G_i) \subseteq V_i$.*

Definition 1 A relational structure R is called (m, n)-indivisible if for every partition $R = R_1 \dot\cup \ldots \dot\cup R_m$ there exist integers $\{i_1, \ldots, i_n\} \subseteq \{1, \ldots, m\}$ and an embedding $\sigma : R \to R$ such that $\sigma(R) \subseteq R_{i_1} \cup \ldots \cup R_{i_n}$. The age of R is called (m, n)-indivisible if for every partition $R = R_1 \dot\cup \ldots \dot\cup R_m$, there exist integers $\{i_1, \ldots, i_n\} \subseteq \{1, \ldots, n\}$ such that $Age(R_{i_1} \cup \ldots \cup R_{i_n}) = Age R$.

It follows from the compactness theorem that the (m, n)-indivisibility of the age of R is equivalent to the following statement: For every finite substructure S of R there exists a finite substructure T such that for every partition $T = T_1 \dot\cup \ldots \dot\cup T_m$ there exist $\{i_1, \ldots, i_n\} \subseteq \{1, \ldots, n\}$ and an embedding $\sigma : S \to T$ with $\sigma(S) \subseteq T_{i_1} \cup \ldots \cup T_{i_n}$.

Let R be an (m, n)-indivisible structure. Clearly R is also (m_1, n_1)-indivisible for every $m_1 \leq m$ and $n_1 \geq n$. Also if $m_1 > m > n$ then R is (m_1, n)-indivisible. Thus the only essential case is the $(n + 1, n)$-indivisibility.

Theorem 3 B_n is $(n + 1, n)$-indivisible.

Proof: Let $V_1 \dot\cup \ldots \dot\cup V_{n+1}$ be a partition of the vertices of B_n. Put $W_i = V(B_n) - V_i$. Let $x_1, x_2, \ldots$ be an ordering of $V(B_n)$ and denote by H_m the n-colored graph spanned by $x_1, \ldots, x_m$. For each $i = 1, \ldots, n$, let m_i be such that there is an embedding $\sigma_i : H_{m_i} \to W_i$ which has no extension $\sigma_i' : H_{m_i+1} \to W_i$. This means that for every extension $\sigma_i' : H_{m_i+1} \to B_n$ of σ_i, we must have $\sigma_i'(x_{m_i+1}) \in V_i$. Assume that $x_{m_i+1} \in L_{k_i}(B_n)$ for $i = 1, \ldots, n$.

First we show that $i \neq j$ implies that $k_i \neq k_j$. Assume not. Let G be an n-colored graph consisting of disjoint copies of H_{m_i} and H_{m_j} and a vertex x such that $H_{m_i} \cup \{x\} \cong H_{m_i+1}$ and $H_{m_j} \cup \{x\} \cong H_{m_j+1}$. Here each vertex of G retains the same unary relation considered as a vertex of B_n. Then

$$\sigma_i \cup \sigma_j : G - x \to \sigma_i(H_{m_i}) \cup \sigma_j(H_{m_j})$$

is an isomorphism which can be extended to an embedding $\sigma : G \to B_n$. This is a contradiction, however, since $\sigma(x)$ cannot belong to both V_i and V_j at the same time.

Renaming if necessary, we assume that $k_i = i$ for each i. Now we construct an embedding $\tau : B_n \to W_{n+1}$ as follows. Assume that $\tau(x_1), \ldots, \tau(x_t)$ have been defined. Let $x_{t+1} \in L_i(B_n)$. Using the same argument as above, we can show the existence of a vertex $y \in B_n$ such $\{\tau(x_1), \ldots, \tau(x_t), y\} \cong H_{t+1}$ and $\sigma_i(H_{m_i}) \cup \{y\} \cong H_{m_i+1}$. This implies that $y \in V_i$. Clearly y can be chosen such that $y \notin \cup_{j=1}^n \sigma_j(H_{m_j})$. We put $\tau(x_{t+1}) = y$. This completes the proof of the theorem. $\square$

Theorem 4 The age of B_n^r is $(n + 1, n)$-indivisible.

Proof: We use induction on n. Assume that $n \geq 2$ and let $V(B_n^r) = V_1 \dot\cup \ldots \dot\cup V_{n+1}$ be a partition of the vertices of B_n^r. Let G be an n-colored K_r-free graph. By way of contradiction, assume for every embedding $\sigma : G \to B_n^r$, that $\sigma(G) \cap V_i \neq \emptyset$ for each $i = 1, \ldots, n$. First we construct embeddings $\sigma_i : G - L_i(G) \to B_n^r - L_i(B_n^r)$, $i = 1, \ldots, n$, such that for $i \neq j$, there is no edge between a vertex in $\sigma_i(G - L_i(G))$ and a vertex in $\sigma_j(G - L_j(g))$. By the induction hypothesis, the age of B_{n-1}^r is $(n, n-1)$-indivisible. Since $B_n^r - L_1(B_n^r)$ contains an isomorphic copy of B_{n-1}^r, then there exist $\{i_1, \ldots, i_{n-1}\} \subset \{1, \ldots, n+1\}$ and an embedding $\sigma_1 : G - L_1(G) \to B_n^r - L_1(B_n^r)$ such that $\sigma_1(G - L_1(G)) \subset (V_{i_1} \cup \ldots \cup V_{i_{n-1}}) - L_1(B_n^r)$. Choose $m_1 \in \{1, \ldots, n+1\} - \{i_1, \ldots, i_{n-1}\}$. By the assumption on G, and the independence of $L_1(G)$, there exists a vertex $v_1 \in L_1(G)$ such that, with only finitely many exceptions, every extension $\sigma_1' : G \to B_n^r$ of σ_1 satisfies $\sigma_1'(v_1) \in V_{m_1}$. Now assume that $\sigma_1, \ldots, \sigma_{i-1}$ have been defined. The subgraph $I(\cup_{j=1}^{i-1} \sigma_j(G - L_j(G)) - L_i(B_n^r)$ contains an isomorphic copy of B_{n-1}^r. Thus, by the induction hypothesis, there exists an embedding

$$\sigma_i : G - L_i(G) \to I(\cup_{j<i} \sigma_j(G - L_j(G))) - L_i(B_n^r)$$

such that $\sigma_i(G - L_i(G)) \subset V(B_n^r) - (V_{m_i'} \cup V_{m_i''})$ where $1 \leq m_i' < m_i'' \leq n+1$. Choose $m_i \in \{m_i', m_i''\}$ and a vertex $v_i \in L_i(G)$ such that, with only finitely many exceptions, every extension $\sigma_i' : G \to I(\cup_{j<1} \sigma_j(G - L_j(G)))$ of σ_i satisfies $\sigma_i'(v_i) \in V_{m_i}$. Let H be an n-colored graph consisting of subgraphs $H_1, \ldots, H_n$ together with a copy of G where $H_i \cong G - L_i(G)$ such that for each $x \in G$, if $x \in L_i(G)$, then $H_i \cup \{x\} \cong (G - L_i(G)) \cup \{v_i\}$. Observe again that $H \in \mathrm{Age}(B_n^r)$. Now $\sigma_1(G - L_1(G)) \cup \ldots \cup \sigma_n(G - L_n(G))$ is an embedding of $H_1 \cup \ldots \cup H_n$ into B_n^r. This embedding can be extended to an embedding $\sigma^* : H \to B_n^r$. Clearly $\sigma^*(G)$ is an isomorphic copy of G with $\sigma^*(G) \subset V_{m_1} \cup \ldots \cup V_{m_n}$. Since G was an arbitrary n-colored K_r-free graph, then the age of B_n^r is $(n+1, n)$-indivisible. $\qquad\square$

Corollary 1 *For each n-chromatic K_r-free graph G, there exists an n-chromatic K_r-free graph H such that for every partition of the vertices of H, $V(H) = V_1 \dot\cup \ldots \dot\cup V_{n+1}$, there exists an i with $1 \leq i \leq n+1$ and an embedding $\sigma : G \to H$ with $\sigma(G) \cap V_i = \emptyset$.*

Partitions of partially ordered sets

For the language $L = \{=, \leq, \prec\}$ of three binary relation symbols with equality $=$, the theory S_n says that $\leq$ is a partial order relation and $\prec$ is a quasi order, whose classes are antichains under $\leq$. S_n contains also the sentence $\forall x, y(x < y \to x \prec y)$ and a sentence saying that the longest chain under $\prec$ has length n. An n-colored poset P is a model of S_n. Clearly, the elements of P have n 'levels' under $\prec$. We denote these levels by $L_1(P), L_2(P), \ldots, L_n(P)$ such that if $i < j$ and $x \in L_i(P)$ and $y \in L_j(P)$ then $x \prec y$. It follows, that if $x < y$ and $x \in L_i(P)$ and $y \in L_i(P)$, then $i < j$. Clearly the length, $\ell(p)$, of P satisfies $\ell(P) \leq n$.

With these definitions, we can define the universal poset P_n of length n as the countable homogeneous n-colored poset whose age is the set of all n-colored posets. If A is a finite subset of elements of P_n and $1 \leq i \leq n$, then we define

$$S(A, i) = L_i(P_n) \cup S_1 \cup S_2$$

where

$$S_1 = \cup_{j>i} \cap_{k<i} \{x \in L_j(P_n) : y < x \text{ for each } y \in L_k(A)\},$$

$$S_2 = \cup_{j<i} \cap_{k>i} \{x \in L_j(P_n) : x < y \text{ for each } y \in L_k(A)\}.$$

Observe that $S(A, i)$ is isomorphic to P_n.

Theorem 5 *Let $n = n_1 + n_2 - 1$. Then for every partition of the elements of P_n into two classes $B \cup R$, either $\mathrm{Age}(P_{n_1}) \subseteq \mathrm{Age}(b)$ or R embeds P_{n_2}.*

Proof: Let Q be a finite n_1-colored poset such that there is no embedding $\sigma : Q \to P_n$ with $\sigma(Q) \subset B$. Put $w = \binom{n}{n_1}$ and let $S_1, \ldots, S_w$ denote the n_1-subsets of $\{1, \ldots, n\}$. Let $\beta_i : \{1, \ldots, n_1\} \to S_i$ be a strictly monotone map. For each $i = 1, \ldots, w$ we construct $Q_i \subset Q$, $q_i \in Q = Q_i$, $m_i \in \{1, \ldots, n\}$ and an embedding $\sigma_i : Q_i \to P_n$ satisfying the following:

(a) $\sigma_i = (\alpha_i, \beta_i)$ where α_i is an order embedding of Q_i into P_n with $\alpha_i(Q_i) \subset B$;

(b) $m_i = \beta_i(k_i)$ where $q_i \in L_{i_i}(Q)$;

(c) If $i < j$ then $\alpha_j(Q_j) \subset S(\alpha_i(Q_i), m_i)$;

(d) There is no extension $\sigma_i' = (\alpha', \beta) : Q_i \cup \{q_i\} \to P_n$ satisfying (a), (c) above.

Clearly $|\{m_i : 1 \leq i \leq w\}| \geq n_2$, so that there exists a strictly monotone map $\beta :$ $\{1, \ldots, n_2\} \to \{1, \ldots, n\}$ such that $\beta(\{1, \ldots, n_2\}) \subseteq \{m_i : 1 \leq i \leq w\}$.

We now define an embedding $\tau = (\alpha, \beta) : P_{n_2} \to P_n \cap R$ as follows. Let $x_1, x_2, \ldots$ be an ordering of the elements of P_{n_2}. Assume that $\tau(x_1), \ldots, \tau(x_{t-1})$ have been defined. Let $x_t \in L_k(P_{n_2})$. Choose i such that $\beta(i) = m_i$. Let y be an element of P_n satisfying

(a) $\{\tau(x_1), \ldots, \tau(x_{t-1}), y\} \cong \{x_1, \ldots, x_t\}$;

(b) $\sigma_i(Q_i) \cup \{y\} \cong Q_i \cup \{q_i\}$;

(c) $y \in \cap_{j \leq w} S(\sigma_j(Q_j), m_j) - \cup_{j \leq w} \sigma_j(Q_j)$.

$Q_i \cup \{q_i\}$ and $\{x_1, \ldots, x_2\}$ are in the age of P_n and any singleton $\{z\}$ is in the age of P. $f :$ $y \to q_i$ and $g : y \to x_t$ forms an amalgamation instance. So, there is an amalgam D with $f' :$ $Q_i \cup \{q_i\} \to D$ and $g' : \{x_1, \ldots, x_2\} \to D$ such that $f'f(z) = g'g(z)$ and $f'(\{x_1, \ldots, x_{t-1}\}) \cap$ $g'(q_i) = \emptyset$. Now, $d - f'f(z)$ has an embedding δ into $\cap_{j \leq w} S(\sigma_j(Q_j), m_m) - \cup_{j \leq w} \sigma_j(Q_j)$ such that $\delta/Q_i = \sigma_i/Q_i$ and $\delta/\{x_1, \ldots, x_{t-1}\} = \tau/\{x_1, \ldots, x_{t-1}\}$. Then, because P_n is homogeneous, δ has an extension δ' to all of D. Put $y = \delta' \circ f' \circ f(z)$.

Obviously any such y must belong to R. We put $\tau(x_t) = y$. This completes the definition of τ and the proof of the theorem. □

As a consequence of Theorem 5 we have the following theorem of Nešetřil and Rödl [2].

Corollary 2 (Nešetřil and Rödl [2].) *For every finite posets $P_1, \ldots, P_m$ there exists a finite poset P with $\ell(P) = \ell(P_1) + \ldots + \ell(P_m) - m + 1$ such that for every partition of the elements of P into m classes $V_1 \dot\cup \ldots \dot\cup V_m$, there exists $1 \leq i \leq m$ and an embedding $\sigma : P_i \to P$ with $\sigma(P_i) \subset V_i$.*

Our final result is the analogue of Theorem 4 for posets.

Theorem 6 *The age of P_n is $(n+1, n)$-indivisible.*

Proof: Let $V_1 \dot\cup \ldots \dot\cup V_{n+1}$ be a partition of the elements of P_n into $n+1$ classes. Assume there is an n-colored poset Q such that for every embedding $\sigma : Q \to P_n$ we have $\sigma(Q) \cap V_i \neq \emptyset$ for each i. This implies that for each $1 \leq i \leq n+1$ there exist $Q_i \subset Q$, $q_i \in Q - Q_i$, $m_i \in \{1, \ldots, n\}$ and an embedding $\sigma_i : Q_i \to P_n$ satisfying:

(a) $\sigma(Q_i) \cap V_i = \emptyset$;

(b) $q_i \in L_{m_i}(Q)$;

(c) If $1 \leq i < j \leq n+1$ then $\sigma_j(Q_j) \subset S(\sigma_i(Q_i), m_i)$;

(d) There is no extension $\sigma_i' : Q_i \cup \{q_i\} \to P_n$ of σ_i satisfying (a), (c) above.

We finish the proof by showing that if $i \neq j$ then $m_i \neq m_j$. So assume that $m_i = m_j$ where $i < j$. Consider an element $y \in P_n$ such that $\sigma(Q_i) \cup \{y\} \cong Q_i \cup \{q_i\}$ and $\sigma(Q_j) \cup \{y\} \cong Q_j \cup \{q_j\}$. From conditions (a) and (b) above, we must have $y \in V_i$ and $y \in V_j$. This contradiction completes the proof of the theorem. □

References

[1] R. Fraïssé, *Theory of Relations*, Studies in Logic and The Foundations of Mathematics, North-Holland, 118, 1986.

[2] J. Nešetřil and V. Rödl, Combinatorial partitions of finite posets and lattices — Ramsey lattices, *Algebra Universalis* **19**(1984), 106–119.

VERTEX-TRANSITIVE GRAPHS THAT ARE NOT CAYLEY GRAPHS

Mark E. WATKINS
Department of Mathematics
Syracuse University
Syracuse, New York 13244-1150
U.S.A.

Abstract

It is shown that if a graph is vertex- and edge-transitive of odd valence, has at least 5 vertices, and is not bipartite, then its derived (line) graph is not a Cayley graph. One is thus able to construct a wide variety of vertex-transitive graphs, both finite and infinite, that are not Cayley graphs. The infinite examples may have either one or uncountably many ends, for we show that every 2-ended vertex- and edge-transitive graph has even valence.

1 Introduction

More than once I have wanted to prove some theorem or other about vertex-transitive graphs, failed in the early (and sometimes later) attempts, but proved it for the more manageable special case of Cayley graphs. I then looked about for vertex-transitive "non-Cayley" graphs on which to try my promising conjecture. I would first seize upon the ubiquitous Petersen graph, but since this graph is virtually the "universal counterexample", little could be inferred from that test. The derived (or line) graphs (and their complements) of the complete graphs K_{2n} ($n \geq 3$) would be tried next. However, other than these "folklore graphs", examples of finite vertex-transitive graphs that are not Cayley graphs seem to be known in a hit-or-miss fashion, while the obvious infinite examples have been products of a finite example by an infinite vertex-transitive graph.

Let us henceforth use the acronym VTNCG for a vertex-transitive graph that is not a Cayley graph. The purpose of this note is to prove some results that enable us to expand the above "folklore family" of finite VTNCG's and to construct some more interesting infinite VTNCG's.

In the next section the terminology and notation for this article are presented, as well as several now classical results from the literature required for the sequel. In Section 3 we present our main constructive tool; we prove that the derived graphs of nearly all vertex- and edge-transitive, odd-valent, non-bipartite graphs are VTNCG's. In Section 4 we construct finite VTNCG's, some with the aid of this theorem and some without. In the final section we construct infinite VTNCG's. For graphs with one or uncountably many ends, our theorem is a fertile source of VTNCG's, but we show that it cannot be applied to graphs with exactly two ends by showing that if such a graph is both vertex- and edge-transitive, then it has even valence.

243

G. Hahn et al. (eds.), Cycles and Rays, 243–256.
© *1990 by Kluwer Academic Publishers. Printed in the Netherlands.*

2   Preliminaries

Throughout this article, capital Greek letters will be used exclusively to denote a finite or infinite graph without loops or multiple edges. If Γ is such an object, then $V(\Gamma)$ will denote its vertex set, $E(\Gamma)$ will denote its edge set and $c(\Gamma)$ will denote its complement. If $x \in V(\Gamma)$, then $\rho(x)$ denotes the valence of x. Only *locally finite* infinite graphs will be considered; that is, $\rho(x)$ is finite for all vertices x. If $\rho(x) = n$ for all $x \in V(\Gamma)$, we say that Γ is *n-valent* and write $\rho(\Gamma) = n$.

The connectivity of Γ is denoted by $\kappa(\Gamma)$. When Γ is infinite, we define

$$\kappa_\infty(\Gamma) = \inf \ \{|S| : S \subseteq V(\Gamma); \ \Gamma - S \text{ has at least two infinite components}\}.$$

Clearly $\kappa_\infty(\Gamma) \geq \kappa(\Gamma)$ for all Γ.

Following Ore [15], a *lobe* of Γ denotes a maximal biconnected subgraph of Γ as well as the copy of K_2 induced by an isthmus (separating edge) of Γ. (The overworked term "block", sometimes used in place of "lobe", ought to be reserved for its conventional meaning in the context of imprimitive permutation groups.)

Given Γ, we let $\partial\Gamma$ denote its derived graph (or "interchange graph" in [15]). Thus $V(\partial\Gamma) = E(\Gamma)$ and $\{e_1, e_2\} \in E(\partial\Gamma)$ if and only if e_1 and e_2 contain a common vertex of Γ. (We eschew the term "line graph" for this object since it is inconsistent with all other generally used terminology.)

The symbols Z, Z^+, and Z_n denote, respectively, the set (or additive group) of integers, the set of positive integers, and the cyclic group of order n where $n \in Z^+$.

If G and G' are abstract groups, then $G \simeq G'$ will mean that G and G' are isomorphic in the usual sense. Suppose now that G and G' are groups of permutations on sets V and V', respectively. We say that G and G' are *isomorphic as permutation groups* and write $G \cong G'$ if there exist an isomorphism $F : G \to G'$ and a bijection $f : V \to V'$ such that for all $\alpha \in G$ and all $x \in V$,

$$f(\alpha(x)) = (F(\alpha))(f(x)),$$

i.e., the following diagram commutes:

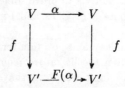

A permutation group G on V is said to be *regular* if it is transitive and all point-stabilizers are trivial; equivalently, for all $x, y \in V$, there exists a unique $\alpha \in G$ such that $\alpha(x) = y$. For any graph Γ, we define its *vertex group* $G_0(\Gamma)$ to be the set of all permutations α of $V(\Gamma)$ such that for all $x, y \in V(\Gamma)$, $\{\alpha(x), \alpha(y)\} \in E(\Gamma)$ if and only if $\{x, y\} \in E(\Gamma)$. The *edge group* $G_1(\Gamma)$ consists of all permutations β of $E(\Gamma)$ such that for all $e, e' \in E(\Gamma)$, e and e' contain a common vertex of Γ if and only if $\beta(e)$ and $\beta(e')$ do.

Proposition 2.1 (Sabidussi [16]) *For a finite graph* Γ, $G_0(\Gamma) \simeq G_1(\Gamma)$ *if and only if* Γ *has at most one isolated vertex and no component isomorphic to* K_2.

We emphasize that in general it is not the case that $G_0(\Gamma) \cong G_1(\Gamma)$. Circuits are the only finite vertex-transitive graphs with this property (cf. [7]). Infinite vertex-transitive graphs with this property include the n-valent trees ($n \geq 2$) and a handful of others. (It is not out of order to speak of infinite handfuls.)

Whitney [24] determined the three small connected graphs Γ for which $G_1(\Gamma)$ does not coincide with $G_0(\partial \Gamma)$. His only vertex-transitive example is K_4 (cf. also [7]). By Proposition 2.1, $G_1(K_4) \simeq G_0(K_4)$, which is the symmetric group on a 4-set; it contains no element of order 6. However, ∂K_4 is the Platonic octahedron, which is also a Cayley graph (see below) of Z_6.

Let G be a group, and let $H \subseteq G$ such that H generates G and $1 \notin H = H^{-1}$. Then the *Cayley graph of* G *with respect to* H, denoted $\Gamma_{G,H}$, has $V(\Gamma_{G,H}) = G$ and

$$E(\Gamma_{G,H}) = \{\{g, gh\} : g \in G \ ; h \in H\}.$$

For each $x \in G$, we define $\lambda_x : G \to G$ by $\lambda_x(g) = xg (g \in G)$. One easily verifies that $\lambda_x \in G_0(\Gamma_{G,H})$ and that the set $\{\lambda_x : x \in G\}$ forms a subgroup of $G_0(\Gamma_{G,H})$ which is a regular permutation group on $V(\Gamma_{G,H})$. In fact, we have:

Proposition 2.2 (Sabidussi [17]) *A necessary and sufficient condition for a graph* Γ *to be a Cayley graph* $\Gamma_{G,H}$ *for some* H *is that* $G_0(\Gamma)$ *contain a subgroup* L *such that* $L \simeq G$ *which is a regular permutation group on* $V(\Gamma)$.

We see that Cayley graphs $\Gamma_{G,H}$ must be vertex-transitive and are $|H|$-valent. They may be edge-transitive or not. *In this article we investigate vertex-transitive graphs that are not Cayley graphs.* Such an object will be called a *VTNCG*.

The notion of an "end" of a graph as formulated by Halin [5] will be useful to us in classifying infinite graphs. The main ideas will be included here for completeness. Let Π_1 and Π_2 be 1-way-infinite paths in Γ, and let us write $\Pi_1 \sim \Pi_2$ if there exists a 1-way-infinite path Π_0 that has infinitely many vertices in common with each of Π_1 and Π_2. Halin easily demonstrates that $\sim$ is an equivalence relation and lets an *end* denote an equivalence class of 1-way-infinite paths with respect to the relation $\sim$. Let $\epsilon(\Gamma)$ denote the cardinality of the set of ends of Γ. If Γ is an infinite, locally finite graph, one could equivalently define $\epsilon(\Gamma)$ as the supremum of the number of infinite components of $\Gamma - S$ as S ranges over all finite subsets of $V(\Gamma)$. Since the relation $\sim$ is respected by graph automorphisms, the automorphisms of Γ permute the ends of Γ.

Proposition 2.3 (Halin [6, Corollary 15], Jung [12, Theorem 1]) *If* Γ *is an infinite, locally finite, vertex-transitive graph, then* $\epsilon(\Gamma)$ *equals 1 or 2 or is uncountable.*

For any notions related to permutation groups not already included, the reader is referred to [25].

Constructions of various examples may require any of three standard graph products of graphs Γ_1 and Γ_2: the *cartesian product* $\Gamma_1 \times \Gamma_2$, the *weak product* $\Gamma_1 * \Gamma_2$, and the *lexicographic* product $\Gamma_1[\Gamma_2]$. (Cf. [18, 19].) The vertex set of each of these products is the cartesian product $V(\Gamma_1) \times V(\Gamma_2)$ of the vertex sets. Let $x_i, y_i \in V(\Gamma_i)$ for $i = 1, 2$. Then:

1. $\{(x_1, x_2), (y_1, y_2)\} \in E(\Gamma_1 \times \Gamma_2)$ if and only if $[x_1 = y_1 \wedge \{x_2, y_2\} \in E(\Gamma_2)] \vee [\{x_1, y_1\} \in E(\Gamma_1) \wedge x_2 = y_2]$.

2. $\{(x_1, x_2), (y_1, y_2)\} \in E(\Gamma_1 * \Gamma_2)$ if and only if $\{x_i, y_i\} \in E(\Gamma_i)$ for $i = 1, 2$.

3. $\{(x_1, x_2), (y_1, y_2)\} \in E(\Gamma_1[\Gamma_2])$ if and only if $[x_1 = y_1 \wedge \{x_2, y_2\} \in E(\Gamma_2)] \vee \{x_1, y_1\} \in E(\Gamma_1)$.

All three products are associative, and the first two are also commutative. It is well-known [18, 19] that a graph produced by any of these products is vertex-transitive if and only if each factor is vertex-transitive. For the categorical product the same holds for edge-transitivity.

3   Main Results

Theorem 3.1 *Let Γ be a vertex-transitive graph that is not bipartite and not K_4. If its derived graph $\partial \Gamma$ is a Cayley graph, then $\rho(\Gamma)$ is even.*

Proof: Without loss of generality, we may assume that Γ is connected. By Proposition 2.2, $G_0(\partial \Gamma)$ contains a regular subgroup M_1. By the remarks following Proposition 2.1, since Γ is not K_4, M_1 is a regular subgroup of $G_1(\Gamma)$.

Consider the function $\Phi : G_0(\Gamma) \to G_1(\Gamma)$ whereby if $\alpha \in G_0(\Gamma)$ and $\{x, y\} \in E(\Gamma)$, then $(\Phi(\alpha))(\{x, y\}) = \{\alpha(x), \alpha(y)\}$. It is immediate from the definition of the vertex group that Φ is well-defined. In fact, Φ is precisely the isomorphism used to prove Proposition 2.1 in [16]. Let us now define $M_0 = \Phi^{-1}[M_1]$. Although M_0 is very unlikely a regular permutation group on $V(\Gamma)$, it nonetheless induces a regular permutation group on $E(\Gamma)$. Thus given $\{x_1, y_1\}, \{x_2, y_2\} \in E(\Gamma)$, there exists a unique $\mu \in M_0$ such that $\{\mu(x_1), \mu(y_1)\} = \{x_2, y_2\}$. To simplify notation, if $\mu \in M_0$ and $e = \{x, y\} \in E(\Gamma)$, then let us understand $\mu(e)$ to denote $\{\mu(x), \mu(y)\}$.

Arbitrarily choose $\{x, y\} \in E(\Gamma)$, and let $X = \{\mu(x) : \mu \in M_0\}$ and $Y = \{\mu(y) : \mu \in M_0\}$. Since M_0 acts transitively on $E(\Gamma)$, we have $V(\Gamma) = X \cup Y$. If $X \cap Y = \emptyset$, then Γ is bipartite. Hence we may pick $z \in X \cap Y$. Define

$$E^+ = \{\{z, u\} \in E(\Gamma) : (\exists \mu \in M_0)[\mu(x) = z \wedge \mu(y) = u\}$$

and

$$E^- = \{\{z, u\} \in E(\Gamma) : (\exists \mu \in M_0)[\mu(x) = u \wedge \mu(y) = z\}.$$

Since M_0 induces regular action on $E(\Gamma)$, we have $E^+ \cap E^- = \emptyset$ while $E^+ \cup E^-$ consists of all edges of Γ incident with z.

If $\mu \in M_0$ and $\mu(z) = z$, then $\mu[E^+] = E^+$ and $\mu[E^-] = E^-$. Conversely, if $e_1, e_2 \in E^+$ (resp., E^-), then there exists a unique $\mu \in M_0$ such that $\mu(e_1) = e_2$; moreover, $\mu(z) = z$.

Hence $|E^+| = |\{\mu \in M_0 : \mu(z) = z\}| = |E^-|$. Finally, $\rho(\Gamma) = \rho(z) = |E^+| + |E^-| = 2|E^+|$ as required. $\qquad\square$

Whenever Γ is edge-transitive, $\partial\Gamma$ is vertex-transitive. Hence

Corollary 3.2 *Let Γ be vertex- and edge-transitive of odd valence. Then Γ is K_4 or bipartite or $\partial\Gamma$ is a VTNCG.*

If Γ is ρ-valent, then clearly $\rho(\partial\Gamma) = 2(\rho - 1)$. Hence the valences of all VTNCG's obtained via Corollary 3.2 are divisible by 4.

4   Some Finite VTNCG's

We have already cited the derived graphs (and their complements) of certain complete graphs as VTNCG's. If $\Gamma = K_{2n}$ $(n \geq 3)$ then $|V(\partial\Gamma)| = 2n^2 - n$, yielding two VTNCG's of valences $\rho(\partial\Gamma) = 4(n - 1)$ and $\rho(c(\partial\Gamma)) = (2n - 3)(n - 1)$, respectively. When $n = 3$, a 6-valent VTNCG and an 8-valent VTNCG, both of order 15 are obtained.

Suppose now that $\Gamma = K_{4n+1}$ $(n \geq 1)$. Then $|V(\partial\Gamma)| = 2n(4n + 1)$, and $\rho(c(\partial\Gamma)) = (2n - 1)(4n - 1)$, which is odd. The graph $c(\partial\Gamma)$ may also be regarded as having as vertices the 2-subsets of a $(4n + 1)$-set, with two vertices adjacent if and only if they denote a pair of disjoint 2-subsets. Viewed in this way, $c(\partial K_{4n+1})$ is clearly edge-transitive since the symmetric group acting on a $(4n + 1)$-set is 4-transitive. Clearly $c(\partial K_{4n+1})$ is not bipartite. Hence, we may apply Corollary 3.2 to the graphs $c(\partial K_{4n+1})$ to yield the VTNCG's $\partial c(\partial K_{4n+1})$; they have $n(2n - 1)(16n^2 - 1)$ vertices and valence $4n(4n - 3)$. When $n = 1$, we obtain the derived graph of the Petersen graph, a third VTNCG of order 15, this one being 4-valent. The next VTNCG in this family is a 40-valent graph on 378 vertices.

The Petersen graph itself is the complement of ∂K_5; it has 10 vertices. No VTNCG has fewer than 10 vertices.

The number of vertices in a VTNCG must be composite, since every vertex-transitive graph on a prime number p of vertices is a Cayley graph of Z_p.

Additional examples of finite VTNCG's include the derived graphs (and their complements) of the familiar graphs listed in Table 1.

Γ	$|V(\Gamma)|$	$|V(\partial\Gamma)|$	$\rho(\partial\Gamma)$
1-Skeleton of the Platonic dodecahedron	20	30	4
1-Skeleton of the Platonic icosahedron	12	30	8
Coxeter graph, cf. ([1, p.241])	28	42	4
Moore graph (cf. [8]) of valence 7	50	175	12

TABLE 1.

If Ω_m denotes the edgeless graph with m vertices and if Λ is any finite vertex- and edge-transitive graph of odd valence, then the lexicographic product $\Lambda[\Omega_{2m+1}]$ is also vertex- and edge-transitive and has odd valence $(2m + 1)\rho(\Lambda)$. If Λ is not bipartite, one considers the derived graphs (or their complements) of these products. For example, $\partial(K_6[\Omega_3])$ is a 28-valent VTNCG with 135 vertices.

If Γ is a vertex- and edge-transitive, non-bipartite graph of odd valence, then the cartesian product of Γ with itself an odd number of times produces another graph with these same properties and so also satisfies the hypothesis of Corollary 3.2.

Certainly finite VTNCG's exist which are not produced by Corollary 3.2. We first present such an example Γ which, interestingly in the light of the alternatives of Corollary 3.2, is also bipartite. It has 20 vertices and is 4-valent but is distinct from the other VTNCG in Table 1 with these same parameters, which is not bipartite.

Let Π denote the Petersen graph. Let $V(\Gamma) = V(\Pi) \times \{0, 1\}$ and let $E(\Gamma) = \{\{(x, 0), (y, 1)\} : \{x, y\} \in E(\Pi)\}$. Thus Γ is the categorical product $\Pi * K_2$. We observe that Γ has girth 6. To each 5-circuit Δ in Π there corresponds a 10-circuit in Γ on the vertex set $V(\Delta) \times \{0, 1\}$; all 10-circuits in Γ are of this form.

Now suppose that Γ is a Cayley graph $\Gamma_{G,H}$. Clearly G is non-abelian, for otherwise for some $a, b \in H$, the identity $aba^{-1}b^{-1} = 1$ would define a 4-circuit in Γ. There are three non-abelian groups of order 20 (cf. [2, p.134]): the dihedral group D_{10}, the K-metacyclic group $\langle x, y : x^{10} = x^5 y^2 = xyxy^{-1} = 1 \rangle$, and the ZS-metacyclic group $\langle x, y : x^5 = y^4 = 1 ; y^{-1}xy = x^2 \rangle$.

The set H must be of either the form $\{a, b, c\}$ where $a^2 = b^2 = c^2 = 1$ or the form $\{a, a^{-1}, b\}$ where $b^2 = 1$ and $a^2 \neq 1$. We can now eliminate the latter two groups, because in both of these cases, no element of order 2 belongs to any 3-element generating set.

There remains only the case where G is $D_{10} = \langle a, b : a^{10} = b^2 = (ab)^2 = 1 \rangle$ (to fix the notation), and we now assume that $V(\Gamma)$ has been properly identified with D_{10}. If H contains only one element of order 2, then without loss of generality we may take $H = \{a, a^{-1}, b\}$ as in the above presentation of D_{10}. But then the relation $(ab)^2 = 1$ determines a 4-circuit in Γ, giving the same contradiction as in the abelian case. Hence for some $0 \leq i < j < k \leq 9$, H is of the form $\{ba^i, ba^j, ba^k\}$ with the stipulation that not all of i, j, and k are even, for then H would fail to generate D_{10}. Let us write $r = ba^i$, $s = ba^j$, $t = ba^k$. Some two of i, j, k are of the same parity — let us say w.l.o.g. — j and k. Then $st = a^{k-j}$ has order 5 and the vertices 1, s, st, sts, $(st)^2, \ldots, (st)^4 s$ determine a 10-circuit Δ in Γ. The automorphism λ_r (see §2) of Γ interchanges Δ with a disjoint 10-circuit Δ' while fixing each of the 10 edges of the form $\{g, rg\}$. Corresponding to Δ and Δ' are a pair of disjoint 5-circuits of Π joined to each other by five edges, each such edge $\{x, y\}$ corresponding to a pair $\{\{(x, 0), (y, 1)\}, \{(x, 1), (y, 0)\}\}$ of edges joining Δ and Δ' in Γ. However, no automorphism interchanging a pair of disjoint circuits in Π fixes each of the five edges joining them. Hence λ_r is not an automorphism of Γ, giving a contradiction.

The above graph has representation also as the generalized Petersen graph $(10, 3)$. Its automorphism group is presented in [3].

An interesting family of VTNCG's was brought to my attention several years ago by Richard Weiss. For a prime p, $\mathrm{PSL}(2, p)$ denotes the group of 2×2 matrices over Z_p modulo its center. This group has order $p(p^2 - 1)/2$ (cf. [9, p.191]). For each $p \equiv \pm 1$ (mod 16), Weiss constructed a vertex-transitive graph Γ_p having $p(p^2 - 1)/48$ vertices and $G_0(\Gamma_p) \simeq \mathrm{PSL}(2, p)$. (In addition, $G_0(\Gamma_p)$ acts primitively on $V(\Gamma_p)$ and possesses certain other highly symmetric properties (see [23, Satz 6.2].) Since $\rho(\Gamma_p) = 3$, these graphs are not constructible via Corollary 3.2.

To see that Γ_p is not a Cayley graph, we first note that the order of any vertex stabilizer in $G_0(\Gamma_p)$ is $|G_0(\Gamma_p)|/|V(\Gamma_p)| = 24$. If Γ_p were a Cayley graph, then the regular subgroup L of Proposition 2.2 (above) would have index $[G_0(\Gamma_p) : L] = 24$. One verifies, however, by Dickson's Theorem [9, p.213] that $\mathrm{PSL}(2,p)$ contains no subgroup of index 24 for any prime $p \equiv \pm 1 \pmod{16}$.

The smallest graph in this family is Γ_{17}. It has 102 vertices. Weiss notes that it is the only graph in this family with the property that $G_1(\Gamma_p)$ acts primitively on $E(\Gamma_p)$.

Another infinite family of finite VTNCG's is due to Godsil [4].

5 Some Infinite VTNCG's

Following Proposition 2.3, we divide our analysis of infinite, locally finite VTNCG's into three subsections according to the number of ends that they possess.

A. $\epsilon(\Gamma)$ *is uncountable.* A point of departure is the class of graphs in which every vertex is a separating vertex. In [13], H.A. Jung and I characterized the infinite, locally finite graphs Γ with $\kappa_\infty(\Gamma) = \kappa(\Gamma) = 1$ subject to the condition that $G_0(\Gamma)$ act transitively, regularly, and primitively, respectively. In [22], I characterized infinite, locally finite Cayley graphs of connectivity 1. Each of these four characterizations involves conditions (ranging from easily-stated to highly technical) concerning the distribution of lobes incident with each vertex, namely: how many lobes are there of each isomorphism class? What global (should I say "lobal"?) properties does each lobe possess? How many orbits does each lobe have? etc. These characterizations are, of course, in accord with following flow of implications:

$$\text{regular} \;\rightarrow\; \text{Cayley}$$
$$\searrow$$
$$\text{vertex-transitive}$$
$$\nearrow$$
$$\text{primitive}$$

and none of these implications is reversible. The characterization in the case of Cayley graphs is particularly complex and will not be repeated here. The point to be made is simply that VTNCG's with $\kappa_\infty = 1$ and uncountably many ends do indeed exist.

Here is a recipe for constructing VTNCG's with ϵ uncountable and $1 < \kappa < \infty$. Let Λ be any biconnected, non-bipartite, vertex- and edge-transitive graph of odd valence other than K_4, and let m be a positive integer. Let Γ_0 be the empty graph and let $\Gamma_1 = \Lambda$. For each $k \geq 1$, form Γ_{k+1} by identifying each vertex in $V(\Gamma_k)\backslash V(\Gamma_{k-1})$ with one vertex in each of $2m$ disjoint isomorphic copies of Λ. The resulting graph

$$\Gamma = \bigcup_{k=0}^{\infty} \Gamma_k$$

satisfies $\kappa(\Gamma) = 1$, has uncountable many ends, and is vertex-and edge-transitive. Moreover, $\rho(\Gamma) = (2m+1)\rho(\Lambda)$, which is odd. By Corollary 3.2, $\partial\Gamma$ is a VTNCG. Clearly $\kappa_\infty(\partial\Gamma) = \rho(\Lambda) \geq \kappa(\Lambda) > 1$.

B. $\epsilon(\Gamma) = 2$. Our first order of business in this subsection is to show that Corollary 3.2 is inapplicable here; its hypothesis is never satisfied by graphs with exactly two ends. If Σ is a two-ended, vertex-transitive graph, a $\kappa_\infty(\Sigma)$-subset T of $V(\Sigma)$ such that $\Sigma - T$ has two infinite components will be called a κ_∞-*separator*, and the infinite components of $\Sigma - T$ are called *fragments*. Certainly if $|T| < \kappa_\infty(\Sigma)$, then $\Sigma - T$ has only one fragment. Our first lemma was obtained jointly with Imrich.

Lemma 5.1 *Let Σ be a vertex-transitive graph with $\epsilon(\Sigma) = 2$, and let $\tau \in G_0(\Sigma)$ have infinite order. Then there exists a $\kappa_\infty(\Sigma)$-set of pairwise-disjoint, τ-invariant, 2-way-infinite paths that span Σ.*

Proof: By [6, Theorem 9e], Σ contains an m-set of pairwise-disjoint, τ-invariant, 2-way-infinite paths $\Pi_1, \ldots, \Pi_m$ and an m-set $T \subseteq V(\Sigma)$ that meets each path Π_i such that each (of the two) infinite components of $\Sigma - T$ contains a 1-way infinite subpath of each Π_i ($i = 1, \ldots, m$). It follows that $\kappa_\infty(\Sigma) \leq |T| = m$. Let $x \in V(\Sigma)$ and choose a $\kappa_\infty(\Sigma)$=separator S that contains x. Again by [6, Theorem 9e], there exist $p, q \in Z$ such that S separates $\tau^p[T]$ from $\tau^q[T]$. Hence S meets each path Π_i. It follows that $m \leq |S| = \kappa_\infty(\Sigma)$, and x must lie on some Π_i.   □

By [14, Theorems 5.13 and 5.6], every two-ended vertex-transitive graph possesses automorphisms of infinite order, and so the hypothesis of Lemma 5.1 always holds non-vacuously. If Σ is such a graph, a $\kappa_\infty(\Sigma)$-subset T of $V(\Sigma)$ such that $\Sigma - T$ has two infinite components will be called a κ_∞-*separator*, and the infinite components of $\Sigma - T$ are called *fragments*. Certainly if $|T| < \kappa_\infty(\Sigma)$, then $\Sigma - T$ has only one fragment.

Lemma 5.2 *Under the hypothesis of Lemma 5.1, if Π is a τ-invariant path, $\alpha \in G_0(\Sigma)$, and T is a κ_∞-separator, then T and $\alpha[\Pi]$ have a unique common vertex.*

Proof: As in Halin's proof [6, Theorem 9], the path Π is the union of two 1-way infinite paths belonging respectively to the two ends of Σ. Hence if T is a finite set such that $\Sigma - T$ has two fragments, then T meets Π. If $|T| = \kappa_\infty(\Sigma)$, then since Π is just one of $\kappa_\infty(\Sigma)$ pairwise-disjoint, 2-way-infinite paths traversing T, it follows that T, and hence $\alpha^{-1}[T]$ meet Π in a unique vertex. Hence T meets $\alpha[\Pi]$ in a unique vertex.   □

Lemma 5.3 *If a vertex- and edge-transitive graph Σ has $\epsilon(\Sigma) = 2$, then every κ_∞-separator is an independent vertex set.*

Proof: Let T be a κ_∞-separator, let $e \in E(\Sigma)$ have both endpoints in T. Let Π be a path in Σ with the properties assured by Lemma 5.1, and let $e' \in E(\Pi)$. Since Σ is edge-transitive, $\alpha(e') = e$ for some $\alpha \in G_0(\Sigma)$. But then $\alpha[\Pi]$ meets T in at least two vertices, contrary to Lemma

For $x, y \in V(\Sigma)$, the distance (in Σ) between x and y is denoted $d(x, y)$. For any graph Γ we let

$$N(\Gamma) = \{\nu \in G_0(\Gamma) : \text{ there exists } d_\nu \text{ such that } d(x, \nu(x)) \leq d_\nu \text{ for all } x \in V(\Gamma)\}.$$

By [14, Lemma 5.9], $N(\Gamma)$ is a normal subgroup of $G_0(\Gamma)$. In case Γ is vertex-transitive and 2-ended, it was shown [14, Theorem 5.10(e)] that $N(\Sigma)$ has index at most 2 in $G_0(\Sigma)$.

Theorem 5.4 *Every vertex- and edge-transitive graph Σ with $\epsilon(\Sigma) = 2$ has even valence.*

Proof: By Lemmas 5.1 and 5.2 we select a 2-way-infinite path Π that meets each κ_∞-separator in a unique vertex. Let us fix notation:

$$V(\Pi) = \{x_i : i \in Z\} \text{ and } E(\Pi) = \{\{x_i, x_{i+1}\} : i \in Z\}.$$

Let T be a κ_∞-separator such that $x_0 \in T$ and let F_1, F_{-1} be fragments (with respect to T) such that $x_i \in V(F_i)$ for $i = \pm 1$. Let us also fix α in the stabilizer of x_0. We first show:

(5.5) $$\alpha(x_i) \in V(F_i) \ (i = \pm 1) \Leftrightarrow \alpha \in N(\Sigma)$$

and

(5.6) $$\alpha(x_i) \in V(F_{-i}) \ (i = \pm 1) \Leftrightarrow \alpha \in G_0(\Sigma) \backslash N(\Sigma).$$

By Lemma 5.2, x_0 is the unique vertex common to T and $\alpha[\Pi]$. Since automorphisms induce permutations of ends (see §2 above), $\alpha[\Pi]$ contains infinitely many vertices of each of F_1 and F_{-1}. Hence $\alpha(x_1)$ and $\alpha(x_{-1})$ lie in different fragments, and (5.5) and (5.6) are equivalent.

Suppose $\alpha(x_i) \in V(F_{-i})$ for $i = \pm 1$. Then $\alpha(x_i) \in V(F_{-1})$ for all $i \in Z^+$. Because Σ is locally finite, F_{-1} contains vertices from $\alpha[\Pi]$ at arbitrarily large distances from T; specifically, for any given n, there exists $i \in Z^+$ such that $d(x_i, \alpha(x_i)) > d(T, \alpha(x_i)) > n$. Thus $\alpha \in G_0(\Sigma) \backslash N(\Sigma)$.

On the other hand, if $\alpha(x_i) \in F_i$ $(i = \pm 1)$, then by a similar argument, $\alpha(x_i) \in V(F_1)$ for all $i \in Z^+$. By [14, Lemma 5.4], there exists a finite upper bound n for the diameters of all the sets

$$V_{i,k} = \{y \in T \cup V(F_i) : d(x_0, y) = k\} \text{ for } i = \pm 1, k \in Z^+.$$

Each set $V_{i,k}$ contains at least one vertex from each of Π and $\alpha[\Pi]$. For all $j \in Z$, both x_j and $\alpha(x_j)$ belong to the same set $V_{i,k}$. For any given $y \in V(\Sigma)$, we may pick x_j in the same set $V_{i,h}$ containing y. It follows that $\alpha(x_j)$ and $\alpha(y)$ belong to the same set $V_{i,l}$. Putting these facts together we obtain

$$d(y, \alpha(y)) \leq d(y, x_j) + d(x_j, \alpha(x_j)) + d(\alpha(x_j), \alpha(y)) \leq 3n.$$

Thus $\alpha \in N(\Sigma)$, and we've proved (5.5) and (5.6).

Let us now suppose that Σ has odd valence. By [14, Lemma 5.4(a)], Σ has linear growth, which is certainly subexponential, and which implies by the main result in [20] that Σ is *1-transitive*, (that is, given $\{u_1, v_1\}, \{u_2, v_2\} \in E(\Sigma)$, there exists $\varphi \in G_0(\Sigma)$ such that $\varphi(u_1) = u_2$ and $\varphi(v_1) = v_2$). In particular, there exists $\alpha \in G_0(\Sigma)$ such that $\alpha(x_0) = x_0$ and $\alpha(x_1) = x_{-1}$. By (5.6), $\alpha \in G_0(\Sigma) \backslash N(\Sigma)$.

Suppose $y \in V(F_1)$ is any neighbor of x_0. Pick $\beta \in G_0(\Sigma)$ such that $\beta(x_0) = x_0$ and $\beta(x_1) = \alpha(y)$. Since $\alpha^{-1}\beta(x_1) = y$, (5.5) implies that $\alpha^{-1}\beta \in N(\Sigma)$. Hence $\beta \in G_0(\Gamma) \backslash N(\Sigma)$, and by (5.6), $\alpha(y) = \beta(x_1) \in V(F_{-1})$. We have shown that $\alpha[V_{1,1}] \subseteq V_{-1,1}$.

Symmetrically one can prove $\alpha[V_{-1,1}] \subseteq V_{1,1}$. By Lemma 5.3, $\rho(\Sigma) = \rho(x_0) = |V_{1,1}| + |V_{-1,1}| = 2|V_{1,1}|$.   □

The above theorem fails if $\epsilon(\Gamma) \neq 2$. Both the regular hexagonal mosaic and the lexicographic product of the 3-valent tree with the edgeless graph Ω_3 are vertex- and edge-transitive and have one and uncountably many ends, respectively. Their respective valences are 3 and 9.

In the light of Theorem 5.4, we cannot exploit Corollary 3.2 to produce 2-ended VT-NCG's and must therefore resort to other methods. In [21], Uebelacker considered a generalized product $\Gamma_1 \circ \Gamma_2$ of graphs Γ_1 and Γ_2 as a graph having $V(\Gamma_1 \circ \Gamma_2) = V(\Gamma_1) \times V(\Gamma_2)$ with $E(\Gamma_1 \circ \Gamma_2)$ determined by the groups $G_0(\Gamma_1)$ and $G_0(\Gamma_2)$. These products are restricted to those that contain the cartesian product $\Gamma_1 \times \Gamma_2$, are contained in the lexicographic product $\Gamma_1[\Gamma_2]$, and generally have the property that $\Gamma_1 \circ \Gamma_2$ is vertex-transitive whenever both Γ_1 and Γ_2 are. In the following theorem, the graph product $\circ$ will be presumed to have all of these properties.

Theorem 5.5 *If Θ is a finite VTNCG and Δ denotes the 2-way infinite path, then $\Delta \circ \Theta$ is a VTNCG and $\epsilon(\Delta \circ \Theta) = 2$.*

Proof: The graph $\Gamma = \Theta \circ \Delta$ is vertex-transitive by definition. Moreover, $\epsilon(\Gamma) = 2$. (Cf. [14, Theorem 5.12] and [11, Proposition 3.1].)

Let us write $V(\Gamma) = \{(n, x) : n \in Z; x \in V(\Theta)\}$, and let Θ_n be the subgraph of Γ spanned by $\{(n, x) : x \in V(\Theta)\}$ for each $n \in Z$. Thus $\Theta_n \cong \Theta$ for all $n \in Z$. By [21, Theorem 2], $G_0(\Delta \times \Theta) \leq G_0(\Gamma) \leq G_0(\Delta[\Gamma])$. By [18, 19], the sets $V(\Theta_n)$ are blocks of imprimitivity for both $G_0(\Delta \times \Theta)$ and $G_0(\Delta[\Theta])$ and hence also for $G_0(\Gamma)$.

Suppose that $G_0(\Gamma)$ were to contain a regular subgroup G, and let $x \in V(\Theta)$. Then for each $y \in V(\Theta)$, there exists a unique $\alpha_y \in G$ such that $\alpha_y(0, x) = (0, y)$. Hence $\alpha_{y|V(\Theta_0)} \in G_0(\Theta_0)$, and the set of these restrictions as y ranges over $V(\Theta)$ forms a regular subgroup of $G_0(\Theta_0)$. By Proposition 2.2, Θ_0 is a Cayley graph, contrary to assumption.□

The following example illustrates that if the condition $\Delta \times \Theta \subseteq \Delta \circ \Theta$ is relaxed in Theorem 5.5 then $\Delta \circ \Theta$ may well be a Cayley graph.

Example 5.6 Let $G = \{(n, x) : n \in Z; x \in Z_5\}$ with the group operation

$$(m, x) \cdot (n, y) = (m + n, 3^n x + y),$$

where the second coordinate is, of course, computed modulo 5. Let $H = \{(0, 1), (0, 4), (1, 0), (-1, 0)\}$ and let Γ be the Cayley graph $\Gamma_{G,H}$. A portion of Γ is shown below in Figure 1, where the broken lines at top and bottom are to be identified as though Γ were wrapped around an infinitely long cylinder. Note that for each $n \in Z$, the subgraph induced by $V(\Theta_n) \cup V(\Theta_{n+1})$ is a copy of the Petersen graph Π, a finite VTNCG, and that Γ contains a product of Δ with Π but does not contain $\Delta \times \Pi$.

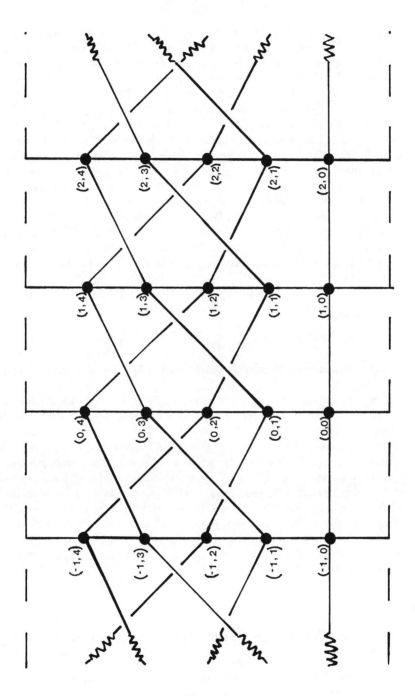

Figure 1

Question: *Do there exist 2-ended VTNCG's not constructible via Theorem 5.5?*

C. $\epsilon(\Gamma) = 1$. Two methods for producing 1-ended VTNCG's will be discussed here. Of lesser interest perhaps is to let Γ be an infinite-ended, odd-valent, vertex- and edge-transitive graph with $\kappa(\Gamma) = 1$, as constructed in ¶A of this section. We form the cartesian product Γ^{2m+1} $(m \geq 1)$ of Γ with itself an odd number $2m+1$ of times. The graph Γ^{2m+1} is, of course, vertex- and edge-transitive and of odd valence $(2m+1)\rho(\Lambda)$, but $\kappa_\infty(\Gamma^{2m+1}) = \infty$ and $\epsilon(\Gamma^{2m+1}) = 1$. By Corollary 3.2, $\partial\Gamma^{2m+1}$ is a VTNCG, and it has only one end.

Our second method is in effect to generalize the dodecahedron of §4 by considering those infinite tilings of the plane where, for given integers $m \geq 1$ and $s \geq 3$, exactly $2m + 1$ s-sided regions meet at each vertex. Such an infinite tiling is always constructible if (a) $m \geq 1$ and $s \geq 7$ or (b) $m \geq 2$ and $s \geq 4$ or (c) $m \geq 3$ and $s \geq 3$. These tilings are vertex- and edge-transitive and odd-valent. Moreover, s is odd, if and only if they are not bipartite. In this case the tilings satisfy the hypothesis of Corollary 3.2; their derived graphs are $(4m)$-valent, 1-ended VTNCG's. When s is even, their derived graphs are not VTNCG's. For suppose that Γ is an r-valent tiling such that all regions have $s = 2t$ edges on their boundary. Then one straightforwardly verifies that $\partial\Gamma$ is the Cayley graph $\Gamma_{G,H}$ of the group

$$G = \langle a_1, a_2 : a_1^r = a_2^r = (a_1 a_2)^t = 1 \rangle$$

with respect to the generating set

$$H = \{a_i^j : i = 1, 2; j = 1, 2, \ldots, r - 1\}.$$

Necessary and sufficient conditions upon r and t for the finitude of G can be found in [2, p.67].

We conclude by showing that no other tiling satisfies the hypothesis of Corollary 3.2. Since these graphs are clearly 3-connected, their planar imbeddings are unique, and the set of elementary circuits that define the "regions" of any and all planar imbeddings is well-defined. (See, for example, [10] for a proof of Whitney's Theorem [24] which is generalizable to infinite graphs.) If Γ is a vertex- and edge-transitive, 1-ended planar graph, then one may associate with Γ a sequence $r_3, r_4, r_5, \ldots$ of parameters such that for all planar imbeddings of Γ, each vertex is incident with exactly r_i i-regions, i.e., regions whose boundary is an i-circuit. Since Γ is edge-transitive, there exist j, k such that for every edge e, the two regions whose boundaries contain e are a j-region and a k-region. Thus $\rho(\Gamma) = r_j + r_k$, and $r_i = 0$ for all $i \neq j, k$. About each vertex, the regions in rotational order must be alternately a j-region and a k-region. Thus $r_j = r_k$, implying that $\rho(\Gamma)$ is even whenever $r_i \neq 0$ for more than one value of i.

References

[1] J. A. Bondy and U. S. R. Murty, *Graph Theory with Applications*, MacMillan Press Ltd., London, 1976.

[2] H. S. M. Coxeter and W. O. J. Moser, *Generators and Relations for Discrete Groups*, 2nd Edition, Springer-Verlag, Berlin, 1965.

[3] R. Frucht, J. E. Graver, and M. E. Watkins, The groups of the generalized Petersen graphs, *Proc. Camb. Phil. Soc.* **70**(1971), 211–218.

[4]   C. D. Godsil, More odd graph theory, *Discrete Math.* **32**(1980), 205–207.

[5]   R. Halin, Über unendliche Wege in Graphen, *Math. Annal.* **157**(1964), 125–137.

[6]   R. Halin, Automorphisms and endomorphisms of infinite locally finite graphs, *Abh. Math. Sem. Univ. Hamburg* **39**(1973), 251–283.

[7]   F. Harary and E. M. Palmer, On the point-group and line-group of a graph, *Acta Math. Acad. Sci. Hungar.* **19**(1968), 263–269.

[8]   A. J. Hoffman and R. R. Singleton, On Moore graphs with diameters 2 and 3, *IBM J. Res. Develop.* **4**(1960), 497–504.

[9]   B. Huppert, *Endliche Gruppen I*, Springer-Verlag, Berlin, 1967.

[10] W. Imrich, On Whitney's Theorem on the unique embeddability of 3-connected planar graphs, in: *Recent Advances in Graph Theory, Proceedings of the Symposium held in Prague, June 1974*, (Editor Miroslav Fiedler), Academia Praha, Prague, 1975, pp. 303–306.

[11] W. Imrich and N. Seifter, A note on the growth of transitive graphs. To appear.

[12] H. A. Jung, A note on fragments of infinite graphs, *Combinatorica* (3) **1**(1981), 285–288.

[13] H. A. Jung and M. E. Watkins, On the structure of infinite vertex-transitive graphs, *Discrete Math.* **18**(1977), 45–53.

[14] H. A. Jung and M. E. Watkins, Fragments and automorphisms of infinite graphs, *Europ. J. Combinatorics* **5**(1984), 149–162.

[15] O. Ore, *Theory of Graphs*, Providence, R.I., 1962.

[16] G. Sabidussi, Loewy-groupoids related to linear graphs, *Amer. J. Math.* **76**(1954), 477–487.

[17] G. Sabidussi, On a class of fixed-point free graphs, *Proc. Amer. Math. Soc.* **9**(1958), 800–804.

[18] G. Sabidussi, Graph multiplication, *Math. Z.* **72**(1960), 446–457.

[19] G. Sabidussi, The lexicographic product of graphs, *Duke Math. J.* **28**(1961), 573–578.

[20] C. Thomassen and M. E. Watkins, Infinite vertex-transitive, edge-transitive, non-one-transitive graphs, *Proc. Amer. Math. Soc.* **105**(1989), 258–261.

[21] J. W. Uebelacker, Product graphs forgiven subgroups of the wreath product of two groups, I, *J. Combin. Theory.* **17**(1974), 124–132.

[22] M. E. Watkins, Les graphes de Cayley de connectivité un, in: *Problèmes Combinatoires et Théorie des Graphes, Proceedings of the International Colloquium on Graph Theory at Orsay, France, July 1976*, C.N.R.S., Paris, 1978, pp. 419–422.

[23] R. M. Weiss, Kantenprimitive Graphen vom Grad drei, *J. Combin. Theory Ser. B* **15**(1973), 269–288.

[24] H. Whitney, Congruent graphs and the connectivity of graphs, *Amer. J. Math.* **54**(1932), 150–168.

[25] H. Wielandt, *Finite Permutation Groups*, Academic Press, New York, 1964.

INDEX